Patent Transactions in the Life Sciences

A Global Guide to Agreements in the Sector

Consulting Editor **Paul England**

Consulting editor
Paul England

Managing director
Sian O'Neill

Commissioning editor
Katherine Cowdrey

Editorial services director
Carolyn Boyle

Production manager
Neal Honney

Group publishing director
Tony Harriss

Patent Transactions in the Life Sciences: A Global Guide to Agreements in the Sector
is published by
Globe Law and Business
Globe Business Publishing Ltd
New Hibernia House
Winchester Walk
London SE1 9AG
United Kingdom

Tel +44 20 7234 0606
Fax +44 20 7234 0808
Web www.globelawandbusiness.com

Printed and bound by CPI Group (UK) Ltd, Croydon, CR0 4YY

ISBN 9781909416284

Table of contents

Introduction

Paul England
Taylor Wessing

1. The growth and diversification of life sciences

A combination of high research costs, complex regulatory requirements and dwindling opportunities for developing small-molecule targets is what is primarily blamed for the difficulties encountered by some branded pharmaceutical companies in recent years when trying to bolster the number of candidates entering and progressing through their product pipelines. At the other end of the pharmaceutical product life cycle, many of the 'blockbuster' drugs that have dominated the market in the last decade are now seeing generic competition. Indeed, many generics (often called 'super-generics') are moving into products based on their own research and development, including the adaption and reformulation of existing drugs for new or improved applications.

As a result of these pressures, some companies are moving away from reliance on the traditional small-molecule pharmaceutical business model and are attempting to broaden the nature of their products. This diversification includes:

- branded companies creating their own lines of branded generic products, and/or buying generic companies;
- developing biological products, which are seen as presenting new opportunities and greater difficulty for competitors to copy;
- moving into medical devices, which are valuable not only as innovative products in themselves but also as a means to extend exclusivity on existing products; and
- developing borderline products – for example, health foods and supplements – which may fall outside the more complex regulatory requirements applying to medicinal products.

Further difficulties for firms have, of course, come from the worst economic downturn in living memory, bringing with it cost pressures and reduced revenues. These factors have turned the following into attractive options:

- acquiring small research companies with just one or two promising drug candidates, rather than conducting early-stage development in-house;
- divesting certain business operations around particular drugs or research areas while retaining equity and intellectual property rights, thereby creating smaller, more focused, teams dedicated to particular product goals, and removing costs from the balance sheet; and
- cooperation and collaboration with third parties to share and exchange resources and expertise on drug production.

Figure1: Patent application filings in the European Patent Office by sector, 2004 to 2013

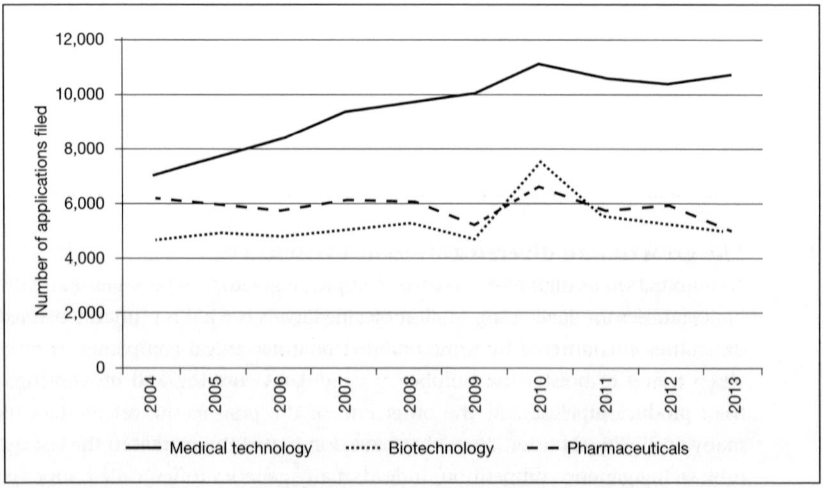

This diversification of the traditional, small-molecules industry at the large scale has helped sustain the life sciences sector. Indeed, it is the many models and subject areas of life sciences activity that may have maintained it through the downturn. This is the view of a report prepared by Battelle in 2012,[1] which observes that a primary reason for the resilience of the bioscience industry through difficult economic conditions is its diversity. The opening paragraph of the report's executive summary states:

At a time when the global economy struggles to recover from a severe recession and uncertainty remains regarding future economic growth, bioscience industry development is generating significant attention both globally and [in the United States]. A number of recent studies have detailed the mounting global competition for biosciences industry development as both developed and developing nations seek to grow and advance in this high-wage, high-growth industry…

A primary reason for the resiliency of the bioscience industry is the diverse set of markets it serves….

To the extent that the number of filings for patent applications can be taken as an indicator of the wealth of innovation that has been taking place and that may develop into products in the coming years, figures recorded by the European Patent Office (EPO) and shown in Figure 1[2] indicate that despite the economic crisis that took hold in 2008, applications for pharmaceutical and biotechnological material have remained steady over the period. Moreover, and reflecting the diversification process described above, 'medical technology'[3] filings, which include those for

1 Battelle/Bio, "State Biosciences Industry Development 2012" (hereinafter the Battelle Report).
2 See European Patent Office, *Facts and Figures 2014.*
3 'Medical technology' is a field in the WIPO technology concordance table, which links the International Patent Classification (IPC) symbols with 35 fields of technology.

medical devices, have soared by almost 50% over the past 10 years. In fact, medical technology was the single biggest class of applications filed in 2013.

Of course, innovation is not restricted to Europe and the United States. For example, although it has been known for many years as a centre of generic drug production, India is now seeing a number of novel drug research projects.[4] In particular, in June 2013 Zydus Cadila launched India's first home-grown new chemical entity, Lipaglyn, onto the market to treat diabetic dyslipidemia. And in October 2013 the Zydus group announced that it had gained US FDA (Food and Drug Administration) approval for Phase I clinical trials for ZYDPLA1 – a next-generation, long-acting DPP-4 inhibitor for the treatment of type-2 diabetes. Other Indian companies such as Glenmark Pharmaceuticals, Sun Pharma and Biocon are anticipated to develop their own novel medicines in the next few years.

Furthermore, new levels of deal making in the sector in 2014 are attributed to a desire for efficiency, productivity and the need to reshape businesses after the sluggish period of the last few years.[5] In the United Kingdom alone – and quite apart from the speculation about a mega-merger between AstraZeneca and Pfizer – acquisition interest has focused on Shire and Allergan as potential targets and Bayer, Reckitt Benckiser, Valeant and Stryker as potential purchasers; Novartis has purchased GlaxoSmithKline's oncology business; Merck & Co is auctioning off its consumer healthcare unit; and companies such as Circassia and Horizon Discovery have had notably successful listings.

With this time and money being spent and the potential commercial rewards at stake, how do those making the investment and depending on the success of a life sciences business ensure that the innovations they are relying on at the centre of their businesses provide a return for them? Furthermore, in such a competitive sector, what is to ensure that the technology being developed by one enterprise is not being appropriated by another? The answer comes from the EPO statistics quoted above: where there is innovation, particularly when it has a prospect of commercial application, there is intellectual property, and in particular patents.

2. The importance of patents

Patents enable the patent owner to have exclusive use of the technology covered by the patent for a fixed period of years. The exact nature of the acts that can be excluded depends on whether the patent is for a product or a process, and the country in which it is registered. For example, for a UK patent or European patent designating the United Kingdom the following apply:

- Anyone who in the United Kingdom and without the permission of the patent holder makes, disposes of, offers to dispose of, uses or imports the product, or keeps it for disposal or otherwise, is an infringer of the patent;
- If the patent is for a process, then using that process or offering it for use in the United Kingdom without the permission of the patent holder will be an infringement if it can be shown that it was obvious to a reasonable person in

4 See Amy Kazmin, "Indian pharmaceutical groups shed copycat image", *Financial Times*, July 22 2013.
5 Andrew Ward, "Pharma deals reach new level of intensity", *Financial Times*, May 1 2014.

the circumstance that its use there without the consent of the owner would be an infringement of the patent; and

- Where the patented invention is a process, a third party will infringe if, without permission from the patent owner, the third party disposes of, offers to dispose of, uses or imports any product obtained directly by means of that process, or keeps any such product whether for disposal or otherwise.

Additionally, in the United Kingdom and many other countries, a patent can be used to prevent a third party from supplying or offering to supply certain means that relate to an essential element of the invention for putting that invention into effect.

A commercially successful, patented technology can therefore corner the market to the extent that competitors cannot use the same technology. And, although patents cannot guarantee that an innovation will be commercially successful, they do ensure, if properly managed, that the rewards of a successful invention flow to the patent owner. Hence, as the Battelle Report states, the industries in the biosciences sector are amongst those with the highest levels of patenting. This, says the report, is a key to their competitive advantage in the marketplace.

This is further explained by another report, published in January 2014 by University College London's School of Pharmacy and co-written with the London School of Economics (hereinafter the UCL Report).[6] The UCL Report examines patient interests in medicines and biomedical innovation, and the extent to which these are served by intellectual property rights – in particular, patents. As the UCL Report points out, the market value placed on medicines is not a reflection of simply the cost of raw materials and manufacture, but of the scientific challenges and material costs of their development. These include not only the preclinical developments and clinical trial costs of successful drugs but also those for drug candidates that never made it to market. These costs all having been considered, the UCL Report supports the view that it is not an exaggeration to estimate the costs of developing an innovative medicine to market to be in excess of $1 billion. The price of medicines reflects these costs.

Consequently, when the exclusivity afforded by patent protection is lost, competitors whose costs are limited largely to raw materials and manufacture can offer the same drug at a small percentage of the price (in the small-molecule drugs market, this is often less than 10%). Hence, as the UCL Report notes, patents do not just enable rewards for innovations that are already available; patents are also important to secure the willingness of investors to continue to put large sums of money towards innovations for the future.

Acknowledging that the high drug costs that the patent system enables are controversial, the UCL Report nonetheless states in summary that:

> *... some commentators question the value of intellectual property rights (IPRs) such as pharmaceutical patents because they increase the prices of recent therapeutic developments. But without intellectual property rights private investment in high risk*

---

6 Mari Lundeby-Grepstad and David Tordrup (LSE), and Tinas Craig and David Taylor (UCL School of Pharmacy), "Patients' needs, medicines innovation and the global public's interests", January 2014.

biomedical research would be very unlikely to take place. This would almost certainly have negative 'knock on' effects on public funding for fundamental research.

Despite continuing criticism of the application of patent rights to medicines, they remain a fact of life and their importance for commercial investment in the life sciences continues. Indeed, and again controversially, with the advent of the World Trade Organization (WTO) Agreement on Trade-Related Aspects of Intellectual Property Rights (TRIPS), concepts of intellectual property have been exported to developing countries, including product and process patents for pharmaceuticals. To follow the example of India, this country signed TRIPS in 1995, foreshadowing the introduction of its Patent Act. Indeed, it is because of this that Zydus has been able to patent the novel drugs mentioned above.

3. Patents for collaboration

But there is more to patenting than the exclusive right to exploit that it affords the patent owner. To return to the Battelle Report, this states that the growth of the biosciences sector can also be attributed to the fast pace of advances in biological sciences and notes that:[7]

the close connections between basic research discoveries, often advanced in academic and federal laboratory settings, and industry product development within biosciences are well documented.

This observation resonates with comments in the recent book *The Second Machine Age*.[8] The authors suggest that the true work of innovation is not necessarily a matter of coming up with something big and new, but recombining technologies that already exist. For this, the authors say, Kary Mullis's Nobel prizewinning polymerase chain reaction (PCR) is a prime example: Mullis had the insight to take well-known techniques in biochemistry and to combine them to create a new and highly valuable research tool. However, the paradigm example in the life sciences sector is probably the decoding of the human genome that was undertaken by the Human Genome Project (a project that PCR had in turn largely made possible). In addition to the medical research directly enabled by the decoding of the genome, this project has seeded the growth of an industry:

... to supply the scientific research community with the tools needed to conduct genomics research and development and associated product development, such as gene sequencers, sample preparation techniques, sample amplification technology, and a range of other analytical tools and technologies. Moreover, a majority of bioscience companies are now using genomics-based tools to advance new product development.[9]

According to the Battelle Report, the US Government has already seen a return of nearly $50 billion in tax receipts from the little more than $10 billion that it has invested in the Human Genome Project.

Arguably, this kind of recombinant innovation can also be seen in the approach being taken by the super-generics mentioned in section 1 above. These are

---

7 Battelle Report, Executive Summary, p1.
8 Erik Brynjolfsson and Andrew McAfee, *The Second Machine Age*, WW Norton & Company, 2014.
9 Battelle Report, p2.

companies originally rooted in a generics model, which are now developing products based on their own research and development aimed at adapting, combining and reformulating off-patent drugs, both for the improved treatment of existing indications and for use in new indications. The commercial driver for these is the promise of patent protection on the product that results. There are also the examples of known research tools, such as phage display and bioinformatics, that can be employed to discover respectively a multitude of new monoclonal antibodies and other biologics.

If the recombinant innovation argument is correct, the more innovations that come into existence, the greater and greater the opportunities created for recombinant technologies – theoretically, an exponential growth becomes possible. For recombinant technology growth to happen in this way, however, there must be collaboration. How is this helped by patenting?

4. The purpose of this book

The answer to the foregoing question is provided by the UCL Report:

> *Innovative potential is closely associated with overall human development and the establishment of settings in which ideas and information can be exchanged safely, with a confident sense that individual and corporate contributions to knowledge will be fairly recognised.*[10]

Patents delineate and record as a property right on a register what the owner's claim to an invention is, allowing it to share that information with the public at large without losing the owner's claim to it. But, importantly for the purpose of commercial activity, this also allows patents to be dealt in with third parties, as with any other property right, or the proprietor to collaborate with third parties to share the technology for the advancement of the interests of both. The fundamental ways in which this is achieved are by patent assignment, by which rights to a patented technology are transferred from one owner to another, and by licence, in which certain rights to the patent can be granted in a controlled way to one or more third parties without ownership itself being transferred. Both allow the proprietor to ask for money by way of exchange. Thus the patent is monetised, without a single product having been sold or, necessarily, developed.

Returning once again to the Indian examples, where this process can be seen in action, it can be noted that Biocon is reported as having development deals with Bristol-Myers Squibb, and Glenmark has licensed patents on promising molecules to Sanofi and Forest Laboratories. They are collaborating with third-party companies to develop products covered by the patented technology without losing control of it, and ensuring that they share in any return.

However, patent licences and assignments are not simple. Typically, where there are patents there are also complex patent law issues to address. For the unlucky few, these may be encountered in disputes; but they are far more often encountered when parties negotiate to strike a mutually beneficial deal for the exploitation of the technology protected by the patents at issue. Licences and assignments also

---

10 UCL Report, p13 (Box 4).

commonly form the core of bigger transactions in the life sciences sector, such as mergers and acquisitions, joint ventures, divestments and development agreements.

This book explains the rights that underpin these life sciences transactions and describes their architecture, features and purpose – from the basics of patents, licences and assignments to the types of deal that commonly feature in the industry at the corporate level. Associated issues of trade secrets, competition rules, securitisation, insolvency and taxation are also outlined.

The book is intended to provide a resource for lawyers, in-house counsel and other professionals advising on transactions in the life sciences industry, particularly where they have global reach. It is designed to provide guidance on the structure and detail of agreements in the life sciences sector that are based on patents and associated rights, covering the practical what, where, why and when of patent transactions in this sector, under the leading legal regimes.

In the life sciences field, patents are the crown jewels of any business; knowing how to use them is paramount, and this book provides an introduction on how to do so.

Looking ahead to the unitary patent

Paul England
Taylor Wessing

1. Introduction

No discussion of patents is nowadays complete without raising the subject of the upcoming 'unitary patent' in Europe. This radical addition to the European patent system is widely anticipated to come into being as early as 2016 and will transform patent ownership and enforcement in those countries to which it applies. This chapter briefly outlines the key features of the unitary patent and its context within the Unified Patent Court (UPC) and the existing (hereinafter 'classic') system of European patents.

2. Context

In the current system, the European Patent Convention provides a centralised application procedure in which the European Patent Office (EPO) grants a bundle of classic patents across Europe. Disputes about patents after grant (with the exception of those patents opposed at the EPO within the first nine months) are adjudicated upon exclusively by national courts.

However, the UPC, when in operation, will create a court in which future unitary patents and classic European patents can be litigated in a largely one-stop shop for the participating EU member states[1] – that is, it will be possible to enforce and revoke a unitary patent simultaneously in all of the territories that it covers in one set of UPC proceedings. The UPC will have a number of divisions across Europe: local and regional divisions making up a Court of First Instance, and a central division split between three sections in Paris, London and Munich. The London section has been assigned responsibility for so-called 'human necessities' and chemistry and metallurgy patents, a class that includes life sciences subject matter. Appeals will be heard by an appeals chamber in Luxembourg.

3. The unitary patent

In contrast to classic European patents, unitary patents, the full name of which is 'European patents with unitary effect', will be single pan-European rights protecting all the EU member states that participate in Council Regulation 1257/2012/EU (hereinafter 'the Regulation') and have ratified the UPC Agreement. At mid-2014, this excludes Italy, Spain and Poland.[2] The patent is intended to represent a key

1 The system of national patents will continue. These will remain exclusively actionable in the national courts.

component of the European single market, but will exist alongside the existing system of classic European patents and national patents. The unitary patent will come into force at the same time as the UPC.

The unitary patent is created by Article 3(1) of the Regulation:

A European patent granted with the same set of claims in respect of all the participating Member States shall benefit from unitary effect in the participating Member States provided that its unitary effect has been registered in the Register for unitary patent protection.

A unitary patent, as further provided in Article 3(2) of the Regulation, is a classic European patent that has been registered as having unitary effect upon grant by the EPO. As mentioned above, this means it has the effect of a single right conferring on its proprietor and exclusive licensees the right to prevent a third party from committing certain infringing acts in all the EU member states covered. The Regulation states that the protection provided across the participating member states must be uniform.[3] Hence, only classic European patents granted with the same claim sets can benefit from unitary status. This, as the recitals to the Regulation state,[4] ensures the fundamental purpose of the unitary patent as a patent of uniform substantive scope of protection across Europe.

The corollary of this breadth of protection is that any change to the status of the unitary patent also takes effect across all the participating member states as a whole, in uniformity, in accordance with Article 3(2) of the Regulation, which states: "It may only be limited, transferred or revoked, or lapse, in respect of all the participating Member States." This presents a great risk that a patentee, faced with a decision as to whether to register for unitary effect, needs to consider the possibility that his patent could be revoked across all the participating member states through one UPC revocation action.

Despite this uniform nature, the Regulation specifically states that a unitary patent may be licensed in respect of the whole or part of the territories of the participating member states.[5] The implication of this provision appears not only to be that licences may be granted to different licensees for different respective participating member states, but that transactions that are based on patent licensing, such as those described in the chapters of this book, will also be possible in respect of the whole or a part of the territories of the participating member states.

The proprietor of a unitary patent may also file a statement with the EPO that he is prepared to grant a licence in return for appropriate consideration,[6] and the Regulation envisages that such a statement should have the consequence of a reduction in renewal fees.[7]

Requests for unitary effect should be filed with the EPO within one month of the date of publication of the mention of the grant in the European Patent Bulletin.[8]

---

2 Italy and Spain did not sign the Unitary Patent legislation. Poland did not sign the UPC Agreement, which has the effect that the unitary patent cannot extend to Poland.

3 Article 5(2) of the Regulation.

4 See Recital 7 and Article 3(1) of the Regulation.

5 Article 8 of the Regulation.

6 Article 8 of the Regulation.

7 Recital 15 of the Regulation, although no such arrangements were in place at the time of writing.

8 See Article 3(2) and Recital 18 of the Regulation.

Hence, if unitary effect is not applied for within this time period, the opportunity to obtain unitary patent status in respect of a classic European patent is lost. The EPO currently states that:

> *The details of the procedure to request and validate a unitary patent, as well as the level of renewal fees and the key under which the fees will be distributed to the participating states, will be prepared by the select committee of the Administrative Council of the European Patent Organisation.*[9]

The EPO will also be responsible for collecting, administering and remitting renewal fees for unitary patents and for keeping a register of unitary patents that will include legal-status information relating to licences, transfers, limitation, revocation or lapse.[10]

4. Classic European patents in the UPC

The UPC will have jurisdiction over classic European patents. Decisions of the UPC in respect of classic patents will simultaneously apply to all designations of the patent in contracting member states who have ratified the UPC Agreement. Such decisions are also, therefore, pan-European in effect. However, when the UPC first comes into force its jurisdiction to deal with disputes concerning classic European patents will be subject to important limitations for a seven-year transitional period.[11] During this transitional period Article 83(1) of the UPC Agreement permits disputes concerning classic patents to be litigated in either the UPC or the national courts of the EU member states of designation that have contracted to and ratified the UPC Agreement.[12]

All classic European patents designated in member states that have ratified the UPC Agreement automatically become subject to the exclusive competence of the UPC after the transitional period has expired, unless they have opted out (see section 5 following).

5. Opt-out of classic European patents

An opt-out removes a classic European patent from the exclusive jurisdiction of the UPC which would otherwise apply to it after expiry of the seven-year transitional period. This is important because a classic European patent that has been opted-out can only be litigated in the national court of its designation, subject to a later withdrawal of the opt-out (see section 5.1 below). A party attempting to revoke the patent must therefore do so in national courts country by county rather than in one action in the UPC. It is currently thought that the opt-out will have effect for the entire life of the patent.[13] And although an opt-out may be withdrawn, once the seven-year transitional period has expired, it may not be opted-out again.

There is an important restriction on the right to opt out. If the UPC is seized with

---

9 EPO website, unitary patent page: www.epo.org/law-practice/unitary/unitary-patent.html.

10 See www.epo.org/law-practice/unitary/unitary-patent.html.

11 This period may be extended by another seven years.

12 This is currently all EU member states except Poland and Spain. Note that unlike the unitary patent, Italy did contract to the UPC and therefore Italian designations of classic European patents are subject to the UPC's jurisdiction.

13 'Patent' includes any supplementary protection certificate based upon the patent.

an action concerning a classic patent before an application to opt-out that patent has appeared on the UPC Register, that action must continue in the UPC[14] and the patent concerned cannot be opted-out. This has the practical effect that patent proprietors or the party responsible for opting out must ensure they opt-out as soon as possible those patents that they would like to take out of the UPC system, to avoid the risk of a revocation action in the UPC.

5.1 Withdrawal of opt-out

Classic European patents that have been opted-out of the UPC system may later have the opt-out withdrawn, under Article 83(4) of the UPC Agreement. Withdrawal of an opt-out for a classic European patent places all the designations of that patent into the exclusive jurisdiction of the UPC. Much like the opt-out, the withdrawal of an opt-out is barred if an action relating to one of the patent designations in question has already been brought before a national court. The danger of opting out for a patent proprietor is therefore that he might lose the opportunity to enforce his classic European patent against an infringer across Europe in the UPC. However, it is currently unclear whether a classic patent that has been litigated in a national court can subsequently have the opt-out withdrawn, or whether it remains only subject to national jurisdiction for the remainder of its term.

5.2 Co-owners of classic European patents

The opt-out procedure is set out in Rule 5 of the Rules of Procedure.[15] This states that the application to opt out must be lodged by the "proprietor" of a granted patent or application. If the application or granted patent is owned by two or more proprietors "all proprietors shall lodge the application". In practical terms, this means that co-owners of a patent must agree between themselves whether they wish that patent/application to be opted-out; and they will need to collaborate on their opt-out strategy before the UPC comes into effect.

5.3 Classic European patent designations with different owners

Rule 5.1(c) of the Rules of Procedure refers to the opt-out application being made in respect of "all of the Contracting Member States for which the European patent is designated". This makes it clear that a classic European patent must be opted-out across all the designated contracting member states, or none at all. Furthermore, the patentee of one classic European patent designation cannot opt-out all designations in circumstances where they have come under different ownership as a result of assignment to third parties. Consequently, agreement must be reached between such different owners. This includes the proprietors of designations of the patent in contracting member states to the UPC Agreement who have not ratified the UPC Agreement at the time of the opt-out. The practical effect is that owners of different designations must collaborate on the issue of opt-out, in similar fashion to co-owners.

---

14 Article 83(2) of the UPC Agreement.

15 The Rules of Procedure are in their 16th draft at the time of writing. See section 6 below for more details.

5.4 The role of exclusive licensees in opt-out

Rule 5 only refers to the consent of "proprietors" being required to opt out. Should an exclusive licensee be treated like a proprietor, who must give his consent to an opt-out application? There is an argument that a party who has bargained for exclusive rights and has the power to enforce them subject only to giving notice to the patentee ought to be able to choose the jurisdiction in which it does so. But there is good reason to argue that the right to decide on opting out from the UPC was deliberately limited to patent proprietors, because ownership can easily be proven based on the appropriate patent register. Extending this right to exclusive licensees would not be practicable, as patent licences are often not registered, and there may be more than one exclusive licence granted under the same patent or designation.

If an exclusive licensee wishes to make the decision on opt-out instead of the patent proprietor, especially in a situation where the exclusive licensee enforces the patent in its own name, the parties can provide for an obligation on the patent proprietor to consent to such a decision in the terms of the licence agreement. This is a matter that requires negotiation before the UPC comes into operation.

6. Current status of the unitary patent and the UPC

A key element of the UPC will be its Rules of Procedure. These provide the detail of how proceedings in the UPC will work for every kind of action that the court is competent to hear (as exemplified by Rule 5 on opt-out). The Rules of Procedure are now in their 16th draft, which was published in March 2014. The draft Rules will now pass to the Preparatory Committee, which was established to deal with practical implementation of the UPC, for further consultation. Although the 16th draft of the Rules of Procedure has clarified many matters, problems still need to be resolved on issues such as the language in which a statement of claim must be lodged. Other details are also under consideration: the highly important question of whether the opt-out of a classic European patent endures for the life of that patent or merely for the period of the opt-out is currently understood to be under consideration by the Committee. However, it is recognised that other details, such as the approach taken by the UPC to the grant of preliminary injunctive relief, and search and seizure, can only be worked out by the UPC judges themselves over time.

A key area of progress overseen by the Preparatory Committee is the recruitment and training of judges. Having requested interest from potential judges at the end of November 2013 (which request drew responses from well over a thousand interested persons), the Committee has now formed a provisional list of eligible judges who can move on to receive specialist training where necessary. The dedicated training centre for this purpose was officially opened on March 13 2014 in Budapest.

In total, 13 ratifications will be needed before the UPC – and with it the unitary patent – can come into force. At the time of writing, Austria, France, Belgium and Sweden have ratified. France's ratification is particularly significant because, as with that of Germany and the United Kingdom, the UPC cannot go ahead without it. The UK's Intellectual Property Act 2014, which provides the necessary powers for UK ratification of the UPC Agreement, has passed into law and it is expected that the United Kingdom will ratify the Agreement by the time of the next General Election

(May 7 2015). Ratification in Denmark is also expected to follow in the second half of 2014, following the approval given by the Danish people to the UPC in a referendum that took place in May 2014.

As regards the establishment of the regional and local divisions of the UPC, Germany has announced that it will host local divisions in Düsseldorf, Mannheim, Munich and Hamburg. These local divisions will be in addition to the Munich branch of the central division. Estonia, Latvia, Lithuania and Sweden have concluded an agreement on the creation of a regional Nordic-Baltic division of the UPC, having its seat in Stockholm.

7. Obstacles that remain

There are still a number of obstacles to the implementation of the UPC and, with it, the unitary patent. IT, staffing and funding issues concerning the UPC remain to be resolved. Moreover, the costs that will apply to those using the UPC and the unitary patent (fee for validation, fees for litigation, fee for the opt-out) have still to be determined.

A potential major obstacle for the UPC and unitary patent is that Spain remains opposed to the new system and two legal challenges await a decision of the Court of Justice of the European Union.[16] Considerable disruption might ensue if Spain's objections are upheld.

Even without these difficulties, the Preparatory Committee has much to do before the UPC can become a reality. Indeed, the Committee has recently acknowledged that it does not expect the UPC to be operational until the end of 2015 at the earliest. As noted at the beginning of this chapter, this is probably optimistic, with 2016 appearing more likely. However, few doubt that the UPC and unitary patent are coming soon and that they will introduce fundamental changes to the European patent system for which those in the life sciences, where patents are so important, must prepare.

16 CJEU cases C-146/13 and C-147/13.

Belgium

Philippe de Jong
Christophe Ronse
ALTIUS

1. Patents, know-how and trade secrets: foundations for transactions

1.1 Patents and supplementary protection certificates

(a) *Legislative basis*

Belgian patent law and Belgium's law on supplementary protection certificates (SPCs) are essentially governed by four legislative instruments:

- The Belgian Patent Act of March 28 1984; this law has most recently been updated by the Law of January 10 2011 to align Belgian patent law with the Patent Law Treaty of June 1 2000 and the Act of November 29 2000 revising the European Patent Convention, which is due to enter into force in the foreseeable future;
- The European Patent Convention of October 5 1973, as revised;
- Regulation (EC) 469/2009 of May 6 2009, concerning the SPC for medicinal products; and
- Regulation (EC) 1610/96 of July 23 1996, concerning the creation of an SPC for plant protection products.

In the future, the following legal instruments will also play an important role in the Belgian patent landscape, once they have entered into force:

- Regulation (EU) 1257/2012 of December 17 2012, implementing enhanced cooperation in the area of the creation of unitary patent protection;
- Regulation (EU) 1260/2012 of December 17 2012, implementing enhanced cooperation in the area of the creation of unitary patent protection with regard to the applicable translation arrangements; and
- The International Agreement on a Unified Patent Court.[1]

Under all of these legal instruments, patents provide their proprietor with a temporary exclusive 'negative' right to prevent third parties that have not obtained the proprietor's authorization from commercially exploiting the subject matter covered by the patent. Patents do not, however, provide the patent proprietor with a 'positive' right to exploit the subject matter of the patent.

Under Belgian patent law, the geographical scope of both a purely Belgian patent

1 See www.unified-patent-court.org/images/documents/upc-agreement.pdf.

and the Belgian part of a European patent is limited to Belgium. Under the aforementioned Regulation (EU) No 1257/2012 on unitary patent protection, once implemented, it will also be possible to bestow a unitary effect on a European patent with effect in all participating EU countries.

As is the case in most other countries, the exclusive right of the patent proprietor lasts 20 years from the patent's application date. Under the aforementioned European Regulations on SPCs it is possible to extend this term of protection, but by five years at most. With respect to medicinal products, a further six months' extension is possible if the criteria of Article 36 of Regulation 1901/2006 of December 12 2006 on medicinal products for paediatric use are fulfilled.

Belgian patents and Belgian parts of European patents are obtained by filing an application with the competent patent office, subsequent to which the application will undergo a formal and (for European patents) substantive examination which, if all applicable conditions are met, will lead to the grant of the patent. The patent is renewed by duly paying the yearly renewal fees. In order for a European patent to be validated in Belgium, it is necessary to file a translation of it in one of the national languages (Dutch, French or German) within the three months after publication of the grant of patent.

Provided that all annual fees are paid, a patent will remain in force until the expiry of its term of protection, unless it is renounced by its proprietor, or declared null (partially or completely) by a competent court of law, or revoked partially or completely in an administrative procedure by the European Patent Office (EPO); in respect of European patents, this may also take the form of a centralised limitation.

(b) Infringement

Under Belgian law, a patent can either be directly or indirectly infringed.

With regard to direct infringement, Article 27(1) of the Belgian Patent Act confers on the patent owner the right to prevent all third parties not having his consent from manufacturing, offering for sale, putting on the market or using a product which is the subject matter of the patent, as well as from importing or stocking such product for those purposes. Furthermore, with regard to process patents, the patent owner has the right to prevent any third party from using the patented process and from offering the process for use on Belgian territory. Finally, the patent owner can also prevent third parties from offering for sale, putting on the market, using or importing or stocking for those purposes the product obtained directly by a process which is the subject matter of the patent.

According to the majority of Belgian case law, a patent is considered directly infringed if all essential elements of the adduced patent claim can be found in the allegedly infringing product or process.[2] In the absence of a literal infringement and in line with the Interpretative Protocol for Article 69 of the European Patent Convention, the Belgian courts apply the so-called 'function–way–result' test to determine whether the patent could alternatively be infringed under the doctrine of equivalence.[3]

In addition to acts of direct infringement, a patent can also be indirectly

2 Court of Cassation, February 3 2012, IRDI 2012/3, 259.

infringed. Article 27(2) of the Belgian Patent Act provides that the patentee has the right to oppose the offer or supply, on Belgian territory and to persons other than those entitled to put the patented invention into effect, means relating to an essential element of the invention for use of the patented invention on Belgian territory, if the supplier knows or if it is clear from the circumstances that these means are suitable for putting, and intended to put, the invention into effect.

1.2 Who owns the patent?

According to Article 8 of the Belgian Patent Act, the right to a patent vests in the inventor or his successor in title. If two or more persons have made an invention independently of each other, the right to a patent belongs to the person whose patent application has the earliest date of filing. Under this first-to-file regime, the applicant is assumed to be entitled to the patent. The same applies to the Belgian tier of a European patent, pursuant to Article 60 of the European Patent Convention.

Where a patent application is filed by an unauthorised third party (having misappropriated the invention or otherwise being in breach of a legal or contractual obligation), Article 9 of the Patent Act allows the inventor or his successor-in-title to claim (full or, as the case may be, partial) title to the patent or patent application.

In an employer–employee relationship, Belgian case law roughly distinguishes between three types of inventions when it comes to entitlement to the patent covering that invention. First, there are so-called 'service inventions', which are made by employees in execution of their employment contract and using the means provided by the employer. Service inventions are to be distinguished from 'independent inventions', which are made by employees outside the scope of their employment and using their own means. Inventions which do not qualify under either type, because they are made with the means of the employer or are related to the employment, are labelled 'mixed' or 'dependent' inventions. According to case law and the legal doctrine, the employer has the exclusive right to service inventions, whereas independent inventions vest in the employee.[4] On dependent/mixed inventions, legal doctrine is divided: some argue that the rights belong with the employer, and others favour entitlement by the employee. In older case law, several solutions were deemed possible: joint property (in the case of an intellectual contribution from the employer), employee entitlement with an obligation to compensate the employer, or vice versa.[5]

If an employee-inventor was mainly employed in Belgium, the same rule applies to European patents pursuant to Article 60(1) of the European Patent Convention.

The above is without prejudice to different contractual arrangements made between an employer and an employee in respect of mixed inventions. The same contractual flexibility applies to contracts with self-employed inventors.

---

3 Ghent Court of Appeal, September 26 2011, IRDI 2012/1, 56; Liège Court of Appeal, February 21 2012, IRDI 2012/3, 266.

4 Brussels Labour Court, January 14 2011, JTT 2011, 1098, 217; Brussels Court of First Instance, September 12 2008, 03/7621/A and 04/7215/A, *Arcelormittal Liège Research v Cold Plasma Applications AO*, not reported.

5 Ghent Labour Court, May 18 1987, TGR 1987–88, 89.

Within this context, it is also necessary to take into account the possible incidence of specific regulations of the (private or public) university, institution or public administration where the inventor is active. For example, Article 169*ter* of the Flemish Decree of June 12 1991 on Universities provides that the right over inventions made by salaried academic personnel in the framework of their research activity is vested exclusively in the university and that the same rule applies for non-remunerated researchers (eg, PhD students) if such is stipulated in a written contract.

Finally, co-ownership of patents is governed by Article 43 of the Belgian Patent Act. Absent any differing contractual arrangement, every co-owner has the right to exploit the invention and to take any provisional measures. However, co-owners cannot award licences or start infringement actions without the other co-owners' consent. Unless provided otherwise by contract, co-owners are assumed to hold equal shares in patent ownership. If a co-owner wishes to assign his part of the patent, the other owner has a right of first purchase.

1.3 Inventor compensation

There is no legal right of an inventor to be remunerated for the value of his invention by his employer or, as a matter of fact, by any third party who benefits from it. Given some legal uncertainty, particularly in respect of so-called mixed inventions, this matter is usually dealt with on a contractual basis.

The inventor has a right to be mentioned as such on the patent title (Article 12 of the Belgian Patent Act; respectively Article 62 of the European Patent Convention).

1.4 Know-how and trade secrets

There is no specific legislation in Belgium on the protection of confidential know-how and other trade secrets; hence there is also no uniform legal definition for these concepts. Although this provision has no direct effect in Belgium, reference is usually made to Article 39(2) of the Agreement on Trade-Related Aspects of Intellectual Property Rights (the TRIPS Agreement), which provides for the protection of information that:

- is secret in the sense that it is not, as a body or in the precise configuration and assembly of its components, generally known among or readily accessible to persons within the circles that normally deal with the kind of information in question;
- has commercial value because it is secret; and
- has been subject to reasonable steps under the circumstances, by the person lawfully in control of the information, to keep it secret.

As to know-how in particular, reference is also made to Article 1(i) of Regulation (EC) 772/2004 (providing for a block exemption for technology transfer agreements), where 'know-how' is defined as a package of non-patented practical information, resulting from experience and testing, which is secret, substantial and identified.

At the national level, a distinction must be drawn between manufacturing secrets the illegitimate disclosure of which by employees and former employees is sanctioned by criminal law under Article 309 of the Belgian Criminal Code, and confidential information in general. In this respect, the Belgian Supreme Court has ruled in a 1943 judgment that manufacturing secrets comprise:

> *technical data which, in contributing to the realisation of operations put in place in a factory to obtain a certain product, are of such nature as to provide to the manufacturer technical advantages and which ensure a competitive superiority over his competitors so that the manufacturer obtains an economic benefit by not disclosing the information to his competitors.*[6]

On November 28 2013 the European Commission proposed a draft directive "on the protection of undisclosed know-how and business information (trade secrets) against their unlawful acquisition, use and disclosure".[7] The definition of a 'trade secret' under the proposal is the same as the one given to 'undisclosed information' under Article 39(2) of the TRIPS Agreement. The objective of the proposal is to improve the legal protection of trade secrets against misappropriation, by providing for civil law remedies akin to those foreseen in the Enforcement Directive.

1.5 Employees, consultants and protecting know-how

As trade secrets and know-how are not expressly mentioned as a specific property right in the legislation applicable for Belgium, the baseline for trade secret protection consists of the holder's efforts to keep the information secret. That is not to say that no remedies are available against the misappropriation of know-how or other confidential information.

As mentioned in section 1.4 above, Article 309 of the Belgian Criminal Code provides that it is prohibited for individuals working or having worked in a factory to disclose manufacturing secrets to third parties in a deceitful or malicious way. Furthermore, and regardless of the intellectual property rights which can be claimed in respect of any work or invention by employees, Article 17(3)(a) of the Belgian Act on Employment Agreements prohibits employees from disclosing, either during their employment or after the termination thereof, "trade secrets, business secrets or secrets in respect to personal or confidential matters which [they] may obtain in the framework of [their] employment". Former and current employees should thus refrain from disclosing trade secrets entrusted to them by their employer, including any unpatented inventions.

In all other situations where one cannot rely on a specific statute or regulation, such as in the case of self-employed contractors, it falls to the proprietor of the know-how or confidential information to provide adequate protection for it on a contractual basis. Third parties to whom the information is divulged in breach of the said contractual provisions will only be tortiously liable if they wrongfully interfere in the contractual relationship – that is, where they are aware, or should have been aware, of the contractual obligations of the party disclosing the information and they nevertheless continue their relationship with the latter.

Where the misappropriation of confidential information and the use thereof

6 Court of Cassation, September 27 1943, Pas 1943, I, 358.
7 COM (2013) 813, procedure 2013/0402 (COD).

causes or threatens to cause prejudice to an economic operator, the latter can furthermore rely on Article 95 of the Belgian Trade Practices Act, which provides for a general prohibition of unfair trade practices.

Finally, the Belgian Constitutional Court has acknowledged, in a case concerning access to the file of a government entity which contained confidential information, that the right to protection of confidential business information can also be derived from the fundamental right to privacy – that is, through Article 8 of the European Convention on Human Rights (ECHR).[8]

2. Context and features of patent transactions

2.1 The principal dealings in a patent

(a) Patent assignments

Pursuant to Article 44(2) of the Belgian Patent Act, a patent assignment must necessarily be in writing. Any such assignment has to be notified to the Belgian Intellectual Property Office.

An assignment does not affect the prior rights of third parties such as licensees, except when the ownership of the patent was successfully claimed by means of an entitlement action in accordance with Article 9 of the Belgian Patent Act. In the latter case, Article 10 provides that prior licences lapse by means of registration of the change of ownership. However, prior licensees who have applied the invention on Belgian territory prior to the change (or have made substantial preparations to do so) can seek a licence from the new patentee provided their use or preparations were in good faith. That licence must be granted for a reasonable period of time and on reasonable terms.

The assignment can only have effect in relation to third parties once it has been registered by the Patent Office, except when that third party was aware of the assignment. As mentioned above, if a co-owner wishes to transfer his share in the ownership, his co-owner has a right of first refusal pursuant to Article 43(2) of the Belgian Patent Act.

Apart from the requirement that the assignment be in writing, the contractual freedom of the parties will prevail as to the contents of the agreement and the parties' rights and obligations (eg, in terms of warranties and indemnities granted by either party). In the absence of any specific contractual arrangements, the assignor will be bound by the obligations imposed by the Belgian Civil Code relating to the sale of an asset – in particular, to guarantee the peaceful possession of the title by the assignee.

(b) Patent licences

Pursuant to Article 45(1) of the Belgian Patent Act, a patent licence must be in writing. A licence can be exclusive or non-exclusive. A licence (and any subsequent transfer) must be notified to the Belgian Intellectual Property Office in order to be opposable

8 Constitutional Court, September 19 2007, RABG 2008, 382.

by third parties. Similarly, the transfer of a licence to another party must be in writing and, to be opposable by third parties, must be registered by the same office.

Here too, the contractual freedom of the parties will prevail as to the content of the licence agreement and the extent of the rights and obligations granted to either party. A typical licence agreement will contain provisions on the scope of the licence (exclusive, non-exclusive, or otherwise limited in territory or scope), the term and termination of the licence, the remuneration due to the licensor (upfront fee and/or royalty), the obligations of the licensee to exploit and/or to sub-license the licensed patent(s) and to report to the licensor, the obligations of the licensor in respect of the prosecution and maintenance of the patent(s), the cooperation between parties in a case of infringement by third parties or where infringement proceedings are started against the licensee by a third party, the legal status of improvements made by either parties, etc.

The Belgian Patent Act (Article 52(2)) provides that, unless the parties agree otherwise, only an exclusive licensee can start infringement proceedings, and then only after having (in vain) requested the patentee to do so. However, a licensee (exclusive or not) will be entitled to join infringement proceedings in order to claim its own damages.

2.2 Patent rights in corporate transactions

(a) *Mergers and acquisitions*

Obtaining patent/know-how ownership: In a share deal, a company as a whole is acquired through the purchase of its stock. The documents transferring the stock allow the purchaser to indirectly become the owner of the assets, including patent rights and know-how. The patent rights and know-how remain with the acquired company. In a business sale or purchase (an 'asset deal'), the business is purchased through acquiring its assets individually. An asset deal under Belgian civil law requires that each asset is transferred and that these transfers should be made enforceable against third parties, subject to the conditions imposed by the *lex specialis* applicable to the particular assets – which, as stated in section 2.1 above, for patents is the special rule governing a patent transfer, as provided in Article 44 of the Belgian Patent Act. A transfer agreement must therefore be in written form, on penalty of invalidity (Article 44(2)), and can only be enforced against third parties upon registration of the transfer in the Belgian Patent Register (Article 44(6)). However, prior to registration, the transfer can also be enforced against third parties acquiring rights after, but with knowledge of, the transfer (Article 44(6)).

However, if all the assets can be qualified as a branch of activity or a universality of assets, Article 760 and following of the Belgian Company Code provide a special procedure to avoid fulfilling the transfer formalities for each individual asset. Under this procedure, all assets, including patents and know-how, are transferred automatically by law and are enforceable from the date of the notarial deed of transfer, or beforehand if the third party knew of the transfer prior to publication. Even so, this provision should not prevent the parties from duly registering the transfer in the Belgian Patent Register.

In a case of merger or fusion, Article 683 of the Belgian Company Code requires the observance of special enforceability conditions for patent transfers, being the registration conditions provided in Article 44 of the Patent Act.

Finally, the transfer of know-how in corporate transactions (and the enforceability thereof) is not specifically regulated by Belgian law and will thus be a matter of contract.

Patent/know-how due diligence: In performing a due diligence on the patent portfolio of the transferor, the transferee is advised to obtain a complete list of all the patent rights of the transferor as well as related information covering the following:

- whether the transferor's patent rights concern granted patents or mere applications and accordingly the expectation of a grant;
- their geographical scope of protection, in the sense of the difference between a national, European or international patent and the number of designated states;
- the (remaining) duration of their protection;
- whether opposition is pending to any of the European patents;
- details of any national infringement or invalidity actions filed or announced by third parties against the patents;
- whether European patents are included which benefit from a presumed validity;
- whether all fees have been paid in due time;
- proof of the identity of the patent proprietor;
- details of any agreements with third parties in relation to the patents, such as licence or R&D agreements; and
- whether the transferor owns all of the patents required for the operation of the transferred business (eg, for the production of a certain medicinal product) or, if not, whether the transferor has all the necessary rights to use such required intellectual property, and whether these are fully in force and freely assignable.

The transferee will also need disclosure of the transferor's know-how, as well as a description of every (exclusive) right granted to others in respect of this know-how and any confidentiality agreements (or lack thereof) governing the granting of such rights.

Patent/know-how warranties and indemnities: Through warranties, the transferor is able to reassure the transferee of the 'perfect shape' of his patent rights and know-how that are subject to transfer. As such, he can warrant that he has provided a true and complete disclosure of information on the transferred patent rights/know-how, to be enclosed in a disclosure letter. He can warrant against licence agreements with, or infringement/invalidity proceedings against, third parties. He can also make assurances that he has valid title to all the necessary intellectual property required for the operation of the transferred business. If the transferee succeeds in proving that the opposite is true and has caused him loss after the agreement entered into force, he will be entitled to damages on the basis of a contractual breach of warranty.

Through indemnities, the transferor promises to reimburse fully the transferee for a specific loss, the cause of which is described beforehand – for example, arising from findings from the due diligence. They also enable the reimbursement of losses to occur where a warranty could not give rise to damages because the transferor had disclosed against the warranty.

It is possible to include a *de minimis* provision, reserving title to damages only from a minimum amount of loss upwards and possibly with a cap on maximum damages.

(b) Financing

Relevant exchanges in the jurisdiction: Exchanges in the Belgian jurisdiction are available on the regulated Euronext Brussels market and the Euronext Brussels NV/SA market for derivative products. The available Multilateral Trading Facilities are published on the website of the Authority for Financial Services and Markets,[9] including "Alternext", "Vrije Markt / Marché Libre", "Publieke Veilingen / Ventes Publiques", "Trading Facility", "Easynext", "MTS Belgium", "MTS Denmark" and "MTS Finland".

Preparation and audit of patents prior to flotation: Prior to any flotation, a bidder needs to straighten out its patent assets to ensure the 'good shape' of its patent rights and thereby help to attract future shareholders. A patent audit process enables identification of the strengths and weaknesses of the patent assets, and of the issues which are responsible for adversely affecting their value. This preparation includes enforcing patent rights against licensees through the collection of due royalties.

Listing rules applicable to patents: A prospectus, subject to approval by the Authority for Financial Services and Markets, must include all the information necessary for future shareholders to reach a properly informed decision on the purchase of the business, in accordance with Article 3 of the Royal Decree of October 31 1991 on the Prospectus to be Published in a case of Public Emission of Securities. The decree contains overviews listing the minimum information required in the prospectus, which includes disclosing all the risks of the business, such as the risk of not obtaining the necessary funding to cover the costs of obtaining, maintaining and enforcing the patent rights required for the operation of the business, or the risk of infringing the patent rights of third parties, leading to costly patent litigation (see Scheme A, Chapter IV). The prospectus must also include a complete list and detailed description of all the patent/know-how assets and licence titles (Scheme A, Chapter V).

Ongoing obligations: The ongoing obligations for listed companies are listed in a Royal Decree of November 14 2007. This decree includes periodic information obligations, such as disclosing the annual report in which the business risks are described – which can possibly include the risks related to patent assets.

9 At www.fsma.be.

(c) *Joint ventures*

Depending on the particular transaction for which a joint venture is established, the partaking partners could decide to contribute the required patent rights as they stand into the newly created separate legal entity of the joint venture, instead of making a cash contribution. Another option is to transfer or license the intellectual property to the joint venture separately from the contribution. The transferee or licensee will respectively receive a compensation fee or royalties. When the patent rights are invested as a contribution in, or transferred to, the joint venture, the latter will become the owner of these patent rights. In such a case, this transfer will also have to be registered with the Belgian Patent Register. Unless otherwise agreed, any new rights flowing from these patent rights will then equally be owned by the joint venture.

In contrast, where a partner merely licenses the required patent rights to the joint venture, the latter has a mere user right, and it might be appropriate to regulate the use and ownership of any new rights flowing from adaptations made to the licensed inventions. As the owner of the patent rights is normally responsible for enforcement, it is furthermore appropriate to provide for a contractual framework allowing the joint venture to take action in any case of infringement by third parties.

Upon termination of a joint venture, the licence agreement will as a rule also terminate. Unless it is specifically agreed so beforehand – for example, in return for specific compensation to the other partner – the progress of the winding-up cannot guarantee that the transferred or contributed patent rights automatically return to their former proprietor.

(d) *Divestment*

A patent audit could point out which patents it is financially more attractive to transfer (divest) or to license out. If the patented technology is in high demand in the market, it is clearly better to grant licences, thereby expanding the inflow of royalties. If not, a transfer could be more appropriate, accompanied by sufficient warranties and indemnities limiting the liabilities of the seller in respect of those aspects which are likely to affect the value of the assigned patent rights.

The transfer of know-how needs to be accompanied by contractual non-disclosure obligations, the breach of which gives rise to damages, without prejudice to the application of specific legal provisions as described above.

(e) *Consortium agreements*

Particularly when setting up a consortium, it is of great importance to stipulate as much as possible beforehand, to avoid uncertainties afterwards. When the project of the partners leads to shared results for which a patent application can be filed, unless otherwise agreed the parties principally share the ownership of the forthcoming patent rights with all legal consequences involved (Article 43 of the Belgian Patent Act). However, most often the parties will contractually deviate from this principle – for example, through assigning the patent rights to the party whose core business is most closely related thereto. The parties can then create a patent pool by cross-licensing each other's patents. The party already owning the patents for the technology required to perform the project could also agree to assign or license this technology to the lead exploitation party.

More than with normal patent transactions, there is the risk that the consortium may constitute a cartel prohibited under Article 101 of the Treaty on the Functioning of the European Union. However, under certain conditions, R&D consortiums can fall under the group exemption from such prohibition, as provided for in EU Regulation 1217/2010.

2.3 Particular business transactions

(a) Material transfer agreements

A material transfer agreement (MTA) is an agreement that governs the transfer of physical research material from one entity to another. In addition to certain regulatory requirements that may apply – for example, in the case of access and benefit sharing[10] – such agreements will have to comply with the above-mentioned legal provisions on the assignment of patents to the extent that certain rights are transferred from one entity to the other.

(b) Commercial research and development agreements

Research and development (R&D) contracts are not regulated by law in Belgium and their features thus mainly depend on the parties' agreement. For pre-existing patent rights, the parties will grant limited cross-licences to each other. Patent rights resulting from an R&D project are usually shared property of all the parties who substantially contributed to the creation of the invention. However, the parties could – and most often will – contractually stipulate otherwise, assigning the ownership to one party and granting licences to the other. As mentioned, EU Regulation 1217/2010 governs the competition exemption for such agreements resulting in cartels.

(c) Contract manufacturing agreements

The party launching a medicinal product onto the market could engage another party for manufacturing this product (often in lower-cost countries). Either the manufacturer owns all the required intellectual property and know-how for manufacturing, so that a mere purchase or distribution contract will be put in place with the party launching the product, or the latter owns the required intellectual property and provides the appropriate licences to the manufacturer. In view of the tax benefit possibly resulting from royalties, as discussed below, a Belgium based licensor would be well advised to charge royalties to the manufacturer.

(d) Product distribution agreements

Article 28(2) of the Belgian Patent Act provides for the exhaustion of patent rights in respect of any product put on the Belgian market by the patent owner or with his assent. The principle also applies to any products marketed in the European

10 See, eg, Article 12(4) of the International Treaty on Plant Genetic Resources for Food and Agriculture.

Economic Area (ie, the European Union and the European Free Trade Area (EFTA) combined). Nevertheless, when patented products are distributed, it is appropriate to provide for a contractual licence for the distributor in respect of those additional acts it is entitled to carry out as a distributor.

(e) ***Co-promotion agreements***

When two or more companies decide to promote a given product jointly under a given brand, they may conclude a co-promotion agreement. Co-promotion agreements relating to patented products equally must include the necessary cross-licences in order to allow the parties to carry out their contractual duties.

(f) ***Clinical trial agreements***

The organisation of clinical trials in Belgium is governed by the Law of May 7 2004 on Experiments on Human Persons. Where the trials relate to as yet unpatented subject matter, it is necessary to provide for sufficient contractual safeguards so as to ensure confidentiality in relation to the participating physicians and patients.

Apart from the exemption discussed in section 3.1 below, it is generally considered by legal doctrine (in the absence of case law) that Phase I and II, and even Phase III, clinical trials will benefit from the 'experimental use' exemption under Article 28(1)(b) of the Belgian Patent Act. In accordance with this provision, any acts carried out on or with the patented subject matter are exempted when they are performed for scientific purposes. The fact that the entity involved is a commercial undertaking is in itself not sufficient to bring the relevant activities outside of the scope of the exemption.

3. Issues associated with patent transactions

3.1 Regulatory law and compliance

The commercialisation of medicinal products in Belgium is governed by the Medicinal Products Law of 25 March 1964, which implements EU Directive 2001/83 concerning the Community Code for Medicinal Products for Human Use. Article 6*bis* of this law contains provisions on marketing authorisations and data exclusivity which are essentially similar to those set out in the directive. It is noteworthy in this respect that the so-called 'Bolar' exemption is limited in Belgium to generic medicinal products, products with well-established medicinal use (applied through the so-called 'bibliographic procedure') and so-called 'biosimilars'.

It is a generally accepted principle that any regulatory decision in respect of market authorisation, pricing or reimbursement is without prejudice to any intellectual property rights. However, Belgian social security legislation provides for exemptions from a number of mandatory price decreases for those medicinal products for which the active ingredient(s) are still patent-protected.

3.2 Commonly disputed matters

Disputes in relation to patent transactions seldom reach the Belgian courts and even less often result in published decisions. Only the Commercial Courts located in the district of one of the five Courts of Appeal are competent to decide on disputes relating to patents, including disputes on patent licences or assignments. In recent times, there have been various disputes in respect of patent licences, especially inventions by employees.[11] In respect of patent licences, the disputes usually relate to the payment of royalties (see section 3.7 below for a recent example). In the case of employee inventions, the disputes are mostly about the extent to which rights have been transferred and the amount of compensation that an employee should get from his employer for his invention.

3.3 Competition rules

Competition law – whether Belgian or European – cannot affect the patent in itself if it has been lawfully obtained, nor the mere enforcement thereof. However, the exercise of such a right may be considered as illegal if the applicable conditions under competition law are met.

In this respect, Article IV.4 of the Belgian Code of Economic Law confirms that the prohibition on restrictive practices contained in Article IV.1(1) (the Belgian equivalent of Article 101(1) of the Treaty on the Functioning of the European Union) is not applicable to agreements that benefit from an EU block exemption. Moreover, it extends the effect of the EU block exemptions to situations where trade between member states is not affected, but which would benefit from an exemption where the trade between the member states would have been affected.

Article IV.5 of the Code of Economic Law provides that "Belgian" block exemptions can be adopted by Royal Decree, but so far no such Belgian block exemption regulation has been issued.

3.4 Securitisation of patent rights

A securitisation is a type of structured financing in which a pool of assets is transferred to a special purpose vehicle (SPV) that issues securities. The securities which are issued can be debt instruments, equity instruments or a combination of both. The issued securities will provide cash to the SPV in order to pay for the assets transferred to the SPV. This technique allows the owner of assets to receive cash for its assets which would otherwise not yield any income. If the arrangement is structured properly, the initial owner of the assets can ensure that he keeps the level of control or use of the assets that he deems appropriate for his purposes. If, for instance, he wants to use patents for this purpose, he can do so, if organised properly; but he will have to remunerate the SPV for its services, possibly in the form of a licence fee. The SPV can use this money to pay interest or a dividend to the securities' holders.

The securitisation technique can be used for patents. If patents are used for securitisation, each patent individually must be transferred to the SPV. All formalities for such a transfer will have to be complied with, as described earlier. This shows

11 Court of Cassation, October 18 2013, C.11.0719.F/2, *Meura, Boccard v Berewtec, Anheuser Busch Inbev*; Brussels Labour Court, January 14 2011, JTT 2011, afl 1098, 217; Ghent Labour Court, November 16 2009, TGR-TWVR 2010, 134.

immediately the downside of patent securitisation: these deals are relatively cumbersome, fees for banks and other service providers will have to be paid, and there are the registration costs related to the transfer of each patent. Also, a contractual arrangement will have to be put in place so as to allow the initial owner to continue to exploit the patents and to generally provide for the further administration and enforcement of the patents.

3.5 The effect of insolvency

In principle, a party's insolvency does not automatically end its (licence) contracts. Two exceptions apply, however. First, the parties can include a contractual clause stipulating that the contract does come to an end as a result of a party's bankruptcy. Secondly, contracts that have been entered into because of the counterparty's identity and/or specific qualities (so-called '*in tuitu personae* contracts') automatically come to an end if the counterparty goes bankrupt. Whether or not a specific licence contract is seen as an *in tuitu personae* contract will have to be judged on a case-by-case basis.

Apart from these two exceptions, a bankruptcy receiver can in principle decide to stop fulfilling the bankrupt company's duties under the contract. As a result, the counterparty can claim that the contract has to be terminated and can introduce its (possible) claim for damages in the bankruptcy.

If a patent holder has been declared bankrupt, the bankruptcy receiver will normally try to sell the patent on the market. The question is nonetheless debated as to whether the receiver can terminate a licence agreement with a view thereto. Some doctrine and jurisprudence state that the receiver cannot end the licence agreement merely because the patent is more valuable or more easily transferrable without the licence agreement.

If the licensee has been declared bankrupt, the bankruptcy receiver will most probably decide to stop executing the contract. If the licence agreement is continued – for example because the bankruptcy receiver wishes to continue the business – the licence fees that become due after the bankruptcy will have to be paid by the bankruptcy receiver in primary order.

3.6 Taxation

In 2007, Belgium introduced the patent income deduction (PID), which allows an 80% tax deduction of qualifying patent income (Articles 205/1–205/4 and 236*bis* of the Income Tax Code). Hence, qualifying patent income is currently taxed at an effective tax rate of only 6.8%. The deduction is available for Belgian companies or Belgian branches of foreign companies. Income resulting from the sale or transfer of co-ownership of patent rights is excluded from the deduction.

The PID covers patents (as defined under Belgian and international intellectual property law) or SPCs. There are two possibilities:

- The patented product or process is developed by a Belgian company or branch. In that case, the patent has to be the result of the Belgian company's own R&D activities in a research and development centre in Belgium or abroad (but belonging to the Belgian company or branch); or

- The patent is acquired or licensed from another party and is further developed by the Belgian company or branch. The Income Tax Code requires the further development of the patented products or processes in an R&D centre of the company or the branch in Belgium or abroad, either fully or partly. It is not necessary that this further development leads to a new patent. If the company has not developed the patent itself but has acquired or licensed the patent from a third party (and further improved the patented invention), depreciation on the patent(s) acquired or payments made to other licensors for licensed patents must be deducted from the basis of the PID. The R&D expenses incurred in relation to the further development of the licensed patents are also deducted from the PID basis. In cases of improved patents, the deduction is limited to the added value resulting from the said improvements.

The company or branch can use the patent for its own manufacturing (even through a third party toll manufacturer) or its own services, or it can license the patent to other parties. The calculation of the PID is, however, different in each case. In the first instance, the PID is calculated on a 'deemed royalty' basis – that is, an amount equal to the at arm's length royalty that the company or branch would receive from an unrelated party using the patent for its activities. In the second instance, the PID amounts to 80% of the patent income received (eg, royalties). The income has to be (re)calculated at arm's length.

To benefit from the PID, the company or establishment has to demonstrate the existence of an R&D centre which qualifies as a 'branch of activity'. This means that the R&D centre should be an entity or a division of an entity that is able to operate autonomously. As from the tax year 2013, this condition does not apply to small and medium-sized enterprises (to be evaluated on a consolidated basis).

3.7 Hot topic

There is just one 'hot topic' that should be mentioned in relation to patent law in Belgium. On June 14 2012, the Belgian Court of Cassation issued a ruling which opens the door to claims for the retroactive reimbursement of already paid licence fees if the licensed patent is amended or revoked. The decision was rendered in the context of an annulment appeal against an arbitral ruling, but it has been written in such broad terms that it seems to affect substantive patent law more generally.

At the heart of the court's reasoning was that when a patent is limited following a decision from the EPO's Technical Board of Appeal, this limitation has retroactive effect (Article 68 of the European Patent Convention). In the case at hand, the licensee had already – well before the limitation of the patent by the EPO – argued that some of his activities were not covered by the licensed patent. On that ground, he refused to pay the contractually agreed licence fee. As a result of this dispute, the licensor terminated the licence agreement. Sometime thereafter, the EPO's Technical Board limited the previously licensed patent. The parties then ended up in arbitration proceedings about whether or not the licence fee that the licensee had previously refused to pay was still due.

The Court of Cassation sided with the licensee and found that, because of the effect of Article 68 of the European Patent Convention, the licensee should no longer pay the earlier agreed-upon licence fees. The court thereby disregarded a provision in the Belgian Patent Act which states that the (partial or entire) annulment of a patent does not (normally) affect already executed agreements. This reasoning of the court is based on the premise not only that Article 68 actually governs the legal effects of the retroactivity it prescribes, but also that there is a conflict between Article 68 and this national (Belgian) provision. In our opinion, neither of these assumptions is correct. By not recognising this, it seems that the court has created a dangerous precedent. Not only does it allow licensees to stop paying contractually agreed licence fees whenever the licensed patent is revoked or amended to such an extent that the licensed product is no longer covered by that patent, but it might also very well – and on the basis of the same arguments – be used to claim reimbursement of already paid licence fees without any margin of discretion for the court adjudicating the matter. That would, of course, create great legal uncertainty among licensors.

Canada

Noel Courage
Bereskin & Parr LLP
Rory Cattanach
Kevin Fritz
Wildeboer Dellelce LLP

1. Patents, know-how and trade secrets: foundations for transactions

1.1 Patents and supplementary protection certificates

(a) Overview

Canada has a utility patent to protect inventions such as machines, articles of manufacture, processes and compositions of matter. There is no utility model or innovation patent. There is also no design patent to protect shape and appearance, but there is a similar right, called an industrial design registration. Canadian patents provide the exclusive right to make, construct, use and sell an invention.

A patent has a maximum term of 20 years from filing date.[1] There is no patent term extension at present, but a maximum two-year extension has been proposed for the future in the Canada–Europe free trade agreement signed in 2013. The extension would be based upon Canadian regulatory delay in approving the drug.

Maintenance fees are payable annually from the second anniversary of the patent application filing date. It is important to ensure that maintenance fees have been paid up to date when doing due diligence for a transaction.

The Canadian approach to patent claim construction is based on the UK purposive construction approach.[2]

Novelty and inventiveness law is set by the Patent Act (Canada) and has recently been considered by the Supreme Court of Canada.[3] There is a one-year grace period to file a patent application in Canada after a public disclosure by the inventor, or by someone who obtained knowledge of the invention from the inventor. Canada requires absolute novelty with respect to third-party disclosures.

An inventor must also be able to demonstrate utility of the invention or show a sound prediction of utility.[4] In recent years, courts have begun to focus on what is promised by the patent and whether the promises are fulfilled (the 'promise doctrine'). This has been contentious and created a fair amount of uncertainty as a result of invalidation of some drug patents for failure to fulfil the promise of the patent.

---

1 Patents filed before October 1 1989 have a patent term of the longer of 20 years from filing date or 17 years from issuance: Patent Act, RSC 1985, c P-4.
2 *Free World Trust v Électro Santé Inc* 2000 SCC 66, [2000] 2 SCR 1024; *Whirlpool Corp v Camco Inc*, 2000 SCC 67, [2000] 2 SCR 1067.
3 *Apotex Inc v Sanofi-Synthelabo Canada Inc* 2008 SCC 61, [2008] 3 SCR 265.
4 *Apotex Inc v. Wellcome Foundation Ltd* 2002 SCC 77, [2002] 4 SCR 153.

There is some controversy in Canada over patentable subject matter – in particular, diagnostic inventions that allegedly represent mere natural correlations, and drug dosing regimens (eg, use of X mg of drug per kg body weight per day). Methods of medical treatment are not patentable,[5] but medical 'use' claims are patentable. Isolated unicellular organisms are patentable, but transgenic higher life forms (genetically modified plants and animals) are not.[6] Generally, transgenic higher life forms are protected by patenting the methods and reagents used to make them, as well as methods of use of the higher life forms.

There is no duty to disclose prior art in the Canadian Patent Office, unlike in the United States.[7] Providing prior art to a Canadian Examiner may, however, facilitate prosecution in a case of patent infringement. Nor is there any US-style continuation-in-part patent practice in Canada. Voluntary divisional applications should only be filed for claims that are novel and inventive over the patent, in view of the risk of double patenting.

Many applicants avoid claiming 'small entity' status when filing patent applications, because of the risks of invalidity if fees are paid at the wrong rate.[8]

(b) ***Litigation***

Canada does not have a specialised patent court. Most patent lawsuits are brought by way of action in the Federal Court of Canada.

A patent is presumed valid. The presumption is not strong, since all that is required is some evidence to rebut the presumption and put the validity in play. The onus is on the party trying to invalidate the patent to prove, on a balance of probabilities, that the patent is invalid.

An infringer is liable to pay reasonable compensation for its activities before patent grant.[9] After patent grant, the infringer is liable for either damages or an accounting of profits (ie, disgorgement of infringer's profits). The accounting of profits is an unusual remedy in that the patent owner can take the infringer's profits arising from the infringement, regardless of whether the patent owners suffered any damages. There are no treble damages for wilful infringement. Punitive damages may be requested, but are infrequently awarded and are relatively insubstantial compared with the United States. There is no file wrapper estoppel in Canada. File wrappers may not be used for claim construction during litigation. There is also no doctrine of equivalents in Canada. Fraud on the Patent Office has had very limited viability to date as a ground to challenge a patent.

1.2 Who owns the patent?

Ownership is derived from inventorship. Inventorship is determined by reference to conception and the reduction of the invention to a practical shape.[10] Persons testing

---

5 *Tennessee Eastman Co v Commissioner of Patents* [1974] SCR 111.

6 *Harvard College v Canada (Commissioner of Patents)* 2002 SCC 76, [2002] 4 SCR 45; *Canadian Intellectual Property Office, Manual of Patent Office Practice*, c 12.05.03, 17.02.01a.

7 *Corlac Inc v Weatherford Canada Ltd* 2011 FCA 228.

8 Claiming small entity status refers to paying reduced government fees for small organizations. For the risks, see: *Barton No-Till Disk Inc v Dutch Industries Ltd* 2003 FCA 121.

9 *Baker Petrolite Corp v Canwell Enviro-Industries Ltd* 2002 FCA 158.

inventions are not necessarily inventors. The failure to mention a co-inventor in a patent application is typically not considered an untrue 'material allegation' resulting in a patent's invalidity.[11]

In employment arrangements, the employee will generally own the invention unless hired to invent.[12] Employment contracts can ensure employer ownership. Typically, the contract should be in place before the employee begins work, so as to ensure that it is enforceable. Past consideration is a concern when assignment of invention clauses are signed after employment begins.

Independent contractors will generally own an invention, irrespective of whether they were hired to invent. Companies should have intellectual property (IP) ownership agreements with contractors.

1.3 Inventor compensation

There is no statutory obligation for an employee to be compensated by his employer for the value of his invention.

1.4 Know-how and trade secrets

Most intellectual property in Canada, such as patents, trademarks and copyright, arises from federal statutes. Trade secrets (confidential information) fall under provincial jurisdiction. Canadian provinces have not enacted trade secrets legislation; this means that trade secrets law is a patchwork of common law court decisions in most provinces, with occasional legislation that touches on trade secrets. Quebec has basic provisions on trade secrets in its civil code.[13]

Canada has federal and provincial access-to-information laws that permit third parties to access limited contents of a new drug submission (NDS) file.[14] These laws do not, however, permit disclosure of confidential information to third parties. In this context, 'confidential information' is defined as information that is not generally known to the public, that confers an economic benefit on the owner, and that has had reasonable efforts made to maintain its secrecy.

Documents containing confidential information must be redacted by Health Canada or held back in their entirety.[15] Health Canada notifies the brand name company prior to disclosing NDS contents.

Remedies for trade secret misappropriation include breach of confidence and, potentially, breach of contract.[16]

1.5 Employees, consultants and protecting know-how

In an employment context, trade secrets are typically protected by contractual

---

10 *Supra*, note 4.

11 *Ibid.*

12 *Techform Products v Wolda* (2001), 56 OR (3d) 1 (CA).

13 Civil Code of Québec, LRQ, c 64, 1991, s 1457, 1458, 1612.

14 Access to Information Act, RSC 1985, c A-1; Freedom of Information and Protection of Privacy Act, RSO 1990, c F31.

15 *Merck Frosst Canada Ltd v Canada (Health)* 2012 SCC 3, [2012] 1 SCR 23.

16 *Lac Minerals Ltd v International Corona Resources Ltd* [1989] 2 SCR 574; *Cadbury Schweppes Inc v FBI Foods Ltd* [1999] 1 SCR 142.

confidentiality obligations. Restrictive covenants (non-compete or non-solicit agreements) may be used. Their scope (time, geography etc) must be reasonable and limited because courts often will not enforce clauses that unreasonably restrain a former employee from earning a living.[17] In some cases, higher-level employees may be subject to fiduciary duties that continue to exist after employment has ended. In company–company dealings, confidentiality agreements and restrictive covenants are common tools to protect trade secrets.

These laws vary from province to province. For example, in Ontario, a non-competition contract clause will often not be enforceable, since a non-solicitation clause is considered sufficient to protect the employer.[18]

2. Context and features of patent transactions

2.1 The principal dealings in a patent

(a) *Patent assignments*

Assignments of inventions typically follow the form used in other countries.

Canada's Patent Act requires registration of assignments and exclusive licences.[19] The Canadian Patent Office does not requisition assignments as a condition of receiving the patent, so assignments are often treated as optional by applicants.

Registration is recommended in order to have priority over subsequent purchasers of rights in the patent application that may not have notice of the earlier assignment. To be registrable at the Canadian Patent Office, an assignment should cover Canadian rights and be evidenced – for example, by a witness signature.

For a co-owned patent, either co-owner may freely assign its interest in the patent without obligation to, or the consent of, the other.[20]

(b) *Patent licences*

Canadian licences can give the right to make, construct, use and sell a patented invention. The basic legal terms in a Canadian licence, including the various approaches to royalties, are typically similar to that of a licence used in other common law countries.

Co-owned patents may present special licensing issues. For a co-owned patent, a co-owner may not license its interest in the patent to a third party without the consent of the other co-owner,[21] and such a licence will be invalid as diluting the rights of the co-owner. A co-owner may otherwise exploit the patented invention freely of the other.

---

17 *Woodward v Stelco Inc* (1998), 80 CPR (3d) 319 (ON CA).
18 *Ibid.*
19 Patent Act, RSC 1985, c P-4, s 51.
20 *Forget v Specialty Tools of Canada Inc* (1995), 62 CPR (3d) 537 (BCCA.).
21 *Ibid.*

2.2 Patent rights in corporate transactions

(a) *Mergers and acquisitions*

Share purchase and asset purchase: A critical decision in the buying or selling of any business is whether the transaction should be structured as a share purchase or an asset purchase. This may be a significant consideration where the purchaser's interests are focused primarily on the vendor's patents or other intellectual property.

While share purchases are commonly more favourable to the vendor from a tax perspective in Canada, extensive vendor representations and warranties and due diligence are often required. Since asset purchases involve a more focused acquisition, it follows that the required representations and warranties will be circumscribed accordingly.

Due diligence: Regardless of the structure, effective due diligence should consider the two key issues of ownership and validity.

A prudent purchaser will seek to verify the ownership of any patent prior to closing the transaction. A search of the registered title documents on file with the Canadian Patent Office should be done on every patent included in the purchase transaction. This search should also identify any security interests registered against a patent.

The purchaser should also conduct searches against the vendor under the Personal Property Security Act (or its provincial equivalent) in the province where the vendor's principal place of business is located, as well as in each province where it carries on business. A registration under the Personal Property Security Act against the vendor could impact on its ability to transfer good title to its assets to the purchaser. In an asset transaction, the purchaser should ask, from any entity (a 'registrant') which has a registration against the vendor under the Personal Property Security Act, for a waiver that releases the assets being purchased from any claims by that registrant. A notification of discharge of the security interest should be submitted to the Canadian Intellectual Property Office (CIPO) prior to closing the transaction, for any security interests registered at CIPO against any patents being assigned. In the case of a share purchase transaction, a purchaser would typically require that all registrations under the Personal Property Security Act or at CIPO be discharged prior to closing. It is also important, when representing a purchaser, to ensure that unqualified representations and warranties are obtained from the vendor on ownership, and that there is clear title to the patents in question.

Furthermore, a purchaser should confirm in its due diligence process that the patent asset is still in force. A search at CIPO will reveal whether a patent has been invalidated or has lapsed; legal databases of court judgments may also be searched. Situations that could impact validity include (but are not limited to) a court judgment rendering the patent invalid, or a failure to pay the required annual maintenance fees. Reviewing the file history will also identify any reissue, re-examination or disclaimer of the patent.

Representations, warranties and indemnities: A purchase agreement involving a business with patents and other intellectual property will usually have extensive representations and warranties on the IP. A vendor would typically represent and warrant that:

- each person who has worked on the IP has validly assigned all rights to the vendor;
- all patent applications have been filed and are in good standing;
- all IP owned by the vendor is valid, subsisting and enforceable;
- the vendor is the sole and exclusive owner of the IP and it is free and clear of all encumbrances;
- the company has complied with all legal requirements with respect to its patents; and
- none of its patents has been involved in an interference, reissue, re-examination or opposition proceeding or other proceeding challenging the validity or enforceability of the patent.

Typically, one of the most contentious representations is one which states that the patent does not violate or infringe upon the rights of any third party. It is virtually impossible for a vendor to know for sure if this statement is absolutely true. Thus, the vendor will want to qualify this representation (along with certain other representations and warranties) with "to the best of the vendor's knowledge", thus transferring some of the risk to the buyer. With this qualification, if a third party comes forward after closing the transaction with an infringement claim of which the vendor had no knowledge, the purchaser will suffer the loss as opposed to being able to recover any losses from the vendor. It is important in the purchase-and-sale agreement, therefore, to define exactly what "to the best of the vendor's knowledge" actually means – that is, to identify in the agreement the individuals or positions who/which constitute the vendor's knowledge.

An often contentious section in a share or asset purchase agreement is the indemnity section. An indemnity acts as a contractual remedy that provides recourse for the indemnified party upon the occurrence of certain specified events, such as a breach of a party's representations, warranties or covenants. Although many representations and warranties in Canada only survive the consummation of the transaction for 12 to 24 months (which means that the purchaser could only sue for a breach under the vendor's indemnity for as long as the representations and warranties survive), a purchaser will often negotiate either a longer survival period for the indemnities pertaining to the vendor's intellectual property or an indefinite survival period. Likewise, although a vendor will often be successful in limiting the amount of damages that a purchaser can claim in the event that it suffers damages as a result of a breach of a representation or warranty to an amount less than the purchase price, the purchaser will often not accept any limitation on the amount of damages that can be claimed in such circumstances.

(b) Financing

All companies that intend to issue their securities to raise money in Canada will be subject to the securities laws of the provinces within which their investors reside, as well as the rules and regulations of the stock exchange upon which their securities are listed, if any. Generally speaking, a company issuing securities must do so pursuant to a prospectus filed with one or more provincial securities regulatory authorities. However, in certain prescribed circumstances, a company wanting to issue securities may rely on one of the various available exemptions from the prospectus requirements. Such a distribution is commonly known as a 'private placement'.

National Instrument 45-106 of the Canadian Securities Administrators outlines the various exemptions from the prospectus requirements under which a company may issue securities. The most common exemptions relied upon are:

- the accredited investor exemption (can only issue securities to sophisticated investors);
- the private issuer exemption (cannot be listed on a stock exchange and can only issue securities to family, friends, business associates, etc); and
- the minimum investment exemption (the investor must invest a minimum of C$150,000, thereby indicating a certain level of investor sophistication).

It should be noted that securities regulators in Canada are currently reviewing these exemptions.

Regardless of whether a company decides to raise money by way of public offering or private placement, due diligence is an important consideration in any financing. An agent or investor will conduct due diligence on all aspects of the company, including its patents, patent applications and other IP. Aside from conducting searches regarding ownership, validity and non-infringement (freedom-to-operate) of patents, investors may request a legal opinion from the company's counsel attesting to such.

Listing rules: In Canada, most public companies either trade on the Toronto Stock Exchange (TSX) or the TSX Venture Exchange (TSXV). The difference lies in the growth stage of the listed company. The TSX is a senior equity exchange, whereas the TSXV is a venture capital marketplace for emerging companies. When undertaking a financing, a company will have to be sure to comply with the rules of the stock exchange upon which its shares are listed.

Many technology companies will initially seek a listing on the TSXV. To be listed on the TSXV as a Tier 1 (the TSXV's premier tier, reserved for its most advanced issuers) technology company, the company must have either C$5 million of net tangible assets or C$5 million in revenue. If the company does not generate revenue, a two-year management plan indicating there is a reasonable likelihood of generating revenue within two years may be an acceptable alternative. In order to be listed as a Tier 2 technology company, the company must have either C$750,000 of net tangible assets, C$500,000 in revenue or C$2 million in arm's length financing. Similarly, if no revenue exists, a two-year management plan, as described above, may suffice. The TSXV has additional listing requirements that must be met at the time of listing.

In order to list on the TSX, Canada's senior stock exchange, the requirements are even more substantial, including a minimum of C$10 million in cash (a majority of which has been raised under a prospectus offering) and a market capitalisation of at least C$50 million.

Disclosure requirements: A company that sells securities to the public by way of a prospectus offering will be required to file either a long-form or short-form prospectus, depending on whether the company is already a reporting issuer (ie, a publicly traded company). If the company is not a reporting issuer and is undertaking an initial public offering, a long-form prospectus will be required. Conversely, if a company is already a reporting issuer in a Canadian jurisdiction, it may be eligible to file a short-form prospectus. Regardless of form, a prospectus must provide "full, true and plain disclosure of all material facts relating to the company and the securities which are proposed to be distributed".[22] This would include all material facts related to the company's patents and other IP.

A company which issues securities in reliance on a prospectus exemption is not required to have a disclosure document like a prospectus that would require full, true and plain disclosure of all material facts. Indeed, companies in Canada frequently complete significant private placements without using an offering memorandum or other disclosure document. However, a Canadian company will, at a minimum, enter into a subscription agreement with the purchaser, which usually includes extensive representations and warranties on the company's business and assets, including, if applicable, its patents and other IP. Representations and warranties given by a company on its patents and IP would be very similar to those given in a share or asset purchase agreement, as set out in section 2.2(a) above.

Ongoing disclosure: In addition to the initial offering disclosure requirements, a public company in Canada (ie, a reporting issuer in one or more Canadian provinces) will be subject to ongoing disclosure obligations. Reporting issuers in Canada are required to file annual and interim financial statements with the provincial securities regulatory authorities, and these must be accompanied by a discussion and analysis by the company's management of the results and a certification of the disclosure by the company's chief executive officer and chief financial officer. In addition, companies listed on the TSX must file an annual information form (not applicable to companies listed on the TSXV) which must provide extensive disclosure on the company and its business.

A reporting issuer in Canada is also required to disclose all material changes to its business and operations. The Securities Act (Ontario) defines a 'material change' as a change in the business, operations or capital of the issuer that would reasonably be expected to have a significant effect on the market price or value of any of the securities of the issuer and includes a decision to implement such a change made by the board of directors of the issuer or by senior management of the issuer who believe that confirmation of the decision by the board of directors is probable.

The TSX and the TSXV require that a listed issuer disclose 'material information' concerning its business affairs forthwith upon the information becoming known to

22 National Instrument 41-101 – General Prospectus Requirements.

management or, in the case of information previously known, forthwith upon it becoming apparent that the information is material. 'Material information' is stated to be any information relating to the business and affairs of a company that results, or would reasonably be expected to result, in a significant change in the market price or value of the listed securities of the company. Material information consists of both material facts and material changes relating to the business and affairs of the listed company. The TSX and TSXV thus consider that 'material information' is a broader term than 'material change' since it encompasses material facts that may not entail a material change as defined in the securities laws of Ontario. Consequently, a reporting issuer will be required to disclose new information, or changes associated with its patent assets or its IP, if such information or change would be considered material.

A privately held company in Canada is not subject to the same disclosure requirements as public companies. The company's incorporating statute would typically require it to provide its shareholders with annual audited financial statements unless the company's shareholders unanimously agree to waive the audit requirement. Other than that, a private company in Canada would only be required to disclose matters that it has a contractual obligation with its shareholders to disclose, typically under a shareholders agreement or a subscription agreement.

(c) Joint ventures

Companies may enter a joint venture agreement or incorporate a joint venture company. In either situation, intellectual property considerations could be a key factor in the negotiations. Issues that the parties might have to address in a joint venture agreement include:

- the division of responsibilities, contributions and ownership;
- the development of intellectual property for use in the joint venture; and
- ownership and use of the developed IP.

Specific intellectual property issues for joint ventures could include those relating to existing patents and single-party inventions, as described next.

If one company contributes technology with an existing patent to the joint venture, the company may want to ensure that it maintains its rights and ownership in the patent rather than assigning the patent to the joint venture. The company could grant the joint venture company a licence to use the technology for the benefit of the joint venture. The specific terms of the licence would have to be negotiated by the parties. For example, a licence may be exclusive or non-exclusive, be for a specific use and/or time period or be applicable only in a specific territory. The licensing party may want to retain the right to use the IP for research or other purposes, provided such uses do not compete with the business of the joint venture. In addition, the parties should provide for what happens to the patents or other IP if the joint venture is terminated. Most likely scenarios would have the patent and other IP going back to the party who originally contributed it to the joint venture, potentially including related IP in improvements.

It is not uncommon for companies involved in a joint venture to apportion various tasks to different parties. If new intellectual property is to be developed by just one of the joint venture partners, the joint venture agreement should set out who owns the resulting IP. This is especially true if the invention can be adapted to a different venture or project. Typically, the joint venture would either own the single-party IP or have an exclusive licence to exploit it. The developing party may have the right to use the IP for research purpose or for other purposes that do not compete with the business of the joint venture. In the alternative, both parties may have the right to use the IP for non-commercial purposes.

(d) Divestment

Companies may wish to divest themselves of their patent assets for a variety of reasons. While patent assets can be a source of significant value, the associated costs might outweigh the benefits for a company. In other instances the intellectual property underlying a patent may no longer be a core part of the company's business, resulting in a decision to divest. Moreover, a research and development (R&D) company will often develop IP which it will then sell or license to a third party to commercialise in exchange for a one-time payment and/or an ongoing royalty.

If a company elects to divest certain patent assets, it must determine whether to do so by way of assignment or licence. Licensing allows the licensor to leverage the value of the patent assets and is usually favoured by companies which would like to maintain their intellectual property portfolios. A licensed patent is capable of generating revenue through one-time licence fees and associated royalties but carries with it some risks, including:

- the exposure to potential litigation from end users of the technology;
- the negative effect on the value of the patent that a licence agreement may have; and
- changes in market conditions that may affect the value of the patents.

Both these divestment options could require extensive representations and warranties similar to those discussed above, and thorough due diligence by the purchaser or licensor as the case may be.

2.3 Particular business transactions

Canada's laws tend to be similar to other major industrialised nations with respect to basic patent agreements.

There is no Canadian counterpart to the US Bayh–Doyle Act, which gives universities ownership of inventions.[23] At Canadian universities, one may have to consider a patchwork of inventions policies, common law, employment contracts and collective agreements.

For clinical trial agreements, the Canadian Institutes of Health Research (CIHR) is developing a model clinical trial template agreement (mCTA) in an effort to simplify and expedite negotiations between sponsors, trial sites and principal investigators for Phase II and III single and multisite studies.

---

23 An Act to Amend the Patent and Trademark Laws, Pub L 96-517, codified at 35 USC §§ 200–212 [the Bayh–Dole Act].

3. Issues associated with patent transactions

3.1 Regulatory law and compliance

Health Canada is the primary government regulator for drug approvals. The marketing authorisation for a drug is called a Notice of Compliance (NOC).[24]

Canada has specified a period of eight years of data protection for innovator drug submissions.[25] Eligibility of a drug for data protection is decided by Health Canada upon review of the drug submission. Protection is limited to drugs that qualify as an 'innovative medicine', which is one that has not previously received an NOC. Minor changes to a previously approved drug do not qualify for additional terms of data protection, such as a new formulation, polymorph, salt or enantiomer.

The eight years of data protection operates as follows. For the first six years after an innovative drug receives its approval, a generic company may not file an abbreviated new drug submission (ANDS) that refers to an innovative medicine. Generic companies filing an ANDS after six years can have Health Canada review safety and efficacy but may not receive an NOC until after the eight-year period expires.

Canada has regulations linking the issuance of an NOC to a review of patent issues, namely the Patented Medicines (Notice of Compliance) Regulations.[26] Conventional Canadian patent litigation typically involves a brand name company suing a generic company for patent infringement after the generic product has been launched. The linkage regulations can trigger a process that allows a brand name company to apply to the court for an order prohibiting issuance of an NOC to the generic company until the patent expires. In order to use this procedure, a brand name company must first have an eligible product, formulation or use patent that is directly pertinent to the approved product. The patent must then be submitted to Health Canada within strict time limits, along with a request that it be added to the Health Canada Patent Register. If the patent is accepted by Health Canada for the register, any generic company ANDS relying on that brand name company's approved drug as a reference product must be accompanied by a notification to the brand name company. The notification document is called a Notice of Allegation (NOA) and it addresses patent issues for the relevant patent on the Health Canada Patent Register. If the generic company wants to get its marketing authorisation before the patent expires, the NOA will assert that the generic company is not infringing the patent or that the patent is invalid. The brand name company then has a limited time period to apply to the court for a prohibition order, trying to block the generic company from receiving its marketing authorisation. The court proceeding is commonly called 'an NOC proceeding'. It is a powerful tool to maintain market exclusivity because the generic company is automatically blocked from receiving its marketing authorisation until after the court decides in the NOC

24 Food and Drugs Act, RSC 1985, c F-27.

25 Data Protection Regulations: Food and Drugs Act, RSC 1985, c F-27, s 30(3); Food and Drug Regulations, CRC, c 870, s C.08.004.1. A six-month extension of data protection term may be available if clinical trials are conducted towards paediatric use of an innovative drug.

26 Patented Medicines (Notice of Compliance) Regulations, SOR/93-133.

Proceeding whether the generic company's allegations of non-infringement or invalidity are reasonably justified.

Canada has price controls on patented medicines.[27] There is a Patented Medicines Prices Review Board (PMPRB) which reviews prices of patented medicine where there is any Canadian patent related to a medicine 'by the slenderest thread'. Even patents for formulations and processes that are not being commercially used can trigger price reviews. Pharmaceutical companies often voluntarily consult with the PMPRB prior to product launch in order to discuss price. If an agreement on price is not reached, the PMPRB can trigger an administrative price review hearing. The hearing decision can be appealed to a court.

3.2 Commonly disputed matters

There is no frequent litigation arising from patent transaction agreements. The most litigated patent issues relate to the NOC proceedings, described in section 3.1 above. The generic product must clear brand name company patent barriers in order to get its generic drug marketing authorisation.

3.3 Competition rules

Competition law in Canada is governed by the Competition Act,[28] and enforced by the Competition Bureau. The Competition Act applies to three main areas of business conduct that can harm competition: coordinated conduct among competitors, abuse of dominance, and mergers. Price fixing, market allocation, and output restriction agreements among competitors are criminal offences, subject to a defence for restraints that are ancillary to a broader and otherwise legitimate agreement. The Competition Act provides for civil remedies for other agreements between competitors, as well as for abuse of dominance and mergers. Generally speaking, remedies can be ordered where conduct lessens or prevents competition substantially – that is, where the conduct creates, enhances or maintains market power.

In most cases, the mere exercise of an intellectual property right is immune from scrutiny. The mere exercise of an intellectual property right means the exercise of the owner's right to use, and to exclude others from using, the intellectual property. Where there is something more than a mere exercise of intellectual property rights, the conduct may become reviewable.[29] In the context of patents, potentially reviewable conduct could include the transfer, licensing, or agreement to use or enforce patent rights, or the merger of firms that collectively own a portfolio of complementary patent rights.

The exercise of patent rights is most likely to attract sanction under the Competition Act's criminal provisions where there is an agreement that pairs the

---

27 *ICN Pharmaceuticals Inc v Canada (Staff of the Patented Medicine Prices Review Board)* [1997] 1 FC 32 (CA).
28 RSC 1985, c C-34.
29 *Eli Lilly and Co v Apotex Inc* 2005 FCA 361; *Molnlycke AB v Kimberly-Clark of Canada Ltd* (1991), 36 CPR (3d) 493 (FCA); *Canada (Competition Act, Director of Investigation and Research) v Warner Music Canada Ltd* [1997] CCTD No. 53; Competition Bureau, *Intellectual Property Enforcement Guidelines* (2000); Competition Act, RSC 1985, c C-34, s. 79(5).

exercise of patent rights with criminally prohibited conduct. For example, an agreement between competitors to cross-license patents that also contains an agreement allocating markets could be an offence, unless the ancillary restraints defence is made out.

There has been no reported enforcement litigation arising from reverse payments in patent litigation settlements between brand name and generic companies. Since reverse payments tend to delay entry by a competitor, there is a risk that they will offend the Competition Act's conspiracy provisions. Their lawfulness may depend on whether the ancillary-restraints defence applies in the circumstances. The Competition Bureau has yet to issue any guidance on reverse payments. There are no notification requirements for reverse payment settlements.

An exercise of patent rights may also attract sanction under the civilly reviewable conduct provisions. For example, a civil remedy may be sought in a case where competitors entered into a patent pooling arrangement that creates, enhances or maintains market power by hampering the ability of other firms to compete against the members of the patent pool.

Patent issues are also of concern in the realm of mergers, as the collective patent rights of the merged entity may create, enhance or maintain market power. There have been several occasions where the Competition Bureau has required, as a condition of approving a merger, the divestiture of patent rights by the merged entity.

The Competition Act also allows the Federal Court to grant special remedies where the mere exercise of intellectual property rights has harmed competition. These remedies include the forced licensing of patents and the revocation of patents. To date, these provisions have never been successfully invoked.

The Competition Bureau has stated that it will only seek special remedies where the following criteria are met:

- there has been an undue restraint of trade or lessened competition;
- the holder of the IP is dominant in the relevant market; and
- the IP is an essential input or resource for firms participating in the relevant market. This third criterion will only be met if a refusal to license IP is stifling further innovation.

The Competition Bureau observes that the preconditions for a remedy are quite stringent, and that a remedy would be appropriate in only very rare circumstances.[30]

There is limited judicial consideration of the interaction between the Competition Act and intellectual property rights. The Competition Bureau published intellectual property guidelines in 2000, and released draft updated guidelines in April 2014.[31]

3.4 The effect of insolvency

Historically, there was concern over the fate of licensees upon a licensor's insolvency

30 Competition Bureau, *Intellectual Property Enforcement Guidelines* (2000) at 9.
31 Competition Bureau, *Intellectual Property (Draft Update)* (April 2014).

or bankruptcy. Absent any clear jurisprudence, there was ambiguity whether to treat licences as mere contracts to use, or as representing an interest in the underlying property.

Canadian courts were divided in their rulings. In 1999, the Ontario Superior Court held that the T Eaton Company was able to repudiate a licence and service agreement because it was merely a contract to use the company's trademark.[32] Conversely, a British Columbia court held, in the case of *Re Erin Features No 1 Ltd*,[33] that a contract granting exclusive marketing rights did indeed transfer property rights. Further, a licensee's rights can be extinguished by operation of a vesting order. If a purchaser buys intellectual property from an insolvent debtor and a court grants a vesting order, the licensee may only be able to pursue a claim against the resulting proceeds.

The Bankruptcy and Insolvency Act (Canada)[34] and the Companies' Creditors Arrangement Act (Canada)[35] have since been amended to resolve the major issues surrounding the treatment of licences upon bankruptcy or insolvency. Accordingly, as long as a licensee continues to perform its obligations under a licence agreement, a licensor is unable to "disclaim or resiliate" a licensee's right to use the intellectual property, including any right to exclusive use.

While the amendments seem to address the conflict at face value, in reality there are many concerns still unanswered. For example, the amendments do not address what 'use' entails and whether or not the right extends to ancillary rights of patent licensees, including a right to manufacture, upgrade, promote etc. Licensees and licensors must be cautious with licensing issues in an insolvency situation.

3.5 Taxation

(a) *Development of patents*

For Canadian tax purposes, expenses incurred in developing a patent can generally be divided between expenses of an income nature and those of a capital nature. Expenses of an income nature, such as salaries, are fully deductible on a current basis in the year that they are incurred. In addition, expenses associated with applying for a patent are currently deductible (although an election can be made to amortise such costs over a 10-year period). The treatment of capital expenses will depend on whether a patent is ultimately obtained by the taxpayer. If a patent is obtained, then the patent will be treated as depreciable capital property and included in Capital Cost Allowance Class 44, which allows for a 25% deduction on a declining-balance basis. If a patent is not obtained, then the capital costs associated with its development will be treated as eligible capital expenditures. Seventy-five per cent of an eligible capital expenditure is added to the taxpayer's cumulative eligible capital pool. This pool can be depreciated at a rate of 7% per year for Canadian income tax purposes.

Canada has a tax incentive programme for scientific research and experimental

32 *Re T Eaton Co* [1999] OJ No 5322.
33 8 CBR (3d) 205 (BCSC).
34 RSC, 1985, c B-3.
35 RSC, 1985, c C-36.

development (SR&ED). SR&ED expenses generally qualify for an investment tax credit at a rate of 15%. Canadian-controlled private corporations may be entitled to refundable investment tax credits at a rate of 35% on SR&ED expenses, subject to certain income and expenditure limits. Such refundable SR&ED credits can be an important source of financing for private Canadian corporations.

(b) *Acquisition of patents*

The cost of a patent or of a licence to use a patent will be treated as depreciable capital property and included in Class 44 (as described in (a) above). Alternatively, a taxpayer may elect instead to include this amount in Class 14, which allows for a deduction on a straight-line basis over the life of the patent. This election will be attractive if the patent has a life of less than nine years. Where a portion of the cost of a patent (but not of a licence to use a patent) is determined by reference to its use, a deduction is available for the amount paid for the use in the year. The cost of acquiring the rights to a patent application will be treated as an eligible capital expenditure (as described above). The payment of a royalty for the use of a patent is typically treated as a currently deductible expense in the year.

(c) *Transfers of patents*

Receipts for the disposition of a patent are generally on the capital account unless the taxpayer is in the business of selling patents. Capital gains are taxed at half the rate of regular income. Similarly, half of a capital loss is deductible, but only against capital gains. The granting of an exclusive licence to use a patent (such as for a specific time period or geographic territory) will typically be treated as a disposition on the capital account.

Despite a disposition of a patent being on the capital account, any payment received as a royalty based on the production from or use of that patent will be treated as regular income for Canadian tax purposes and taxed at the highest marginal tax rate. Similarly, amounts received for the granting of a non-exclusive licence to use a patent will be treated as regular income.

(d) *International transactions*

Payments made by a Canadian resident person to a non-resident person for the right to use a patent in Canada will be subject to Canadian withholding tax at a rate of 25% on the gross amount of the payment, unless a tax treaty with the non-resident person's jurisdiction of residence provides for a reduced rate. For example, Canada's treaties with both the United States and the United Kingdom provide for an exemption from Canadian withholding tax on patent royalties so long as the payment was not made in connection with a rental or franchise agreement. Under each of these treaties, the withholding tax rate would be 10% if made as part of such an agreement.

Canadian withholding tax does not apply to an outright purchase of a patent by a Canadian resident from a non-resident, unless the purchase price was dependent on the use of, or production from, the patent in Canada.

(e) ***Sales taxes***

Canada's federal goods and services tax will generally apply to the purchase or licence of a patent in Canada at a rate of 5%. A provincial sales tax may also apply, depending on the province in which the patent is supplied: the provincial sales tax rate ranges from 0% in Alberta to 10% in Nova Scotia. An exemption applies if the patent is acquired as part of the purchase of a business. Such taxes do not apply to the supply of a patent to a non-resident of Canada for use outside Canada.

3.6 Hot topics

Two currently hot topics are worthy of mention here.

Eli Lilly, a pharmaceuticals company, has launched a trade arbitration against the Canadian Government under the North American Free Trade Agreement.[36] In general terms, Eli Lilly contends that Canadian court decisions regarding the minimum standards for utility and the 'promise of the patent' are not in compliance with Canada's trade obligations. The company is requesting $500 million in damages after two of its drug patents were invalidated in court.

Health Canada issued a guidance document on biosimilar approvals[37] (also called 'subsequent entry biologics'). The document refers potential applicants to the detailed guidance of the European Medicines Agency as generally consistent with Canadian principles. Several biosimilars have been approved in Canada. There has been very little biosimilar litigation. Amgen and Teva engaged in litigation that settled in August 2013 on confidential terms. Amgen had used a filgrastim-related patent to try to block Teva from obtaining its marketing authorisation.

36 *Eli Lilly & Co v Government of Canada* (NAFTA trade arbitration).

37 Health Canada, Health Products and Food Branch, *Guidance for Sponsors: Information and Submission Requirements for Subsequent Entry Biology* (SEBs).

China and Hong Kong

Joanna Chan
Lewis Ho
Monique Lee
Dechert LLP

---

1. Patents, know-how and trade secrets: foundations for transactions

1.1 Patents and supplementary protection certificates

The current version (third amendment) of the Patent Law of the People's Republic of China (PRC) was promulgated in 2009, and a fourth amendment is currently under consideration and may be passed in the near future. This would be in keeping with the country's typical eight-year cycle of amending its patent law, which first came into force in 1984.

It should be noted that while Hong Kong and Macao are special administrative regions (SARs) under the sovereignty of China, patents granted in mainland China do not automatically confer protection in Hong Kong and Macao, which have their own set of intellectual property laws and administrative procedures. However, there is currently no primary patent review procedure in Hong Kong, and patents can only be granted on the basis of patents granted by one of three designated patent offices, namely the State Intellectual Property Office of China (SIPO), the United Kingdom Patent Office and the European Patent Office.

(a) Types of patents

There are three types of patent in China: invention, utility model and design. Invention patents protect a product or a process. In comparison, utility model patents only protect the functional structure of a product (but not a method), and no substantive examination is performed. Design patents protect aesthetic shape, pattern, and colour for a product.

The terms of a patent vary: an invention patent expires 20 years after application date, while utility model patents and design patents have a term of 10 years from application. The PRC patent laws do not provide for the extension of patent terms or supplementary protection certificates. Renewal fees are payable annually.

(b) Language of patent application

Under the PRC's Patent Law and its implementing rules, documents submitted for patent applications must be written in Chinese. Foreign applicants, including individuals, enterprises or organisations not domiciled in China, must appoint a locally registered patent agent to represent them. However, Patent Cooperation Treaty (PCT) applications filed with the SIPO may be written in English.

(c) ***Non-patentable subject matter***

Additionally, the Patent Law provides that inventions pertaining to the following subject matter shall not be patentable:

- scientific discoveries;
- rules and methods for performing intellectual activities;
- methods for the diagnosis or treatment of diseases;
- substances obtained by means of nuclear transformation;
- animal and plant varieties (except those new plant varieties protected pursuant to the Regulations on the Protection of New Plant Varieties, which came into effect in 1997);
- inventions contrary to social morals;
- inventions that are detrimental to the public interest; and
- inventions that rely on genetic resources, the procurement or utilisation of which contravenes any law or administrative regulations.

(d) ***Specific patentability issues in life science***

The PRC's Patent Examination Guidelines set out the considerations in deciding patentability on specific categories of inventions. Restrictive provisions on patentable subject matter can be circumvented by careful drafting of the patent application in adherence to the guidelines, provided that it satisfies the four main criteria for patentability (ie, novelty, inventiveness, practical applicability, and not being detrimental to the public interest – see further (e) below). As such, under the prevailing law, human embryonic stem cells and preparations are still not patentable.

(e) ***Validity***

Four general patentability requirements apply to the grant of patents in China.

- novelty;
- inventiveness (non-obviousness);
- practical applicability (usefulness); and
- not being detrimental to the public interest (not against public policy).

According to the Patent Law:

- 'novelty' means that the invention or utility model does not form part of the prior art, nor has an application been filed before the date of filing with the SIPO relating to identical subject matter disclosed in patent application or documents published after the filing date;
- 'inventiveness' means that the invention has prominent substantive features and represents notable progress as compared with the prior art; and
- 'practical applicability' means that the invention can be made or used and can produce effective results.

Under the Patent Examination Guidelines, 'detrimental to the public interest' means exploitation or use of a patent in a way that "may harm the interests of the public or society, or may affect the normal order of the State and society".

(f) *Patent infringement*

Under the Patent Law, with regard to direct infringement of a patent a person who manufactures, uses, offers for sale, sells, or imports a patented product without the permission of the patentee infringes the patent. The use of a patent process, or the use, offer for sale, sale or importation of products obtained directly from such a patented process without the permission of the patentee also constitutes an act of infringement.

With regard to indirect (contributory) infringement, the scope and applicability of this concept is not readily defined under the law or by the Supreme People's Court. In other words, there is no statutory provision governing indirect infringement under PRC patent laws; the Supreme People's Court has merely established that if more than one party is involved in using an infringing component to manufacture a product, all such parties will be jointly liable for infringement. In subsequent unofficial speeches given by Supreme People's Court judges, it has been asserted that indirect infringement is not a straightforward issue and that the court was still waiting for the right time to unify requirements for indirect infringement.

Thus, different courts have set out different requirements for indirect infringement. For example, in a judicial opinion published in 2001, the Beijing High People's Court defined 'indirect infringement' as an act of a party which, although not constituting direct infringement of another person's patent rights on its own, intentionally induces, entices or abets others to exploit the right-owner's patent, thereby resulting in an act of direct infringement.

Factors that have been taken into account by PRC courts in relation to establishing infringement are as follows:

- The only purpose of the infringed product is to contribute to the infringement of the claimed method/product.
- There is a subjective intent by the defendant to induce others to practise the claimed method (eg, by providing instructions on how to use its products in a way that infringes the claim).
- If the accused product merely contains the structural features in a claim but does not practise those features as required by the claim, the accused product would not fall within the scope of the claim and should not be found to be infringing (but in that case, the claim is a product claim that contains both structural and installation features).
- To be liable for indirect infringement, direct infringement (against another party) must be found first.

According to the Patent Law, the limitation to bringing an action for infringement is two years from the date on which the patentee or any interested party knew or should have known about the infringement. However, the limitation is only triggered when the infringement activities stop and does not apply when there is ongoing infringement.

(g) *Bifurcated system for determining validity and infringement*

Invalidity and infringement proceedings relating to patents are handled by separate authorities in China. In general, invalidity proceedings are filed with the Patent Re-examination Board (PRB) within the SIPO seated in Beijing. Patent invalidity decisions of the PRB may be appealed to the Beijing First Intermediate People's Court.

On the other hand, infringement proceedings are commenced at the Intellectual Property Divisions of Intermediate People's Courts throughout China. With the rising number of patent disputes each year, a judicial interpretation issued by the Supreme People's Court in April 2013 stated that Basic People's Courts may be appointed to hear patent disputes in the future when deemed necessary.

Where there are simultaneous proceedings regarding both the validity and infringement of a patent, the court handling the infringement claim has discretion to stay the infringement proceedings upon request of the parties, provided that the invalidity claim is filed in a timely manner after an infringement claim is asserted. However, courts are less likely to stay the infringement for invention patents, which will have undergone substantive prior examination, and it is not uncommon for invalidity and infringement actions for such cases to proceed in parallel.

(h) Confidentiality examination requirement

According to the Patent Law, for inventions made in China an applicant seeking to apply for a foreign patent must first undergo a 'confidentiality examination' conducted by SIPO prior to proceeding with patent filing. Under the Implementing Rules of the Patent Law, an invention 'made in China' refers to an invention in which "the substantive contents of the relevant technical solution are completed within China" – for example, where one of a group of inventors – regardless of his nationality – was located and conducting his work in China at the time of invention.

Upon receipt of a confidentiality examination request, SIPO will examine the invention and notify the applicant if the relevant invention involves 'national security or interest'. If SIPO does not issue any such notification within four months of the request, the applicant may proceed with filing an overseas patent application. However, SIPO may, if deemed necessary, extend that deadline by two months.

If a party files a PCT application with SIPO, it will be regarded as having automatically made a request for a confidentiality examination. If, however, the party has not made a request for a confidentiality examination, the applicant may be required to prove that the invention was not made in China. When questioned by an examiner, failure by the applicant to provide a satisfactory explanation may render the application rejected (although in practice an examiner usually defers to the applicant to comply with the confidential examination requirement). Also, under the Patent Law, failure to comply with the confidentiality examination is a ground for invalidation of a patent in China.

1.2 Who owns the patent?

(a) General rules on ownership

Under the Implementing Rules of the PRC Patent Law, an 'inventor' is defined as any person who makes innovative contributions to substantive features of a patent invention. The Implementing Rules also expressly exclude from being named as inventors those merely involved in work of an auxiliary nature, such as organisational support, or those who facilitate the use of materials or technical means for the invention.

(b) Inventions by employees

The Patent Law classifies employee inventions in two ways: service invention-creation and non-service invention-creation.

Service invention-creation: The term 'service invention-creation' is defined as any invention created or discovered by an employee in the course of employment or mainly using the material or technical means or facilities of the employer. The Implementing Rules further specify a service invention-creation to be one that arises:

- in the course of an employee performing his normal employment duties;
- from execution of any task assigned to him by the employer, whether or not within an employee's employment duties; or
- within one year from his resignation or retirement, where the invention-creation relates to his previous employment duties or tasks assigned to him by his previous employer.

Ownership of service invention-creations usually vests in the employer, who has the right to apply for patent protection. However, given the one-year claw-back period as discussed above, there may be risk where former employers of an employee may claim ownership over service invention-creations if the subject matter is substantially related to the type of work conducted in his previous employment.

Under the PRC's Contract Law, when a company intends to sell a service invention, unless otherwise agreed the inventor-employee has the right of first refusal to acquire the invention. However, it is possible to exclude this pre-emptive right by introducing terms to deal with such situations in the employment contract.

Non-service invention-creation: Ownership of non-service invention creations vests in the employee-inventor, and the inventor can apply for a patent in his own name.

The SPC made a pro-employee ruling in a 2008 case concerning ownership of a medical compound. In that case, the court granted ownership to the employee-inventor on the grounds that the employer's contributions were not made specifically for the research and development of the compound, even though the employer participated in and financially contributed to the clinical testing of the compound in question.

(c) Inventions made by way of consultancy or commissioning arrangements

Under the Patent Law, unless agreed otherwise an invention created or made by an entity in execution of a commission belongs to the entity who was commissioned to make the invention. Further, pursuant to the Contract Law, unless otherwise agreed by the parties the right to apply for a patent on the invention resulting from a commissioned development vests with the developer, where the commissioning party may exploit the patent free of charge.

(d) ***Inventions affecting China's national defence***

According to the Regulation on National Defence Patents, 'national defence patent' refers to a patent of any invention that relates to national defence interests and which is therefore potentially important to the defence of the PRC. It should be kept confidential. However, if the subject matter involves top-secret information of the state or relates to national defence, it cannot be covered by a national defence patent. If, upon examination, an application meets the relevant requirements of the Regulation on National Defence Patents, the National Defence Patent Institution of the PRC (rather than SIPO) has the power to grant a national defence patent right to the applicant.

While a national defence patent may be assigned to a Chinese party upon approval, transfer to a foreign entity or individual is strictly prohibited.

Once a national defence patent is granted, the State Council or People's Liberation Army may submit an application to exploit a national defence patent of another entity.

1.3 Inventor compensation

The Patent Law provides that any entity to which a patent is granted shall provide a reasonable reward to the inventor, and upon exploitation of the patented invention the entity must pay the inventor a reasonable remuneration based on the success of the invention and the economic benefits secured.

The Implementing Rules of the Patent Law further provide that the patent owner (employer) may agree with its employee-inventor or specify in a legally binding employment policy the amount and payment method of such reward and remuneration. Where no prior agreement or legally binding employment policy exists, the following minimum amounts of reward and remuneration shall be payable to the inventor:

- a reward of no less than RMB3,000 (around US$480) for an invention patent (or no less than RMB1,000 for a utility model or design patent) to be provided to the inventor within three months of the publication date of the patent;
- an annual remuneration of not less than 2% of the net profit derived from exploiting the invention or utility patent (or not less than 0.2% for design patents) for the duration of the patent; and
- where the patent is being licensed to a third party, a reward to the inventor of not less than 10% of each licence fee.

In the past, the courts have awarded as much as triple the statutory minimum to employee-inventors. To provide a further basis of default remunerations, the Shanghai High Court issued a guideline in June 2013 to provide some references for the determination of reasonable remuneration on service inventions.

Most notably, the guideline states that remunerations do not have to be money but may be in the form of stocks, options, promotions, salary rises, paid leave, bonuses, etc. Furthermore, a mutually agreed remuneration lower than the statutory minimum would be acceptable provided that there is justification for the arrangement, based on the industry and the circumstances surrounding the research and development (R&D) activities; and if the statutory remuneration is applied, the statutory minimum should be awarded and the court should not entertain any request beyond that level.

The guideline also clarifies that service-invention remuneration is a statutory benefit designed especially for employees of the patent owner. Thus, while contract workers may be entitled to such remuneration, commissioned entities (eg, clinical trial institutions) and their employees may not.

1.4 Know-how and trade secrets

While patent protection comes under patent laws, confidential information and know-how may fall under the ambit of trade secrets laws, which are governed primarily by the PRC's Anti-Unfair Competition Law. The Anti-Unfair Competition Law defines a 'trade secret' as technical or business information that:

- has practical utility;
- may create business interests or profits for its owners;
- is unknown to the public; and
- is maintained in secrecy by the owners.

The protection of trade secrets is also supplemented by provisions in various other laws, such as the Company Law, Labour Law and Labour Contract Law (which establishes liability for directors, managers or other relevant employees on breach of confidentiality and provides that trade secrets may also be protected by a confidentiality agreement between employer and employee).

Trade secrets shared with third parties during contract negotiations are protected by the Contract Law, which provides that trade secrets obtained during negotiation of a contract shall not be disclosed or improperly exploited regardless of whether the contract is concluded or not. Serious trade secret infringement also attracts criminal liability and carries a penalty of up to seven years' imprisonment in addition to the imposition of fines (see below).

The revised Civil Procedure Law, which came into effect on January 1 2013, provides interlocutory injunctions to offer additional protection in trade secret litigation.

(a) Enforcement of trade secrets

Under the Anti-Unfair Competition Law, the following are actionable causes for trade secret misappropriation:

- acquiring a trade secret by theft, inducement, duress, or other illegal means;
- disclosing, using, or allowing others to use the trade secret of another that has been acquired by the above illegal means; and/or
- disclosing, using, or allowing others to use the trade secret of another in breach of an agreement, or a confidentiality obligation imposed by a legal owner.

If a third party acquires, uses or discloses a trade secret that it knows, or should have known, to have been misappropriated in any of the aforementioned ways, the third party is also liable.

The enforcement of trade secret rights against misappropriation is difficult. Preventive measures are therefore critical in the protection of trade secrets and technological know-how. The burden of proof rests with a plaintiff who claims that a defendant has misappropriated its trade secrets. Plaintiffs must collect and submit their own evidence (preferably documentary evidence) to show that:

- the information being misappropriated is a trade secret, for example it is unknown to the public, has economic benefit and practical utility, and the trade secret owner has undertaken specific confidentiality measures;
- what the defendant had used is similar, or substantially similar, to the said information; and
- the defendant has used illegal means to obtain the said information.

Once the plaintiff prevails on a claim of trade secret misappropriation, the courts are likely to grant a permanent injunction. Under the PRC's Criminal Law, serious infringement of trade secrets may be publishable with criminal penalties, triggered by "serious" or "exceptionally serious" losses on the part of the trade secret owner. Specifically, if a party is found to have caused the trade secret owner losses to the extent of more than RMB500,000, the court can impose prison sentences of up to three years and fines for the misappropriation. As to "exceptionally serious" losses – that is, when the amount of losses exceeds RMB2.5 million – a defendant can be imprisoned for three to seven years in addition to fines being imposed.

The local Administration of Industry and Commerce (AIC) has power to initiate administrative investigations for trade secret misappropriation. The AIC can also order the return of confidential and proprietary material and the destruction of goods manufactured with such material if those goods would disclose trade secrets to the public when made available. The local AIC may order the cessation of the misappropriation acts and impose a civil fine of between RMB10,000 and RMB200,000 against the offenders. However, it does not have the power to award damages to the aggrieved party. To claim damages for trade secret infringement, a party must resort to civil litigation and institute legal actions in the people's court under the Anti-Unfair Competition Law.[1]

In comparison, as patents are registered intellectual property rights, the enforcement against patent infringement would generally be much easier than the enforcement against infringement of unregistered rights such as trade secrets – at least with respect to the establishment of ownership.

1.5 Employees, consultants and protecting know-how

As mentioned above, the protection of trade secrets between an employer and its employees or third parties is governed by various statutes, which are briefly introduced below.

With respect to contract research organisations (CROs), not only should

1 Where a defendant's actions have caused a trade secret to become known to the general public, the damages will be calculated according to the commercial value of the trade secret (and this will be determined taking into account research costs as well as the proceeds from, and competitive advantage of, the trade secret).

confidentiality agreements be established between the contracting parties but back-to-back confidentiality agreements between a CRO and its employees should also be obtained, in order to ensure enforceability of confidentiality policies and the subsequent availability of legal redress in the case of any breach.

The following are the relevant provisions under the respective laws in relation to trade secrets:

- Company Law: Directors or managers of a company shall not disclose the company's trade secrets. If they are found to have done so during the course of their employment, and in such a way as to damage the company, such individuals shall be liable for damages.
- Labour Law: Employer and employee can enter into a confidentiality agreement regarding the latter's obligation to the employer's trade secrets as a result of the employment. If an employee breaches or violates a confidentiality obligation and causes economic losses to the employer, the employee shall be liable for damages.
- Labour Contract Law: In addition to the provisions of the Labour Law (see previous bullet point), the Labour Contract Law provides that employers may impose confidentiality obligations on their employees with regard to trade secrets or other intellectual property in the employee agreements.
- Contract Law: Neither party involved in the signing of a contract may disclose, or inappropriately exploit, trade secrets obtained in the negotiation of a contract, whether or not the contract is eventually executed.

(a) Duration of confidentiality obligations

While PRC laws do not specify any duration for which employees' confidentiality obligations should subsist, based on positions taken by judges in trade secret cases the injunction granted must not be permanent but should end if and when the trade secret at least becomes public information. Therefore, employers should adopt a similar stance with regard to their confidentiality policy and incorporate post-contractual obligations accordingly in employment agreements.

In general, even after the performance of a contract (so that termination of the rights and obligations provided in it has occurred), the parties must still observe the principle of honesty and good faith and perform their obligations of notification, cooperation and confidentiality in accordance with industry good practice for the particular transaction under the contract. However, issues may arise during disputes in deciding what constitutes 'industry good practice', and therefore it may be better to adopt the approach for contractual parties that is described in the preceding paragraph for employees, namely to incorporate confidentiality provisions into a (surviving) clause in the contract.

The rule on maintaining confidentiality in relation to technology transfer – at least until the information ceases to be confidential in nature – is more clearly spelled out in transactions involving the assignment or licensing of technology. For example, according to the Regulations on the Administration of Technology Import and Export, "the grantee and grantor of a technology transfer contract shall be obligated to keep confidential information concerning the technology which has not been made public pursuant to the scope and term of the period of confidentiality as agreed upon." The same regulations further provide that the confidentiality obligation will terminate immediately the underlying information is disclosed to the public by a person or party which is not under that confidentiality obligation.

(b) *Statistics and success rates in trade secret disputes*

As to the frequency of trade secret infringements in China, no official statistics regarding the number of cases involving trade secret infringement have been separately published since 2006, because incidents of trade secret infringement are now reported in the category of unfair competition cases. Statistics of cases involving trade secret infringement in China during 1995–2005 are nonetheless shown in Table 1. In 2009, there were 1,282 cases involving unfair competition, out of which 253 are believed to have involved trade secrets. In 2012, the number of unfair competition cases was about the same, at 1,123, but no information on trade secret infringement seems available. All that can be reported is where serious infringement of trade secrets results in criminal penalties: in 2012 there were 49 individuals convicted for trade secret misappropriation according to the statistics released by the Supreme People's Court.

Table 1: Registered numbers of disputes involving trade secrets

Year	No of trade secret disputes
1995	587
1996	700
1997	914
1998	658
1999	645
2000	714
2001	830
2002	271
2003	349
2004	379
2005	500

According to statistics released by the Shanghai Higher People's Court in 2012, only 30% of the trade secret cases reached a final decision (with 70% ending up withdrawn or settled), and infringement was found in less than half of the cases decided, making an overall success rate in trial of less than 15%.[2] Furthermore, trade secret cases are unlikely to result in an injunction, take a relatively long time to litigate, and have one of the lowest percentages of damages awarded in relation to damages claimed in all types of IP litigation. Accordingly, trade secret litigants may be having a difficult time in the civil courts in China. For example, it was not until August 2013 that the Shanghai court granted its very first interlocutory injunction in a trade secret dispute (see (c) below).

(c) *Recent cases*

A newly revised Civil Procedure Law came into effect in the PRC on January 1 2013, and this has provided a legal basis for the grant of interlocutory injunctions. As a result, the Shanghai First Intermediate People's Court granted its first interlocutory injunction in August 2013 to Eli Lilly against its former chief chemist in China, who refused to destroy certain core trade secrets downloaded without authorisation. This shows the commitment of the Chinese courts to the protection and enforcement of trade-secret and confidentiality obligations.

Over at its US headquarters, two of Eli Lilly's former high-level scientists, originally from China, were reported to have conspired with an ex-colleague (now part of the senior management of a competing Chinese pharmaceuticals company) to steal trade secrets that belong to Eli Lilly on nine different research areas over the course of two years (allegedly worth US$55 million). Eli Lilly has filed a suit against these two former employees in the United States. There was also a case where a junior researcher of Wuxi AppTec, the biggest CRO in China, was convicted of stealing samples of two patented proprietary compounds made for a leading multinational pharmaceuticals company and selling them on the internet. However, despite the seriousness of the criminal conduct, the punishment was far from harsh and the employee was merely sentenced to 18 months' suspended imprisonment.

2. Context and features of patent transactions

2.1 The principal dealings in a patent

(a) *Regulations on the import or export of technology*

The assignment or licence of patents and associated know-how is considered to be technology transfer and subject to the relevant regulations and restrictions. It should be noted, however, that technology transfer between Chinese and non-Chinese parties constitutes the import or export of technologies and is regulated by the PRC

2 To justify the low success rate, the Shanghai High Court identified three attributing factors: plaintiffs are ill prepared or don't understand the standards for identifying what is a trade secret; they become concerned about risking a second disclosure of trade secrets during the case; and judges encounter problems with technical aspects of the case which extend the cases beyond the periods normally required for adjudication.

Government with regard to the level of national security, social morality, public interest and other prevailing governmental policies (which are subject to changes from time to time).

Technology import and export is classified into three categories: prohibited, restricted and freely transferable. The Ministry of Commerce (MOFCOM) publishes two catalogues, which together set out the list of prohibited and restricted technologies: the *Catalogue of Technologies Prohibited and Restricted from Import* (last published in 2007; hereinafter the Import Catalogue), and the *Catalogue of Technologies Prohibited and Restricted from Export* (last published in 2008; hereinafter the Export Catalogue). Technologies not included in these lists are freely transferable by default.

The catalogues cover a wide range of technologies, but most importantly, for the purposes of this publication, they cover certain technologies in the pharmaceuticals, chemicals processing, and forestry and agricultural industries. The Import Catalogue currently lists 32 prohibited technologies across 10 industries and 87 restricted technologies across 15 industries, while the Export Catalogue covers a wider range, prohibiting the export of 33 technologies across 20 industries and setting restrictions on 117 technologies covering 32 industries. As such, before entering into any type of cross-border patent transactions, an entity should first consult the two catalogues. Further, companies that intend to commission Chinese entities to conduct R&D should, before entering into any agreement, consider whether the resulting technology would be freely exportable and be capable of being transferred out of China (including, first, a successful application for patent protection outside China).

It should also be noted that, for the purposes of remitting funds, foreign exchange, taxation or customs clearance, the banks or authorities will require the presentation of the licence or certificate of recordal of the relevant technology transfer contract.

Pursuant to the Regulations on the Administration of Technology Import and Export, a licence must be obtained from MOFCOM for the import or export of technology in the restricted category, while freely transferable technologies only require a simple recordal. The grant of a technology import or export licence on restricted technology requires substantive review. The party may request a provisional licence, where the authorities have 30 days to consider whether to grant it in order for the applicant to negotiate the prospective agreement, and a further 10- to 15-day processing time to issue an official licence after furnishing the authorities with the executed agreement and other required documents. Alternatively, a recordal certificate for an agreement concerning freely transferable technologies only takes three days to process.

(b) Recordal of assignment of patents and applications

Patent assignments also require to be recorded with SIPO. Pursuant to the PRC's Patent Law, a written assignment agreement must be executed and recorded with SIPO. The assignment will take effect as of the date of recordal. In practice, if the assignment affects the public interest (eg, it involves the transfer of national security patents or patents involving protected technology from a PRC patentee to a foreign assignee), it may not be approved by SIPO.

(c) ***Recordal of licence of patents and applications***

PRC's Measures on the Registration of Patent Licence Agreements requires an application for recordal of a patent licence to be filed within 90 days from the date the licence agreement becomes effective (which is the signing date, unless otherwise specified). While failure to file a patent licence will not affect the validity of the licence contract, the 90-day rule must be strictly complied with: SIPO might refuse to process the application if the 90 days are exceeded, and might require a new licence contract be signed in conformity with the rule. An unrecorded licence may not be effective against a bona fide third party.

If a patent is subsequently invalidated, the licence fee need not be returned unless the damage to the licensees is caused by the bad faith of the patentee. Further, where a dispute arises over licensing matters, the terms of the duly recorded contracts may confer a higher probative value if adduced in court.

The licence recordation system provides greater protection to licensees as it adds a layer of administrative oversight over matters that might affect the exploitation of the patent. The Measures on the Registration of Patent Licence Agreements set out certain situations in which the authorities may refuse to record a licence agreement, as follows:

- The patent has been terminated or declared invalid.
- The licensor is not the patent holder or the holder with the right to grant a licence that has been recorded in the register.
- A co-holder of the patent at issue proves that the licensing contract is in violation of the law or in breach of another agreement.
- Renewal fees for the patent have not been paid.
- The patent at issue is involved in a dispute.
- A recordal filing has already been made with respect to the same patent licence contract.
- A pledge has been created over the patent, and the creditor to the pledge has not given consent.
- The patent licence contract at issue is in conflict with an existing licence recorded in the register.

Proof of recordation issued by SIPO is also needed where funds are to be remitted overseas as part of the transaction. However, starting on September 1 2013 the State Administration of Foreign Exchange set in place new guidelines and implementation rules, which no longer require a SIPO recordation certificate to be produced if the transaction payments are less than or equal to US$50,000. Nonetheless, the guidelines still recommend that banks should continue to review and request such a certificate where reasonable. In any event, patent assignments and licences must always be recorded with SIPO.

(d) ***The drafting of contract terms***

To be in conformance with the formal requirements of the authorities and to ensure that the patent agreements are valid and enforceable, the matters set out next should be taken into consideration when drafting licence agreements.

Implied nature of licensing agreements: According to the *Judicial Interpretation of the SPC Concerning Some Issues on Application of Law for the Trial of Cases on Disputes Over Technology Contracts*, a publication issued in 2004, a licence may be granted as a sole-and-exclusive licence, an exclusive licence, or a common licence. In a sole-and-exclusive licence, even the patentee-licensor cannot exploit the patent, while in an exclusive licence the patentee-licensor retains the right to exploit the patent (pursuant to contractual stipulations, as agreed by the parties). A common licence is a non-exclusive licence, where the licensor may exploit the patent in the absence of contractual stipulations. Unless otherwise specified, the courts will interpret a licence agreement to be a common licence.

National competition rules: Certain restrictive clauses are prohibited for technology import contracts. These are set out in section 3.4 below.

Obligations of the licensor: Ownership of a patent does not pass until its assignment is approved and recorded by SIPO, while recordal of a licence agreement confers benefits on the licensee. As such, contracts should stipulate the relevant obligations on the licensor.

Payment terms: It should be noted that cross-border transfers of funds (for payment of consideration or ongoing royalties) cannot be made unless accompanied by presentation of the licence agreement and/or the licence/registration certificate issued by MOFCOM and SIPO, depending on the nature and amount of the transaction. Tax benefits and withholding issues are described further in section 3.7 below.

Warranties and indemnities: There are certain statutory warranties which apply to technology transfer contracts. The Contract Law stipulates that "the transferor under a technology transfer shall warrant that it is the lawful owner of the technology provided and that the technology provided is complete, free from error, effective, and capable of achieving the prescribed goals". This provision is also replicated in the Regulations on the Administration of Technology Import and Export for technology import contracts. However, it is currently unclear as to whether this provision is strictly applied in courts and whether it can be disclaimed in a technology transfer contract as in the case of a contract for the sale of goods. Under the Contract Law, a party is liable for damages if it contracts in bad faith, conceals important facts or provides false information causing losses to the other party. According to the same law, the parties may agree on the amount and method of calculation of liquidated damages; however, according to the Judicial Interpretation of the SPC Concerning the Application of the Contract Law issued in 2009, the courts will have discretion to reduce or increase the amount of liquidated damages payable to reflect the actual amount of losses incurred. Specifically, the courts may reduce the amount of liquidated damages if it exceeds the actual loss by 30% or more. Alternatively, if the liquidated damages stipulated are less than the actual loss incurred, the courts may increase the amount of liquidated damages up to the amount of actual loss. That being said, these provisions only apply if PRC law is chosen as the governing law for the contract.

2.2 Patent rights in corporate transactions

(a) *Mergers and acquisitions*

The transfer of the ownership of patents in a merger or acquisition is dependent upon whether the deal is an asset purchase or an equity purchase. In the case of an asset purchase, patent rights are typically automatically transferred with the company to whom they previously belonged (ie, if company A acquires all the assets of company B, company A now holds the rights to company B's patents); however, the legal title of the intellectual property might need to be changed. In the case of an equity purchase, the parties must specify, in the process of negotiating the merger or acquisition, the ownership rights of the patents owned by the parties prior to the merger or acquisition taking place.

The former is preferable where the acquirer would like to avoid hidden liabilities of the target company, and allows more flexibility in the cherry-picking of assets from other less desirable assets. However, acquisitions are regulated by MOFCOM, which can deny an acquiring company the rights to acquire the equity with the transfer of patents (or impose other conditions) as it deems fit to protect State interests and acting in line with its national competition rules. Further, it is not limited to cross-border transactions where a Chinese target is involved: so long as it impacts state or public interests in China, MOFCOM may intervene even if the transaction is offshore or at a global level. An example relevant to the life sciences sector is the Pfizer–Wyeth merger: to prevent Pfizer from gaining nearly half the market share for swine vaccines in China after the merger, MOFCOM imposed a condition for merger clearance and required Pfizer to divest its swine vaccine business and the associated IP rights in China to a third party.

In addition, SIPO may also reject or delay the transfer of certain patents or patent applications if the subject matter involves state secrets (as under PRC law, a patent is not assigned until successfully recorded).

IP due diligence in an asset purchase deal will focus on the target assets and generally be less extensive than in an equity purchase deal. Specifically, the following matters are reviewed:

- ownership and freedom-to-operate issues (especially, identifying any service-invention issues, and investigating the employment history of existing employee inventors associated with the IP assets being acquired – for more information on ownership issues, refer to section 1.2 above);
- whether the patents are still valid and enforceable;
- whether the patents are free from encumbrances;
- whether the patents are subject to disputes, infringed or misappropriated by others; and
- whether the patents are subject to any licences.

Additional considerations for an IP due diligence exercise in an equity purchase deal would also focus on the business model of the target, such as:

- whether the IP portfolio of the target will be adequate for the ongoing development and expansion plans of the business;

- whether the business has the relevant permits and licences to conduct the business activity to which the IP assets relate (ie, manufacturing licences or marketing authorisations from the State Food and Drug Administration for pharmaceutical businesses, etc); and
- protection and management policies of the IP assets.

With regard to the transfer of assets, warranties and indemnities would be as required in typical transactions in other jurisdictions in addition to the formalities required by MOFCOM and SIPO concerning assignments (see section 2.1(d) above).

(b) Financing

In addition to choosing to list on one of the two stock exchanges in China (in Shanghai and Shenzhen respectively), companies usually prefer to be listed on the Hong Kong Stock Exchange (SEHK), which is more transparent and liquid. In Hong Kong, subject to the satisfaction of certain requirements on their financial and operating history, companies may list either on its main board or on the Growth Enterprise Market (GEM), which is similar to the Alternative Investment Market of the London Stock Exchange. A listing application to the SEHK is vetted by the listing division of the SEHK in conjunction with higher-level oversight by the Securities and Futures Commission (SFC), an independent statutory body set up to regulate the securities and futures market of Hong Kong.

While there are no specific listing rules applicable to patents, a company wishing to become listed is required to disclose all IP rights that are material to the business in its prospectus. Further, it should be noted that in practice, in the interest of shareholders, all intellectual property which is disclosed to be material in the prospectus should be held by the listing company or its direct subsidiaries within the listed group. Therefore, in instances where the corporate structure is reorganised in anticipation of listing, any patents held by subsidiaries that are not to be included in the final listed group should be duly assigned back to the holding company or wholly-owned subsidiaries that make up the listed group.

(c) Joint ventures

The Chinese government sets restrictions or controls on foreign involvement in certain industries, where in order to conduct business in China in restricted industries one must partner with or set up joint ventures (JVs) with Chinese partners.

Two types of joint venture are permitted by law in China: contractual joint venture (CJV) and equity joint venture (EJV). The difference is that for EJVs the joint venture partners share all rewards, risks and losses in proportion to their capital contribution, whereas for CJVs profit allocation and other liabilities, rights and responsibilities are subject to contractual agreement. Establishment of CJVs and EJVs is regulated by a separate set of laws and is subject to approval by the relevant authorities. According to the Contract Law, up to 70% of registered capital in foreign invested enterprises (FIEs) may be in the form of non-cash assets, including intangible property such as IP rights. Again, such contributions are subject to government verification and approval.

With a view to driving technological innovation and development, JVs in China are often set up where the foreign partner contributes IP rights or other technological expertise while the Chinese partner contributes local know-how, land use rights, and funding. In practice, a licence of patents may not be considered to be sufficient for the purposes of capital contribution and a full assignment to the JV may be required.

Regarding the distribution of IP rights at the termination of a JV, according to the PRC's Law on Sino-Foreign Contractual Joint Ventures, ownership of a CJV's property upon dissolution may be determined by contract, while according to the Implementing Rules for the Law on Sino-Foreign Equity Joint Ventures the liquidation process of an EJV would be subject to a liquidation committee set up at the announcement of dissolution of the EJV, which would consist of the directors or other nominees of the EJV.

It should be noted, however, that under the latest *Catalogue for the Guidance of Foreign Investment Industries*, as amended in 2011, most types of businesses in the medical and pharmaceuticals industries do not have specific requirements as to the legal structure of the company and may be set up as wholly foreign-owned enterprises (WFOEs), typically owned by an offshore joint venture company incorporated in Hong Kong or the British Virgin Islands. As such, the restrictive operating requirements and potential complications to IP rights ownership in setting up a JV may be circumvented.

(d) Divestment

As seen with the Pfizer–Wyeth merger, MOFCOM can impose divestment obligations upon a party that it must fulfil in order for the divestment to be approved. This safeguards Chinese business interests. Therefore, large-scale transactions in China or cross-border acquisitions may trigger 'red flags' for the Chinese government, and one may have to part with certain IP rights as a trade-off in order to proceed with the transaction.

(e) Consortium agreements

Where an invention is jointly owned by more than one party, unless otherwise stated in an agreement the PRC Patent Law provides that the consent of all co-owners is required for assigning the right to apply for a patent, assigning or withdrawing a patent application, or assigning, abandoning or pledging a patent right.

Unless otherwise agreed between the co-owners, a co-owner may exploit a patent alone without the consent of the other owner(s). A co-owner may also unilaterally grant non exclusive licences to third parties, but all joint owners are entitled to a share of the royalties in these transactions. If such an agreement exists, the agreement takes precedence.

In research collaborations, the parties may license technologies to the other parties. However, the right over any improvements in a technology is vested in the party that made the improvement. If desired, the parties may implement clauses to provide for a license-back mechanism on such improvements (please refer to Section 2.3 for more information on collaboration arrangements). Finally, there may also be potential problems with technology exporting when the collaborative research is brought to fruition, if such technology falls within a prohibited or restricted technology (see section 2.1(a) above).

When selecting a local collaborator, the foreign company should conduct due diligence investigations against the potential local collaborator (eg, whether it is related to a state-owned enterprise) and ascertain whether the potential collaborator has received financial assistance from the Chinese government in carrying out its researches, which may add complications to IP rights ownership and exploitation.

It should also be noted that IP rights ownership requirements under government-funded research projects vary from municipality to municipality. Therefore, it is advisable to review the requirements in the policy under which government funding is granted, in order to ascertain the IP rights ownership position. In some cases, the relevant government funding policy requires co-ownership of IP rights between government and the project undertaker; in other cases, the relevant government funding policy allows the vesting of IP rights solely in the project undertaker.

Where there is a transfer or disposal (including sale, transfer and assignment) of state-owned assets by a state-owned entity or public entity, approval must be obtained from the relevant authorities at different levels, depending on the value of the state-owned assets to be disposed of. For example, any disposal by a public entity of intangible state-owned assets valued over US$1.3 million must be approved by MOFCOM, while the disposal of state-owned assets valued between US$160,000 and US$1.3 million must be approved by the Ministry of Education.

2.3 Particular business transactions

The points in (a) and (b) below are considerations peculiar to contracting in China, regardless of the type of business transaction entered into. Thereafter, subsections (c) to (h) relate to specific business transactions.

(a) Contracting in China – due diligence

Ascertaining the identity of the company: Background searches on the company by its Chinese name and/or enterprise number should be conducted for the purpose of document review. An English name or a phonetic transliteration of the company's Chinese name does not have legal status. To avoid any ambiguities, the business registration number can also be used to verify the identity of the company. Unlike some countries, China does not have a unified public database containing records of all registered businesses in China. However, the local Administration for Industry and Commerce (AIC) for larger cities such as Shanghai or Guangzhou will operate its own online database that is available for public searching.

Verification of authorised signatory and proper company stamp: A company's common seal is often affixed to important documents such as official filings, contracts, operating licences, etc. Under the Contract Law, a party may enter into a contract by either signing, or sealing the contract with a proper company stamp. In practice, if the party is a PRC entity, it would be preferable if it is signed by an authorised signatory (usually the legal representative of a corporation) and also sealed by a company stamp normally dedicated for contract execution (known as a 'contract stamp') if the entity has more than one stamp. Although a company stamp is not definitive proof as to the validity of a contract, its absence should invite further investigation.

Therefore, it might be necessary to verify whether the signatory has legal authority equivalent to that of legal representative, which would empower the signatory to enter into agreements on behalf of the company. It might also be necessary to verify whether a general stamp can be used for sealing contracts, as it may be determinative as to whether an agreement has been properly executed.

While official checks and balances are in place for obtaining a common stamp – where a common stamp may only be produced by officially designated seal carvers and the imprint of which is required to be recorded with the local AIC – fake stamps are not uncommon in China. Because of the level of authority inherent in the common stamp, and the fact that a company would usually only have one copy of its common stamp, the handling of it is usually under strict controls within a company to ensure that it is not subject to unauthorised use and is properly accounted for as it is used in various administrative processes. As such, larger companies may also register a second common stamp – one used strictly for contract signing purposes, the imprint of which would indicate that particular use.

Ownership of IP and the capability to assign rights/grant licences: One of the main objectives in due diligence is to determine whether there is sufficient proof that the assignor / licensor owns all intellectual property rights and know-how associated with the development, commercialisation (including the right to seek regulatory approvals) and manufacturing of subject matter, and to identify any pre-emptive or superseding rights potentially vested with a third party.

Status and pre-existing rights to apply for drug registration: If the prospective agreement involves the right of a contracted party to obtain drug registration, it would be important to identify the status of existing regulatory approval and whether any party other than the contracted party has a preceding right to apply for and be named in the drug registration or clinical trial approval.

(b) *Contracting in China – contract format*

Name of the Chinese contracting party and company seal: For the reasons explained above, the agreement must include the Chinese name (in Chinese script) and the business registration number of the contracting party. The contracting parties must also ensure that the proper stamp has been duly affixed where provided for.

Bilingual agreement: The China Food and Drug Administration (CFDA) and SIPO require that application documents and other material for drug registrations and assignment/license agreements that are submitted to them must be in Chinese. Thus it is always helpful to prepare a bilingual version of any such materials or agreements. However, while the parties may designate English to be the prevailing language, authorities and courts will usually defer to the Chinese version if a dispute arises. To avoid any such discrepancies, it is advisable that the actual agreement be written in English and a simplified bilingual version prepared separately for the sole purposes of submitting the agreement to the authorities as required.

Governing law: Chinese businesses are becoming more receptive to accepting foreign jurisdictions as the governing law for their contracts. Further, for neutrality reasons it has been the case for certain Chinese academic institutions to opt for a 'third party' jurisdiction for dispute resolution. Hong Kong, given its proximity to China and relatively mature and transparent common law legal system, is a popular jurisdiction chosen by foreign and Chinese contracting parties as the governing law of the contracts. However, notwithstanding a choice-of-law clause adopted in the contract, the transfer and validity of PRC patent rights are always governed by Chinese law, while the foreign law may govern any dispute that arises surrounding the contracts. A choice-of-law clause that stipulates other than Chinese law may not be enforceable if there is no 'foreign' element to the contract (eg, both contracting parties and the IP rights are from China).

Companies also favour conducting arbitration in Hong Kong (for instance under the Hong Kong International Arbitration Centre), given the lower costs and cross-border recognition of arbitral awards.

(c) Material transfer agreements

Material transfer agreements (MTAs) govern the rights and obligations between the parties in a transfer of tangible materials (eg, biological materials, such as reagents, cell lines, plasmids and vectors, and chemical compounds) which are generally used for research purposes. The most important terms in an MTA relate to the disclosure of confidential information, rights to existing and potential IP, restrictions on the mode of usage of the IP or the material itself (ie, use for research purposes only, but not for human research nor for use in fulfilling other contractual obligations unless with consent, nor for unlawful purposes etc), publication rights on research results, or warranty against potential liabilities. China-specific considerations include import/exportability of IP for the transfer and potential license-back (see topics related to technology transfer in section 2.1) and observing the import requirements and restrictions for different materials into China. For example, the collection and export of human genetic resources of Chinese subjects require approval from the PRC Human Genetic Resources Administration Office (HGRAO). Further, the import and export of biological materials (eg, plasmids, DNA, cell-lines, bacteria and viruses) also require additional approval from the HGRAO.

(d) Commercial research and development agreements

The licensing of technology is commonly involved in research collaboration arrangements. Typically, improvements are owned by the party that makes the improvements and not automatically 'granted-back' to the other party without some form of remuneration or reciprocation clause. This means that typical automatic ownership of improvements made by the collaborating party is not common in China. In practice, it would be difficult to restrict a Chinese licensee from making improvements to the licensed technology and using such improvements.

To protect the interest of the licensor, ownership must be specified for 'background IP' (ie, which arose before the collaboration) and 'foreground IP' (ie, which resulted from the collaboration). For IP rights licensed to a Chinese party, the scope of the licence must be specified: for example, territory covered, duration, products/services covered, limitations, type and amount of royalties to be paid, and ownership and exploitation rights of any jointly or solely developed IP resulting from the collaboration.

As discussed above, for any service invention made, the employee-inventor must be provided with reasonable remuneration as required by the PRC's patent and contract laws. To clarify the ownership of inventions made by Chinese contractors, the foreign commissioning party must:

- define 'background' IP, 'foreground' IP and jointly developed IP, and the ownership of each;
- specify the ownership of inventions or technological developments in service agreements;
- oblige contractors to make timely disclosure of any new invention or development with a full written description of it, and to make available all work, reports, writings, data and records related to the invention, upon request;
- request employees of CROs and institutions to sign template transfers of inventions before joining the project, and require confirmatory assignment deeds to be executed by employee-inventors of contractors if a particular invention is made;
- insert representations and warranties that the contractors and their employees will be obligated to assign their inventions to the commissioning party, and that they will not enter into agreements with third parties regarding such inventions; and
- review contractors' own compliance with reasonable remunerations for service-invention.

On the other hand, if a foreign party licenses certain inventions from a Chinese entity, it must ensure that the Chinese entity owns and has the right to grant that licence (eg, the technology is a service invention and is not a national asset – particularly if the licensor is a public research institution, hospital or university). If the collaboration involves public institutions such as state-owned hospitals or universities, section 2.2(e) on consortium agreements provides more information on IP issues in transactions of subject matter involving government funding.

In a research collaboration agreement, it is particularly prudent to cover issues such as: the right to apply for patent or drug registration, the sharing and publication of inventions or research data, the right to make patent applications, and the granting of options and licences to third parties.

In addition to the points stated above, in a contract research arrangement, unless specified otherwise, the Contract Law provides that the commissioned IP will belong to the commissioned party. This means that it is important for the contracting parties to clearly state what IP is to be developed and who will own it (see section 1.2

for more details on IP ownership issues, section 1.5 on confidentiality issues and section 2.1 on technology transfer issues when dealing with CROs).

(e) ***Contract manufacturing agreements***

Contract manufacturing arrangements are common in China for its large workforce, relatively inexpensive labour costs, efficiency of production. However, the hassles of obtaining approvals and permits in setting up a manufacturing plant from scratch in China should not be overlooked. It is thus a popular model for the highly regulated pharmaceuticals industry. Outsourcing of drug production has inherent risks as it involves the disclosure of proprietary synthesis and manufacturing processes to an external party, especially considering that drug counterfeiting is a big issue in China. As discussed above, manufacturing know-how and processes are generally not patented but merely protected as trade secrets, which is an IP right that is harder to establish and enforce.

Under CFDA regulations, the production facility in which a drug is to be manufactured is designated in the drug registration application and would undergo extensive inspection by the CFDA authorities to ensure that drugs produced at the site conform to the quality and safety requirements prior to and after the approval of drug registration. For safety and traceability reasons, the approved drug may only be produced by the designated manufacturer named on the drug registration certificate. While it is possible to change the manufacturer, such changes are subject to the CFDA's approval by demonstrating that the drug produced by the new manufacturer is of equivalent quality, and the overall process to establish that fact can be time consuming; furthermore, it requires the cooperation of the previous manufacturer in full disclosure and handing over any know-how related to the manufacturing process. As such, selection of a contract manufacturer requires even more comprehensive due diligence, well beyond general aspects such as production capacity and checking whether the production facility has obtained full and proper permits and is in compliance with the Good Manufacturing Practice standard. Confidentiality and non-compete obligations, as well as provisions on monetary damages upon breach, are also essential parts of a contract manufacturing agreement. Moreover, the agreement should provide for procedures to keep proper logs and records, as well as the ownership of and permitted access to such data.

On top of obtaining proper IP protection over the product itself and the synthesis and manufacturing processes, any specially designed equipment used for the manufacturing process, such as moulds or other tooling, should be patented where possible.

In addition, foreign companies sometimes commission contract manufacturers in China to manufacture drugs solely for export purposes, partly because China is one of the biggest suppliers of pharmaceutical raw materials. While a PRC drug registration is not required for such manufacturing activities, the parties must file an application to record the arrangement with the local office of the CFDA within 30 days after execution of the contract. The local drug control institute will periodically test samples of the drug for quality assurance purposes.

(f) ***Product distribution agreements***

Appointment of a local distributor would help foreign pharmaceuticals companies to overcome challenges arising from language barriers, unfamiliarity with local market demands and problems in logistics due to China's vast geographical spread, and would also help those companies break into the market through a local sales network. In addition to the Contract Law, several laws are related to distribution arrangements in China, including the Anti-Monopoly Law, the Anti-Unfair Competition Law, the Price Law, and the Anti-Corruption Law.

With distribution contracts, one overarching principle to bear in mind is market dominance. A supplier who is deemed to be in a dominant market position would have more legal restraints with respect to its conduct, and this could affect how its contract with a local distributor would be structured. Owing to the complexity of the issue, the specific legal duties imposed on a supplier in a dominant market position are outside the scope of this chapter, but section 3.4 at least gives more information regarding the definition of 'market dominance'. Further advice should be sought if necessary to avoid contracts being in breach of PRC competition laws – although certain conduct is always prohibited regardless of whether the supplier is in a dominant market position, as follows:

- Bundled sales – 'Bundling', namely the sales practice of attaching to a sale a less desirable product as a condition for a buyer or distributor to gain the right to a more desirable product, contravenes the Anti-Unfair Competition Law.
- Resale price maintenance – Fixing the resale price or setting a minimum resale price by a supplier is prohibited under the Anti-Monopoly Law. However, there is no prohibition on setting a maximum resale price for distributors as long as the conditions are not so restrictive as to be construed to be fixing of resale price.
- Discriminatory pricing, discounts or rebates to distributors – Under the Price Law, where products are being traded and the circumstances surrounding such trading are the same among different distributors, the pricing, discounts or rebates offered should be the same for all distributors unless otherwise justified. However, the law is unclear as to what comprises 'justification'.
- Predatory pricing – The supply of goods below cost price for the purpose of excluding competition is prohibited under the Anti-Unfair Competition Law and the Price Law, apart from certain exceptions provided by the National Development and Reform Commission – for instance, where the goods are:
  - overstocked;
  - out of season;
  - near the end of their shelf life or validity period (as in the case for perishable goods); or
  - sold in the process of a lawful liquidation, change of operations or suspension of business operations.

IP issues to be considered in a product distribution agreement include the disclosure of trade secrets and confidentiality obligations by distributor, where the distributor may require the use of certain trade secrets and the trademarks of the supplier in its performance of the distribution. Clauses may be introduced to limit or control the use and dissemination of such information, as well as to provide a separate licence agreement on the use of trademarks by distributors and/or sub-distributors, so as to impose a limitation on the use of trademarks in relation to the sale of the supplier's goods.

In relation to trademark rights, a foreign supplier should not overlook the need to apply for trademark protections before engaging in negotiations with prospective distributors, as the latter may, unbeknown to the supplier, start to use the trademark in furtherance of its business. However, the PRC's Trademark Law provides that where an applicant applies for a trademark on identical products, or similar products that are considered to be the same or similar to a prior unregistered trademark used by another party, and that applicant learns about the existence of the unregistered trademark through a contractual, business or other relationship, the application will be rejected upon opposition by the owner of the unregistered trademark. Nonetheless, the supplier would still need to overcome legal hurdles in order to cancel the illicitly used mark and reclaim its rights in the same for its own use or by its other distributors.

With respect to a distributor's right to enforce trademarks, a distributor-licensee may unilaterally bring an infringement action where a licence for the use of the trademark has been granted on a sole-and-exclusive basis, while an exclusive licensee may jointly initiate an infringement claim with the trademark owner. Thus, under certain circumstances and unless otherwise provided, a distributor-licensee can bring a trademark infringement action without the licensor's authorisation; and in such a case, the rights and interests of the supplier may not be properly accounted for if remedies were granted as a result of the proceedings. The supplier-licensor may also want to impose obligations on the distributor-licensee to notify or to be joint-claimants with the supplier-licensor in bringing legal proceedings for infringement.

(g) Co-promotion agreements

Co-promotion arrangements, where two or more companies combine their sales and marketing resources for the marketing of a product under the same brand, are a relatively new concept in China. Under this type of arrangement, foreign companies may take advantage of the existing sales network of an established local Chinese distributor when marketing their goods. There is less inherent risk in terms of the misappropriation of IP in co-promotion arrangements, because the product has already been fully developed. IP issues arising at this stage may include the ownership of trademarks or know-how arising from the promotion of the product, and the right to audit the sales made.

If a co-promotion agreement is entered into for a drug pending its registration phase, the parties involved should be mindful of the fact that commercialisation rights in the drug belong inherently to the named applicants on the drug registration certificate.

(h) Commissioned research and clinical trial agreements

Clinical trials in China may only be conducted by institutions which have been accredited by the CFDA. Details of accredited clinical research institutions can be found on the CFDA's online database. To conduct clinical trials in China, a clinical trial application must be submitted to the CFDA and include pre-clinical trial data, chemistry, manufacturing and controls (CMC) and the results of clinical-trial tests conducted abroad together with the study's protocol (if available). As the clinical trial application dossier submitted to the CFDA must be in Chinese, many foreign companies will engage contract research organisations to prepare the application and coordinate the clinical trials that follow.

The IP risks associated with clinical trial arrangements in China include the misappropriation of IP and trade secrets, data integrity scandals, and ownership of clinical trial data and the resulting IP.

In addition to clinical data collected, the handling of personal data of patients in China is still at its rudimentary stages; and while guidelines have been published with regard to the handling of personal data, the guidelines are not legally binding. Further, the guidelines do not impose any restrictions on the transfer of data outside China. Therefore, in the meantime, the clinical trial agreement can be drafted in accordance with the foreign contracting party's national laws and regulations, as well as applicable international standards, on the handling of personal data.

Hong Kong has a personal data protection law, namely, the Personal Data (Privacy) Ordinance enacted in 1996. Section 33 of the ordinance, which deals with data transfers abroad, is not yet in force, although the provisions have been drafted. Subject to any further amendments before its enactment, Section 33 in its current form provides that personal data may not be transferred outside Hong Kong except under certain exemptions or if all of the following conditions are met:

- The data is transferred with written consent from the data subject.
- The data transferor has reasonable grounds to believe that the place to which the data is transferred has laws in force which substantially resemble, or serve the purpose of, Hong Kong's Personal Data (Privacy) Ordinance.
- The data transferor has taken all reasonable precautions and exercised all due diligence to ensure that the personal data transferred will not be dealt with in a manner that would contravene the Personal Data (Privacy) Ordinance.

The Hong Kong government has no exact timetable for the implementation of Section 33, but the privacy commissioner for personal data, in an address to the Legislative Council in January 2014, named it as a key strategic focus for the year. A survey of 50 jurisdictions has already been completed to compile a 'whitelist' of jurisdictions which have enacted data protection laws substantially similar to Hong Kong's. It is suggested that provisions be drafted into a Hong Kong clinical trial agreement in anticipation of its coming into force.[3]

Other China-specific considerations for clinical trial agreements will be similar to those discussed for commercial research and development agreements in section 2.3(d) above.

---

3 One may also refer to a fact sheet containing model contractual terms issued by the Office of the Privacy Commissioner for Personal Data in Hong Kong, which is available online at www.pcpd.org.hk/english/publications/fact1_model.html.

3. Issues associated with patent transactions

3.1 Regulatory law and compliance

(a) *Administrative protection of drugs*

China is a relatively late starter in terms of the protection of drugs by patents. As mentioned in section 1.1, China does not provide for the granting of supplementary protection certificates or patent term extensions. Indeed, prior to 1993, the Patent Law did not even provide protection for pharmaceuticals.

Between 1986 and 1993, the CFDA provided special administrative protection on drugs for applicants who fulfilled certain requirements stipulated in the Regulations on the Administrative Protection of Drugs at that time. Upon approval, the applicant would receive a certificate of administrative protection and enjoy certain quasi-patent rights such as for the exclusive production and sales of its drugs and the enforcement of its rights against infringers.

However, the administrative protection scheme was only available to applicants from countries that had concluded bilateral treaties or agreements with the PRC on administrative protection; these included the United States, Japan, Switzerland and the countries of the European Union at that time. Furthermore, the following requirements had to be met for pharmaceuticals for which administrative protection was applied:

- The exclusive right was not available under the Patent Law prior to January 1 1993.
- An exclusive right prohibiting others from manufacturing, using or selling in the country to which the applicant belonged was granted after January 1 1986 but before January 1 1993.
- The drug had not been marketed in China prior to the date on which the administrative protection application was filed.

An application was also subject to examination by the CFDA. Once approved, the duration of administrative protection was seven years and six months, subject to timely payment of annual fees. Further, the applicant had to make necessary arrangements to apply for the relevant permits for manufacturing and marketing of the drug within one year from the date on which the administrative protection certificate was issued, and failure to do so could render the protection void.

In practice, this protection scheme is only useful for those drugs patented before 1993 where the patent is still in effect (ie, drugs which were patented before 1993 but still enjoy exclusivity by virtue of supplementary protection certificates). As such, administrative protection is soon to become a creature of the past if the provisions are not revised.

(b) *Patent exclusivity*

Patents of a drug may be used to provide an exclusive period for its patentees. Drug regulations in China do not provide for a patent linkage system akin to the one in the United States, which requires generic drug applicants to serve notices to the pioneering company on their intention to file an abbreviated new drug application. Under the PRC's Provisions for Drug Registration, a drug registration applicant is merely required to submit patent information and a declaration of non-infringement to the CFDA.

Applications for generic drug may be filed when the patent is about to expire – and generic drug manufacturers are allowed to file drug applications two years before the expiry of the drug patent.[4] However, the CFDA will generally not perform an independent assessment of whether a generic drug infringes existing drug patents and therefore, presumably, will proceed with the review of the generic drug application if the generic applicant declares non-infringement.

The practical effect of patent linkage provisions have been questionable. Specifically:

- Patent linkage information may not be available if the patentee has not filed for drug registration or reported the registration to the CFDA.
- The scope of the patents may be subject to challenge by generic manufacturers.
- The latest amendment of the Patent Law (in 2008) introduced the 'Bolar' exemption from infringement "for purposes of providing information necessary for administration approval" – which forms the basis for the CFDA's currently proposed amendment to the Provisions for Drug Registration to remove the two-year restriction.
- The working practice of the CFDA is to enforce only compound patents but not secondary patents (eg, enantiomer, dosage or method of administration).

(c) Administrative new drug observation period

An advantage that is available exclusively to new drug registrations (but not generic or imported drug registrations) is the 'new drug monitoring period' imposed by the CFDA. The monitoring period is between three and five years from the date of approval. While its intended purpose is to ensure the safety and efficacy of the new drug, such a period is sometimes perceived as an exclusive period granted to the registrant, for instance because the CFDA will not grant approval to other entities to manufacture, change dosage forms on or import the same drug. However, if a third party has obtained approval for a clinical trial before the monitoring period is triggered, its application will proceed as usual and, once approved, will not be barred by the monitoring period granted to the first registrant.

(d) Data exclusivity

According to the Provisions for Drug Registration, the CFDA will not accept subsequent drug applications made by other applicants using the clinical and testing data of the original applicant within six years from the date the original applicant was approved. However, no such data exclusivity applies for independent clinical

4 This two-year requirement may be repealed subject to the passing of the latest draft amendments released by the CFDA in November 2013. Please refer to section 3.8 below for more information on the latest draft amendments and their impact on patent rights.

and testing data conducted by a different applicant. In practice, the working protocol of the CFDA is only to enforce this exclusivity to new chemical entities (NCEs) first applied for in China, which makes it almost impossible for foreign drug makers to benefit from such exclusivity.

3.3 Commonly disputed matters

Some of the most notable cases involving conflicts in patent transfers include the following:

- *Xiamen Dayang Crafts Limited v Xiamen Huanghe Technology Trading Limited* (SPC 2004): The issue in dispute was when a patent licence agreement would be held invalid when it "illegally monopolises technology or impairs technological advancement", which violates Article 329 of the Contract Law. The Supreme People's Court held that such actions require the licensee to accept conditions that are not essential for exploiting the technology, including the purchase of unnecessary technology, services, raw materials, equipment and products, to take on board unnecessary staff, or to unreasonably limit the licensee's freedom of choice in sourcing raw materials, components or equipment. In the quoted case, as part of the licence agreement the licensor required the licensee to purchase certain equipment to build a product line to manufacture the patented product. The Supreme People's Court held that the agreement did not violate Article 239 because the equipment was necessary for the production of the patented product.
- *Wang Xinghua v Heilongjiang First Radio Factory* (SPC 2006): The issue involved whether a patentee may terminate a licence agreement with the licensee without the consent of other non-patentees. In the quoted case, the patentee at the time, along with and representing two other individuals, entered into a licence agreement of a patent with a licensee. The Supreme People's Court held that a subsequent agreement to terminate the licence agreement between the patentee and the licensee was invalid, and held that even though the other two individuals were not patentees at the time of the contract, as a party to the licence agreement their consent was required in order to terminate the licence agreement.
- *Shanxi University v Shanxi Provincial Science and Forestry Environmental Protection Technology Centre* (SPC 1999): This case was about whether a third party to a joint collaboration agreement has standing to invalidate a technology transfer agreement on the subject matter developed. Shanxi University and Shanxi Provincial Environmental Protection Department had signed a joint collaboration agreement to develop certain technology for environmental protection. The latter subsequently delegated the R&D to the Shanxi Provincial Science and Forestry Environmental Protection Technology Centre, which signed a contract to enter into a joint venture with the university. The joint venture was never in fact formed, because of funding issues, but the parties nonetheless proceeded to collaborate. The university subsequently filed two patents in the resulting IP and transferred them to a third party. The Supreme People's Court held that even though the technology centre was not

a party to the initial collaboration agreement, it was engaged in collaboration with the university and so there was a de facto collaboration relationship between the parties. Therefore, the technology centre had a direct stake in the case, and had standing to dispute the transfer of the foreground IP. Further, the technology centre was a joint owner of the patent (because, according to the Contract Law, unless otherwise stated collaborating parties shall jointly own resulting IP) and was entitled to half of the assignment fee.

The Supreme People's Court has designated approximately 80 courts around China (mostly at intermediate level) to be first-instance courts adjudicating IP-related disputes. In addition, the Supreme People's Court has its own special panel for hearing IP cases. Panels hearing patent infringement cases will usually consist of three judges, sometimes assisted by a court-appointed technical expert.

3.4 Competition rules

The Anti-monopoly Law specifically clarifies that it does not govern the exercise of IP rights by business operators. Nevertheless, the exploitation of IP rights pursuant to the IP laws and relevant administrative regulations, to the extent that it eliminates or significantly restricts market competition, would be governed by the Anti-monopoly Law. This could cover the establishment of monopoly agreements, abuse of a dominant market position by business operators,[5] or cooperation by business operators with the effect of eliminating or restricting competition.

Provisions in certain PRC laws also set limitations on monopolistic behaviour of patent owners.

The Regulations on the Administration of Technology Import and Export explicitly prohibit the following restrictive clauses in technology import contracts (a concept reinforced in the Judicial Interpretation of the SPC Concerning Some Issues on Application of Law for Disputes Over Technology Contracts, issued in 2004):

- clauses requiring an assignee or licensee to accept conditions which are not indispensable for the import of a technology, such as purchasing unnecessary technologies, raw materials, products, equipment or services;
- clauses requiring an assignee or licensee to pay fees or assume certain obligations in relation to a patent that has expired or been declared void;
- clauses restricting an assignee or licensee from making an improvement to a technology or from using such improved technology;
- clauses restricting an assignee or licensee from obtaining, from other sources, technologies which are similar to or in competition with the technologies of the assignor or licensor;
- clauses which unreasonably restrict the channels or sources for an assignee or

5 Under the Anti-Monopoly Law, there is a presumption of market dominance if: (i) the market share of one business operator accounts for 50% or more in the relevant market; (ii) the joint market share of two business operators accounts for two-thirds or more in the relevant market; or (iii) the joint market share of three business operators accounts for three-quarters or more in the relevant market. However, a business operator in (ii) and (iii) who has a market share of less than 10% is not deemed to be in a dominant position.

licensee to purchase raw materials, spare parts, products or equipment;
- clauses which unreasonably restrict the quantity, type or price of any products of an assignee or licensee; and
- clauses which unreasonably restrict the export of products produced by an assignee or licensee by using the imported technologies.

The Patent Law allows the administrative grant of a compulsory licence by SIPO under some circumstances (including where the patent owner's conduct is deemed to be monopolistic) so as to mitigate the adverse consequences of monopolistic conduct on fair competition. Specifically, any party may request SIPO to grant a compulsory licence to exploit a patented subject matter when:
- the patentee fails to exploit or sufficiently exploit a patent without justifiable reason at least three years from the grant date; or
- the exercising of the patent right by the patentee is legally determined as an act of monopoly, for the purpose of eliminating or reducing the adverse effects of the act on competition.

3.5 Securitisation of patent rights

Patents can be pledged. PRC patent laws required that the pledge must be in writing and be perfected upon recordation with SIPO – ie, it must be in accordance with the steps stipulated in the Measures for the Registration of a Pledge of Patent Rights, which came into effect in 2010.

Elements which must be included in the pledge agreement in order to be accepted by SIPO include (but are not limited to):
- the name or title and address of the pledgor and pledgee;
- the type and amount of the secured debt;
- the duration of the pledge;
- the particulars to each patent pledged, including the patent number, title, application and grant date; and
- the scope of guarantee of the pledge.

The pledge will be publicly recorded in the patent register. In 2013, patent pledges exceeded US$4 billion in China, a record-breaking amount and representing an 80% increase over 2012.

3.6 The effect of insolvency

China's bankruptcy law does not state explicitly the effects of bankruptcy on IP rights and licences. Enterprises that are not state-owned are governed by civil law and its related regulations. If IP rights belong to a bankrupt company, they should be regarded as assets subject to liquidation.

As provided for by the Enterprise Insolvency Law, intangible assets and other assets may be sold separately in an auction while selling off the enterprise (unless otherwise agreed pursuant to a resolution at a creditor's meeting). As seen in the *Proview v Apple* case, if the IP right is valuable, creditors of a bankrupt company may assert influence in the disposal process.

For intellectual property rights owned by a joint venture, any rights belonging to the non-bankrupt partner should be exempted from liquidation, and the rights' exertion should be influenced by the other owner.

As to the effect of bankruptcy on IP licences, it is not clearly provided for in the statutes. In general, a bankruptcy trustee has the right to decide whether to terminate or continue an unfulfilled contract, and to notify the parties. If no such notice is given in two months, it is deemed that the contract is terminated. In view of the above, and in anticipation of such uncertainty, the parties should specify in the licence contract the arrangement that should apply when one of the parties becomes insolvent.

3.7 Taxation

In China, the main taxes involved in IP transactions are enterprise income tax, business tax and stamp duty. Each is described briefly below, within the context of patents and intellectual property.

(a) Enterprise income tax

Foreign enterprises earning income derived from China are subject to enterprise income tax (EIT). According to the EIT Law and its implementing rules, two different tax rates are applicable to foreign enterprises. 'Foreign' (ie, non-resident) enterprises which do not have any establishment or place of business in China are subject to a reduced EIT rate of 10%, while foreign enterprises with an establishment or place of business in China are subject to the standard EIT rate of 25%. However, if there is a tax treaty signed between China and the region or country in which the foreign enterprise is domiciled, the tax rate according to the treaty prevails. For example, under the Arrangement Between the Mainland of China and the Hong Kong Special Administrative Region for the Avoidance of Double Taxation and the Prevention of Fiscal Evasion with Respect to Taxes on Income, companies incorporated in Hong Kong enjoy the lowest EIT rate with respect to income from royalties paid from China as compared with other countries – currently (2014) 7%.

To encourage technological innovations, China has also adopted income tax incentives for qualified high-tech and new-tech enterprises (HNTEs), where the income tax rate is reduced to 15%. One of the requirements for HNTE certification is that the company must be a local incorporated company continuously conducting R&D activities and transferring its IP into products or services. The company must also obtain ownership of certain proprietary rights for core technologies used in its products or services in China.

(b) Business tax

Income derived from technology transfers, and the related consultation or services provided, are normally exempt from business tax. Therefore, income derived purely from the assignment or licensing of patents and the associated royalty income is exempt from business tax.

However, where the technology is transferred via goods such as samples, special machinery or other tangible goods, the value of such goods is not exempt from business tax. Thus, for example, in a technology transfer in the form of a system consisting of a patented method as well as specialised production machinery, the value of the machinery will be subject to business tax, while the transfer of the patented method will not be taxed. Notwithstanding the foregoing, the value of goods in the form of bacterial strains or new animal or plant varieties which are incidental to the transfer of biological technologies remains exempt from business tax unless sold on a large scale.

If goods subject to business tax are viewed by the competent tax authorities as undervalued, those authorities may independently appraise and adjust the price to reflect the fair value of the goods for taxation purposes.

Royalties or other similar income deriving from trademarks under a technology transfer are subject to business tax.

(c) ***Stamp Duty***

Stamp duty is payable by both parties in a technology transaction. The assignment or licensing of a non-patented technology, and the assignment of patent application rights, are each subject to 0.03% stamp duty. The assignment or licensing of a patent is subject to 0.05% stamp duty.

(d) ***Tax withholding***

In a technology transfer between a foreign and a Chinese company, the EIT and business tax payable by the foreign transferor with respect to the transfer from China will be withheld by the Chinese transferee.

3.8 Hot topics

(a) ***Drug registration in China***

As the healthcare sector in China continues to show strong growth year after year, pharmaceuticals companies now look to China not only as a source for inexpensive labour but as a big market for the sales of their drugs. With this in mind, many Western pharmaceuticals companies are now interested in obtaining marketing approval for their drugs from the CFDA.

As mentioned above, the main legislation governing the registration of drugs in China is the Provisions for Drug Registration. The review process for a new drug is roughly the same as in other jurisdictions: it is a two-stage process involving an application for clinical trial approval (CTA), followed by a marketing approval application (also commonly known as a new drug application in the United States or NDA). A company may register its drug as either a domestic new drug (for which a new drug certificate will be issued) or an imported drug (for which an import drug licence will be issued). Regardless of whether the drug is a new chemical entity or not, a drug is deemed a 'local new drug' if it has not been marketed in China prior to the application and is to be manufactured in China. It will then be able to enjoy market exclusivity by way of the new drug observation period (see section 3.1 above). All drugs that are manufactured abroad will be registered as imported drugs and will not be subject to the new drug observation period.

It should be noted that the regulatory review of drugs varies for biological or chemical entities and any sub-classes thereof, but the details are outside the scope of this chapter.

Finally, we offer below a non-exhaustive list of special characteristics of the Chinese drug regulatory system, and comparative differences with other major jurisdictions where relevant. Thus:

- Review time: The timeline for reviews by the CFDA is significantly longer than other jurisdictions. Drug applications are reviewed on a 'first come first served' basis. However, the CFDA is often backlogged and, in practice, the average waiting time for a CTA can be 18–20 months. Subsequent marketing approval after completion of clinical trials takes a similar amount of time for review.
- Language: All submissions to the CFDA for review, including scientific and technical information, are expected to be made in Chinese. Therefore, it is essential for a foreign applicant to find a competent agent to translate the application dossiers and effectively communicate with the CFDA.
- Format: Unlike the regulators of Europe, the United States or Japan, the CFDA has not adopted the ICH[6] guidelines and does not universally accept the Common Technical Document (CTD) format for drug application submissions. However, the CFDA has taken initiatives towards harmonisation and will accept NDAs made in the CTD format for certain classes of chemical pharmaceuticals.

(b) Draft amendments to the Provisions for Drug Registration

The CFDA released a second draft proposal of the Provisions for Drug Registration in February 2014 with a consultation period concluding in March of the same year. The CFDA has yet to release the results of the consultation, but certain amendments have raised concerns and were generally noted as 'pro-generics' by commentators, further undermining the already lacking protection given to innovative drug makers.

Two significant aspects of the proposed amendments to the provisions are worthy of mention:

- Repeal of the two-year limit for generic applications: The draft repeals Article 19 of the provisions, which states that generic drug applications may only be made when the patent for the said drug is due to expire within two years and that a marketing approval for the generic drug will only be issued upon the expiry of the innovator's patent on the drug. With regard to the marketing approval, the proposed amendments also state that the CFDA may issue marketing approval for the generic drug, but the relevant licence given under the approval will be effective only after the expiry of the innovator's patent. Acceptance of the draft would mean that generic drug manufacturers may start to register and market their drugs sooner, regardless of the patent status of the drug, thus speeding up the lag time in bringing generic drugs to the

6 International Conference on Harmonisation of Technical Requirements for Registration of Pharmaceuticals for Human Use.

market. The CFDA is essentially distancing itself from its role of enforcing intellectual property rights and leaving such matters to be dealt with by the courts via patent laws, further weakening the inadequate patent linkage system.

- Erosion of market exclusivity via new drug observation period: As mentioned earlier in the chapter, domestic new drugs are subject to a monitoring period of up to five years, during which no marketing approvals will be given for the same drug. Specifically, the CFDA will reject any other applications for the same drug during this time unless the applicant has already obtained a CTA for the said drug, wherein the review process will proceed concurrently with that of the first applicant for the drug. Generic drug makers in China are opportunistic and aggressive, and this provision essentially creates a rat race where generic drug makers are racing against the innovative drug makers, sometimes even being ahead of the innovative drug maker's application. The current draft amendments propose to make this race even easier by giving generic makers an earlier start. To appreciate the impact this amendment creates, it should be mentioned that obtaining a CTA is a two-stage process. Upon submission of an application, the administrative department within the CFDA will check to ensure that the dossier is complete and will then issue a formal acceptance, after which the information will undergo substantive technical review before (if all is well) issuance of a CTA. The formal acceptance is only an administrative measure, while the bottleneck for obtaining a CTA lies in the time taken for technical review. Where currently a CTA will be required, a formal acceptance will be sufficient to keep a follow-on application alive under the draft amendments.

Denmark

Klaus Ewald Madsen
Christian Bennet Rosenbæk
Bech-Bruun

1. Patents, know-how and trade secrets: foundations for transactions

1.1 Patents and supplementary protection certificates

According to Section 3.1 of Denmark's Consolidated Patents Act,[1] a Danish patent confers an exclusive right on the patent owner to make, offer, put on the market or use a product, a process or a product obtained by a process, which is the subject matter of a patent. Under Section 3.2 of the same act, a patent also protects against contributory infringement through the supply of means for working out the protected invention.

Under Section 3.3 of the act, the exclusive right conferred by a patent does not extend to:

- acts done for non-commercial purposes;
- acts concerning products put on the market in Denmark or in another country within the European Economic Area by the proprietor of the patent or with his consent;
- acts done for experimental purposes relating to the subject matter of the patented invention;
- acts limited to the subject matter of the patented invention which are necessary for obtaining a marketing authorisation for a medical product for humans or animals; or
- the preparation in a pharmacy of a medical product according to a medical prescription for individual cases, or acts concerning the medical product so prepared.

A supplementary protection certificate (SPC) confers the same exclusive right as the underlying patent, meaning that an SPC in practice works as an extension of the patent but only for the product subject to the SPC.

In accordance with the principle of territoriality, the legal protection conferred by a Danish patent is limited to the geographical borders of Denmark. The protection is extended, however, to include products that have been (legally or illegally) produced outside Denmark and imported into Denmark, if the products have been obtained by a process that has been patented in Denmark (Consolidated Patents Act, Section 3.1.3). These products will constitute a patent infringement in Denmark as if

1 Act 108 of January 24 2012.

they had been produced in Denmark, and they cannot therefore be legally imported into Denmark.

Patent applications must be filed with the Danish Patent and Trademark Office.[2] The application must contain information regarding the invention, including but not limited to a description of the invention, the patent claims being made, an abstract of the description and the patent claims, and the name of the inventor (Consolidated Patents Act, Section 8). Additionally, the applicant must pay an application fee (2014: DKK3,000 – approximately €400).

SPC applications must also be filed with the Danish Patent and Trademark Office. The applicant must provide information such as:

- which product the SPC application concerns;
- the number of the underlying patent and the name of the invention; and
- the date of the first marketing authorisation for the product in Denmark or in the European Union.

A granted patent may be maintained for 20 years from the filing date of the patent application (Consolidated Patents Act, Section 40). A renewal fee must be paid every year beginning after the grant of the patent. An SPC provides additional protection from the expiration date of the patent. The protection period of an SPC is calculated as the period between the filing of the patent application and the date of the first marketing authorisation for the product granted in Europe, reduced by a period of five years; however, the term of an SPC may not exceed five years.

Direct infringement is determined by comparing the claims of a patent with the potentially infringing product. When determining the scope of protection provided by the claims, the description may be taken into account (Consolidated Patents Act, Section 39).

Section 39 of the Consolidated Patents Act materially corresponds to Article 69 of the European Patent Convention, and Danish case law also recognises the doctrine of equivalence as stipulated in the Protocol on the Interpretation of Article 69 of the European Patent Convention, dated October 5 1973. In U.2009.1523H,[3] the Danish Supreme Court expressly recognised the doctrine of equivalence. Infringement was established on the basis of technical equivalence as the defendant's product "utilised the characteristic idea of the patent in a way which was obvious to a person skilled in the art". Accordingly, infringement can be determined even if a product is outside the explicit wording of the patent claims, provided the differences are immaterial and the chosen solutions can be seen as technically equivalent.

A presumption of validity exists for issued Danish patents. However, a patent may be revoked administratively or by a court decision if the patent has been granted without complying with the requirements of novelty, inventive step and protectable subject matter. Additionally, a patent may be revoked if (Consolidated Patents Act, Section 52):

- it has been granted to another person than the person entitled thereto;

---

2 At www.dkpto.org.

3 Supreme Court judgments are published in the *Official Legal Gazette* (In Danish, *Ugeskrift for Retsvæsen*).

- it relates to an invention which is not disclosed in a manner sufficiently clear to enable a person skilled in the art to carry out the invention on the basis of the description; or
- the subject matter extends beyond the contents of the application as filed.

An administrative opposition can be filed against a patent within nine months from the grant of the patent (Consolidated Patents Act, Section 21). The opposition should be based on similar grounds to those stipulated in the bullet list immediately above. After the expiration of the opposition period, a request for administrative re-examination can be filed by the patent proprietor or by a third party (Consolidated Patents Act, Section 53b). As a result of the re-examination, the Danish Patent and Trademark Office may revoke the patent or maintain it in amended or unamended form.

Administrative re-examination and invalidity proceedings before the courts cannot proceed concurrently. Invalidity proceedings take priority and, consequently, re-examination cannot be requested if invalidity proceedings before the courts are ongoing. Furthermore, if invalidity proceedings are instituted while an administrative re-examination filed by a third party is ongoing, the re-examination will be suspended by the Danish Patent and Trademark Office until the invalidity case before the courts has been finally decided upon. If, however, the request for administrative re-examination has been filed by the proprietor of the patent, the administrative re-examination will proceed (Consolidated Patents Act, Section 53b).

In Denmark, a compulsory licence may be granted to persons who have not been able to obtain a licence by agreement on reasonable terms and who may be considered capable of exploiting the invention in a reasonable and proper manner and in compliance with the licence (Consolidated Patents Act, Section 49).

1.2 Who owns the patent?

A Danish patent is granted to the physical person who has made an invention (the inventor) or the inventor's successor in title. According to Danish case law, a person must have made an independent contribution of an intellectual character in order to be recognised as an inventor. In U.2004/1018H, the Danish Supreme Court held that the originator of an idea to a new valve was the inventor, even though the idea far from constituted a finished product and extensive development was needed to create the finished product.

If there is more than one named inventor, the inventors become co-inventors. Danish law provides that co-inventors become joint owners of the invention, which means that the co-inventors have to be in agreement on all material decisions regarding the invention, including whether to file for a patent or otherwise exploit the invention commercially. If a patent is in fact applied for and granted, the co-owners need be in agreement to utilise the patent, including any sale or licensing of the patent (Consolidated Patents Act, Section 1).

In most practical instances, ownership of inventions is automatically transferred from employees to employers. Transfers are regulated by law in the Consolidated Act on Employees' Inventions and in the Act on Inventions at Public Research Institutions.

If an employee has made an invention which he may be considered to have arrived at through his employment, the employer may, provided that the exploitation of the invention falls within the scope of the employer's activities, request that all rights to the invention be transferred to the employer. The same right is enjoyed by the employer even if the exploitation of the invention does not fall within the scope of the employer's activities, provided that the employee's invention relates to a specific task assigned to him by the employer (the Consolidated Act on Employees' Inventions, Section 5). An employee who has made an invention is obliged to notify the employer thereof, and the employer must decide within four months thereafter whether he wishes to acquire the invention.

1.3 Inventor compensation

According to the Consolidated Act on Employees' Inventions, an employee is entitled to reasonable compensation for the transfer of an invention to his employer, unless the value of the invention does not exceed what the employee, in view of his working conditions as a whole, may reasonably be assumed to produce. The employee's specific employment conditions and work assignments, the value of the invention and its importance for the employer must be taken into consideration when determining the compensation.

Employee compensation is normally modest – that is, both in situations where the compensation for inventions has been agreed upon or stipulated in company guidelines, and where compensation is agreed upon on an individual basis. Danish case law is very limited, but in U.2004/1018H the Danish Supreme Court decided that the development of a radiator valve exceeded what a sales consultant was expected to produce. The Supreme Court assessed the appropriate compensation at DKK50,000. In very special circumstances, however, the compensation can be substantial. In a case before the Eastern High Court (B-3233-88), the parties made a settlement in court for DKK2.5 million after an employee had developed a new and cost-saving system for loading containers onto container ships.

According to the Danish Act on Inventions at Public Research Institutions, public institutions are allowed to request that all rights to inventions made through employment at the public institution are transferred to the public institution. In this case, the public institution has an obligation to utilise the invention in the best way possible, and the inventors are entitled to equitable remuneration if the invention is used commercially. The remuneration is calculated based on the income generated by the invention. There is no Danish case law on the subject, but a generally accepted rule is a split of the income of one-third to the inventor and two-thirds to the institution.

1.4 Know-how and trade secrets

An alternative to patenting an invention is to exploit the invention without disclosing the inventive details. In many cases this may not be possible because third parties will be able to work out the invention once they have access to the product (eg, through reverse engineering); but in some cases it may not be possible to figure out how the product/process functions and in such cases protection as a trade secret may be an effective weapon. The legal protection provided for trade secrets and know-how is clearly weaker than for patents in the sense that no exclusive right as such exists, but in practice such protection may be very effective as long as the trade secrets and know-how remain confidential.

Companies are protected by Section 19 of Denmark's Marketing Practices Act against the illegal use of trade secrets by employees and by cooperating or affiliated partners, who have authorised access to the trade secrets. This provision applies even though no specific confidentiality obligation has been agreed upon between the parties.

Sections 263 and 264 of the Danish Criminal Code protect against the theft of trade secrets by people without authorised access to the property containing the trade secrets. Access to the property does not need to be restricted, as long as the property does not give the impression of being public property.

The protection of know-how has developed through case law. A general rule on the scope of protection does not exist, as the scope depends on specific circumstances. According to case law, however, know-how is protected against unauthorised use by former employees, as illustrated by the Danish Supreme Court's decision in U.1980/717H. In this case, three employees founded a company that was supposed to compete against their former employer. The employees planned to use among other things technical knowledge and customer lists from the former employer, but the Danish Supreme Court held that the use of such know-how from the former employer by the newly established company would be illegal.

Case law has also established that an unwarranted cancellation of a contract can result in prohibition on the use of know-how related to the contract, as illustrated by the Danish Supreme Court ruling in U.1955/765H. In this case, a company had paid a licence fee for the rights to produce egg boxes using a specific process (know-how) and purchased dedicated equipment for the production. After four years, the licensee terminated the licence agreement but wanted to continue producing the egg boxes. The Danish Supreme Court held that further manufacturing of egg boxes would be illegal, as such manufacturing would be impossible without using the licensor's proprietary know-how.

1.5 Employees, consultants and the use of non-disclosure agreements

Confidentiality agreements and non-compete clauses are standard in many employment contracts in Denmark. Even without an express confidentiality agreement, employees will be bound by a duty of loyalty. The duty of loyalty prohibits employees from disclosing confidential information and in any other way acting disloyally to the employer. The duty of loyalty applies to employees during the employment, while the duty of confidentiality remains in force for a period of three years after termination of the employment.

Companies often protect their trade secrets and know-how by requiring that third parties sign non-disclosure agreements before the companies initiate any co-operation and enter into agreements which necessitate the disclosure of trade secrets. A non-disclosure agreement can protect any kind of trade secret, know-how or information generally not known, whether oral or written, which is to be exchanged between the parties.

The non-disclosure agreement (NDA) should include definitions of confidential information, time periods, restrictions on use, obligations on the receiving party, remedies such as liquidated damages, etc. In Denmark, NDAs are very common and it is always recommended to enter into an NDA before confidential information is disclosed to a third party. It should be noted, though, that enforcement of NDAs is often very difficult because illegal use of the confidential information may be hard to prove.

2. Context and features of patent transactions

2.1 The principal dealings in a patent

Most inventions are conceived during research at public research institutions and during focused development work carried out by companies. Patent rights are granted to the physical inventor(s). However, patent rights are transferable and they are very often transferred from the inventor(s) to a company with the necessary funds to commercialise the invention. In fact, this is also often the case where an invention has been conceived by a physical person on his own, simply because he cannot afford the further development work that is often required to commercialise the invention.

A patent is treated as a proprietary right, and a patent owner can without limitations sell, license or in any other way transfer a Danish patent to a third party. With regard to securitisation, see section 3.4 later in this chapter.

Patent licensing is used by companies in order to achieve goals such as:

- generating revenue from their technology through license payments;
- increasing their competitors' costs of doing business;
- ensuring that their competitors are not free-riding on the licensor's R&D;
- encouraging growth in a particular industry;
- establishing or improving supplier/customer relationships; and
- expanding the use of the technology into new geographical areas.

Licensees use licensing in order to take advantage of the licensor's innovations, enter into new markets, make up for a lack of R&D in their own firm or avoid litigation.

The transfer of a patent or the grant of a licence must, on request by any of the parties, be entered into the Danish Register of Patents (Consolidated Patents Act, Section 44). Registration is strictly formal and informative and has no legal effect. A licensee may not transfer the right under a licence agreement to others in the absence of an agreement to the contrary (Consolidated Patents Act, Section 43).

A licence to exploit an invention commercially can exist in many forms. A licence can be granted exclusively or non-exclusively,[4] be limited in time or perpetual, be irrevocable, be sub-licensable or otherwise, etc. The term of a patent licence is often the lifetime of the patent, but the term can also be a shorter period.

A licensee cannot apply for an SPC, as only the patent proprietor has this

4 Sole licences and semi-exclusive licences exist but they are not very common in Denmark.

privilege. Thus, the parties to a patent licence agreement should, at the time of contracting, consider the possible future relevance of an SPC, and the agreement should govern the obligations of both parties in this respect. For example, it could be agreed that the patent proprietor must apply for an SPC, while the licensee must cover any expenses incurred by the patent proprietor and on request provide all information necessary for the SPC application.

A non-exclusive licence does not prevent the licensor from exploiting the invention or licensing the patent to others. An exclusive licence prevents the licensor from utilising the patent and from licensing the patent to others within the field covered by the exclusive licence. An exclusive licence can be general or be granted for a defined geographical area, or it can be granted for all fields of application or only for specified fields. Whether an exclusive or a non-exclusive licence is the norm varies depending entirely on the specific circumstances. If nothing has been agreed, the licence is presumed to be non-exclusive.

Normally, a patent licence agreement must govern various legal topics, such as the possible departure from non-mandatory patent legislation, and compliance with competition law, the law of contract and, in international licence agreements, often private international law. Further, a licence agreement must consider topics such as risk allocation, termination, audit, assignment and confidentiality. When entering into licence agreements, the parties often enter into several ancillary agreements relating to the transfer of know-how, trademark licensing and other topics.

Depending on the circumstances and the negotiating power of the parties involved, the allocation of risk for unforeseen circumstances may differ significantly. The seller/licensor will normally seek to add 'knowledge qualifiers' to representations and warranties – for example, wording such as "... to the licensor's best knowledge ...". Such knowledge qualifiers are very important because they may change a warranty from a risk allocation clause to a simple disclosure clause.

Likewise, the seller/licensor may seek to add 'materiality qualifiers' wherever possible. By using the word 'material' throughout the warranties, the seller/licensor limits the effect of the warranties.

A licensor will often request access to audits in order to control the royalty payments and impose adequate payment and penalty provisions (for instance, that the audit must be paid for by the licensee) where licence payments turn out not to be contractual. Another very important point is the limitation of liability. Such limitation clauses will often include a cap on damages amounts, and very often the exclusion of indirect and consequential damages, etc.

Normally, the licensor will provide some standard warranties to the licensee upon entering the agreement. The warranties often include statements that:

- the licensor is the sole owner of the patent;
- the patent is valid and enforceable;
- no third party has challenged or, to the licensor's best knowledge, is considering challenging the validity of the patent; and
- the patent has, to the licensor's best knowledge, not been infringed by any third party.

Payment for patent licensing often takes the form of a lump-sum 'milestone' payment and/or royalties, or combinations thereof. Most commonly, licence agreements set out a payment scheme that includes ongoing royalty payments. Royalties can be calculated in many ways, such as payment per unit, payment via a percentage of net sales or net profits, tiered royalties, etc. Royalty payments vary between products, industries and countries and should always be researched before an agreement is entered into.

Patent due diligence is relevant in most transactions within the life sciences sector, but the scope of the patent due diligence largely depends on the specific patents comprising the contemplated transaction, as well as the type of transaction. Normally, a patent due diligence exercise will include a review of patent validity, any current or potential infringement cases, the estimated strength of patents (including the degree of freedom to operate), licensing terms, period until expiration of patents, etc. The level of detail of the due diligence depends on the transaction, but in general the patent due diligence exercise will be more detailed if a company has just a few patents on which the entire company is based – often the case in the pharmaceuticals sector.

The principle of 'buyer beware' is introduced in Section 47 of Denmark's Sale of Goods Act. According to a literal interpretation, Section 47 does not impose a strict duty of inspection on the buyer; instead, it provides that if, prior to the conclusion of a contract, the buyer has examined the goods or the buyer has, without due cause, failed to examine the goods after a request to do so from the seller or if, prior to the conclusion of the contract, the buyer was given the opportunity to examine a sample of the goods, the buyer may not rely on any lack of conformity that he ought to have discovered by such an examination, unless the seller has acted fraudulently.

Nowadays, some kind of due diligence is performed in almost all transactions, and according to case law this trend might have imposed a stricter duty of inspection on a buyer than what follows from Section 47. In U.1994.698H, the Danish Supreme Court held that a buyer of a company had no remedies against a seller who had provided the buyer with limited but incorrect information regarding the company's economic outlook. Based on the fact that the information was clearly incomplete, the Supreme Court held that the buyer could not uncritically rely on the information and that the buyer should have investigated the company's economic status further. In essence, the principle of 'buyer beware' applies to transactions in Denmark but the extent is unclear.

2.2 Patent rights in corporate transactions

(a) Mergers and acquisitions

In a share sale-and-purchase agreement, the company itself remains untouched even though the shares change hands. Accordingly, no agreements are transferred from the company. In an asset sale-and-purchase agreement, however, the individual assets are sold, which means that agreements will be transferred from seller to buyer.

As a licensee cannot transfer a patent licence in the absence of an agreement to the contrary, a share sale-and-purchase agreement is often preferable if the target company is reliant on patent licences, because no agreements are deemed to be transferred during a share sale-and-purchase agreement. Some licence agreements, however, may contain change-of-control provisions, in which case consent from the licensor is required even in share sale-and-purchase agreements. In an asset sale-and-purchase agreement, the license agreements will be transferred, which in the absence of any agreement to the contrary requires consent from the licensor.

A patent due diligence exercise will normally include a review both of patent applications and of the registration of patents. The patent application history can be used to determine the patent's scope of protection. Additionally, the due diligence will include a review of documentation such as:

- agreements with employees and independent contractors who participated in the development of any products in order to make sure no patent rights have been conferred on a third party;
- agreements regarding the transfer or licensing of patent rights;
- third-party challenges to the validity of a patent; and
- any infringement of a patent.

(b) Financing

In Denmark, stocks are being traded on the NASDAQ OMX Copenhagen, the GXG markets and First North. The 20 biggest Danish companies are traded on the NASDAQ OMX Copenhagen, while smaller listed companies are traded in the GXG and First North markets.

An audit of patents prior to any flotation is made in order to describe material assets and risk, and the audit is similar to the due diligence exercise described in section 2.1 above. No special rules apply to patents, which are treated as any other asset.

Disclosure requirements follow from Denmark's Securities Trading Act and the Danish Executive Order on Prospectus Requirements. The prospectus must contain a description of the issuer's business, including a description of the issuer's reliance on patents and licences (Danish Executive Order on Prospectus Requirements, Section 7).

The issuer has an ongoing obligation to disclose inside information regarding the issuer's business (Danish Securities Trading Act, Section 27). If patents play an important part of the business, the disclosure obligation includes information regarding patents. The relevant information includes clinical test results relevant to patent applications, the issuance of patents, any infringement of patents, and challenges to the validity of patents.

(c) Joint ventures

When establishing a joint venture, the parties must ensure they regulate the ownership of patents, including the patents that each party brings into the joint venture and the patents that might be developed by the joint venture. The parties must also agree on issues such as whether the parent companies can license patents from/to the joint venture, who will enforce the patents that have been transferred to the joint venture, what will happen to the patents if or when the joint venture is terminated, etc. These issues are often regulated through a joint venture agreement.

(d) ***Divestment***

Divestment of a patent portfolio can be a possibility when part of a company's patent portfolio concerns a specific product or business area which for commercial purposes it might make sense to divest. Further, divestment makes sense if a company has patents in a business area which the company will no longer pursue.

(e) ***Consortium agreements***

Consortium agreements resemble joint ventures (see section 2.2(c) above), but consortium agreements are often entered into regarding specific projects.

2.3 Particular business transactions

(a) ***Material transfer agreements***

Material transfer agreements (MTAs) regulate the transfer of research materials to a recipient. The recipient may typically use the materials for his own research purposes, and the agreement specifies the rights and obligations of both the provider and the recipient of the materials. Important issues include:

- the allocation of possible future intellectual property rights;
- the confidentiality of information;
- the right to publish research results;
- specifications for using the material; and
- return of the material to the provider.

(b) ***Commercial research and development agreements***

Research and development (R&D) agreements are often concluded between two or more parties (private companies and/or public research institutions) jointly conducting research and development activities within a specified area. The R&D agreement must address issues such as:

- the ownership of the intellectual property rights developed during the project;
- the use of each party's existing intellectual property rights;
- the allocation of intellectual property rights upon termination;
- confidentiality; and
- competition law (because an R&D agreement might have the potential to restrict competition).

(c) ***Contract manufacturing agreements***

Contract manufacturing agreements are typically used to outsource the production of a component or product to a third party. To guard against liability, the supplier (the contract manufacturer) will usually request that the customer represents and warrants that it owns or controls all the intellectual property rights necessary to manufacture, test, label, package, store and supply the product according to the terms of the agreement, and that the supplier's fulfilment of the terms of the agreement will not infringe the intellectual property rights of any third party.

The customer, on the other hand, can request a confidentiality clause prohibiting the supplier from using any disclosed intellectual property rights in a different manner from that agreed in the agreement between the parties. The contract manufacturing agreement also governs issues such as:

- the manufacturing process to be used;
- that good manufacturing practices (GMP) have to be applied;
- the specifications of the product to be manufactured;
- exclusivity and minimum purchase orders;
- ownership of intellectual property developed during the contract relationship; and
- the right to conduct regular audits, and liability assessments.

(d) Product distribution agreements

A product distribution agreement governs ownership of intellectual property rights in the products covered by the agreement. Furthermore, such an agreement may contain provisions that are customary in distribution agreements, such as exclusivity/non-exclusivity, frequency of price changes, confidentiality, termination periods, etc.

(e) Co-promotion agreements

Co-promotion agreements, where two companies market a single product under one brand name and a single marketing strategy, should at least govern the promotion rights granted, whether any intellectual property rights are transferred, and who will be in charge of infringement cases. The main reason for entering into co-promotion agreements is often to maximise the promotion activity and expand into new geographical markets and, possibly, new customer categories.

(f) Clinical trial agreements

Clinical trial agreements – where a sponsor (often a private company) asks an investigator (often a university) to conduct certain clinical trials – should govern, among other things, the ownership of intellectual property rights developed during the trial. Generally, sponsors want to control all intellectual property rights, as the clinical trials are based on the sponsor's drug/device. The investigator often accepts that the sponsor controls all intellectual property rights, but the investigator might want a licence to use the developed intellectual property rights.

Another important issue is the investigator's right to publish trial data. Key issues here are at what stage the investigator is allowed to publish and to what extent the sponsor can control the content. Furthermore, issues such as confidentiality and indemnification should be addressed.

3. Issues associated with patent transactions

3.1 Regulatory law and compliance

Anyone interested in putting drugs on the market in Denmark must comply with Denmark's Medicines Act and the Order on Marketing Authorisation for Medicinal Products.

A marketing authorisation is necessary in order to sell drugs on the Danish market. The Danish National Health and Medicines Authority issues marketing authorisations for an initial period of five years, with the possibility of indefinite extension. The marketing authorisation application must contain extensive information, including toxicological, pharmacological and clinical results.

No toxicological, pharmacological or clinical results are necessary for generic pharmaceuticals if the original product has been approved in Denmark or somewhere else in the European Union or the European Economic Area for at least eight years. However, a generic pharmaceutical drug cannot be marketed before 10 years have passed since the issuance of the first marketing authorisation for the original product.

In licence agreements or patent transactions, the licensor often warrants that:

- the company has all necessary approvals;
- the company has complied with all legislative procedures; and
- the marketing authorisations are valid and in force.

3.2 Commonly disputed matters

In U.1991.652/2H, the Danish Supreme Court decided on whether a patent licence agreement automatically expired at the date of patent expiry. The patent licence agreement was not limited in time but the licensee had a termination right. The Supreme Court held that the licence period should be determined by the patent licence agreement entered into by the parties. To the extent that the licence agreement did not refer directly to the specific patent, the expiry of the patent did not automatically rescind the licence agreement. Thus, the licensee was obligated to pay the licence fee even though the patent had expired.

In U.2012.3007H,[5] the Danish Supreme Court decided on whether a royalty provision in a licence agreement regarding know-how and design could be set aside as unfair. The licensor had provided know-how and design for a collection of glass beads, and the royalty provision gave the licensee royalties on the sales of all glass beads, whether designed by the licensee or not. After some years, the royalty payments had become much higher than expected by any of the parties and the licensor had started using other designers in addition to the licensee, for which reason the licensor wanted to amend or invalidate the royalty provision. The Supreme Court held the royalty provision to be valid as the licence agreement was entered into by two professional parties and the unexpected development was a normal commercial risk.

3.3 Competition rules

As part of the European Union, Denmark is obligated to comply with the Treaty on the Functioning of the European Union, including Articles 101 and 102 regarding competition rules. A patent confers an exclusive right, which might be problematic in relation to free competition. Thus, competition law is relevant for companies owning or licensing patents.

---

5 This case does not relate to patents but instead know-how and design.

The Competition Act in Denmark governs both agreements that restrict competition and the abuse of a dominant position. Primarily exclusive patent licence agreements may illegally restrict competition, which the licensor should consider before entering into such agreements. Furthermore, some issues related to competition law become relevant if a patent-owning company has a dominant position; in that case, the Competition Act prevents the patent owner from abusing its dominant position – for example, through unfairly high prices – despite the exclusive right conferred by the patent.

Agreements might be exempted from Article 101(1) of the Treaty on the Functioning of the European Union if they fall under the 'block exemption' regulation in Article 101(3) of the same treaty. Block exemption applies if the agreement contributes to improving the production or distribution of goods, or to promoting technical or economic progress, while allowing consumers a fair share of the resulting benefit. The agreement may not impose restrictions on the companies concerned, which are not indispensable for the attainment of these objectives, and may not afford such companies the possibility of eliminating competition in respect of a substantial part of the products in question. Similar exemptions are implemented in Section 8 of Denmark's Competition Act.

3.5 Securitisation of patent rights

A patent is an asset which may be pledged as security – for example, for a loan. The act of perfection is the registration of title in the Registry of Persons at the Danish Registration Court. The pledge of a patent does not restrict the rights of the patent owner, including the right to enter into licence agreements, unless otherwise agreed between the pledgor and the pledgee.

The pledge of a patent can, on request by any of the parties, be entered into the Danish Register of Patents (Consolidated Patents Act, Section 44). Registration is strictly formal and informative and has no legal effect.

3.6 The effect of insolvency

Patents can be the subject of creditor proceedings such as execution or bankruptcy if the patent owner becomes insolvent. In a case of bankruptcy, the patent rights become part of the bankruptcy estate.

According to Section 55 of Denmark's Bankruptcy Act, the bankruptcy estate, represented by the bankruptcy trustee, has the right to decide whether it will become a party to any agreements entered into by the debtor. If the bankruptcy estate decides to become a party to (for instance) a licence agreement, the bankruptcy estate assumes the rights and duties of the debtor and the agreement continues on its normal terms. In this case, the licensee is not affected by the bankruptcy. However, in most cases the bankruptcy estate is keen to dispose of the assets in order to finalise the estate, and therefore the bankruptcy estate often decides not to become a party to any bilateral agreements. In this case, the bankruptcy estate is entitled to terminate any agreement at the usual or a reasonable period of notice, notwithstanding that a longer period of notice or even non-terminability had been agreed between the original parties.

3.7 Taxation

Profits or losses from the sale of patents or other intellectual property rights must be included when calculating taxable income – see Section 40 of the Danish Amortisation and Depreciation Act. The profits or losses are calculated as the difference between the purchase price and the sale price after deduction for amortisation.

Regarding tax, the purchase price of patents may be amortised by the buyer by up to a seventh part of the purchase price each year, or the purchase price may be deducted in full in the year of purchase. The same rules apply for the purchase of patent licences.

Foreign companies are taxable on royalties that originate from sources in Denmark – see Section 2 of Denmark's Corporation Tax Act.

3.8 Hot topics

Until recently, it has been contested in Denmark whether an application for a marketing authorisation for a patented medicinal product constitutes sufficient grounds for a preliminary injunction. In a case before the Danish Supreme Court (case 266/2009), a generic pharmaceutical company T had applied for a marketing authorisation for a drug containing a patented active ingredient. The patent owner N applied for a preliminary injunction in order to prevent T from introducing the generic drug on the market. Both before and during the trial, T declared that it had no current plans for marketing the generic drug. Further, it was explained that the reason why an application for marketing authorisation had been submitted in Denmark at an early stage was that Denmark had been used as the reference country for the decentralised European application procedure.

In these circumstances, the Danish Supreme Court ruled that a marketing authorisation for a patented product did not in itself constitute sufficient grounds for a preliminary injunction unless other actions (such as registration in the DKMA price list) indicated that the holder of the marketing authorisation was actually going to market the product within a short time.

France

Laëtitia Bénard
Allen & Overy LLP

1. Patents, know-how and trade secrets: foundations for transactions

1.1 Patents and supplementary protection certificates

Pursuant to Article L611-10 and following of the French Intellectual Property Code, a patent may be granted, in all fields of technology, for any invention provided that it is new, involves an inventive step and is susceptible of industrial application.

Additional specific provisions are of relevance to patents in the life sciences sector. First, French law expressly excludes from patentability:

- discoveries, scientific theories and mathematical methods;
- surgical or therapeutic methods and diagnosis methods applied to human or animal body;[1]
- inventions the commercial exploitation of which would be contrary to public policy or morality; and
- the human body, at the various stages of its formation and development, and the simple discovery of one of its elements including the complete or partial sequence of a gene.[2]

Secondly, Article L611-11 of the Intellectual Property Code provides that:

- an invention is considered to be new if it does not form part of the state of the art;
- the state of the art is held to comprise everything made available to the public, before the date of filing of the patent application, by a written or oral description, by use, or in any other way; and
- the patentability of any substance or composition, comprised in the state of the art, for use in surgical or therapeutic methods and diagnosis methods applied to human or animal bodies, is not excluded, provided that its use for any such method is not comprised in the state of the art.

However, Article L611-16 of the Intellectual Property Code expressly excludes from patentability the surgical or therapeutic methods and diagnosis methods applied to human or animal bodies. In other words, there are specific criteria for the assessment of novelty for inventions in the life sciences field: indeed, a substance or

1 Article L611-16 of the Intellectual Property Code.
2 Article L611-18 of the Intellectual Property Code.

a composition known for its use outside the life sciences field would be deemed to be new if it has never been disclosed for a use in that field. Also, a new medical use of a substance or a composition previously known for another medical use would equally be deemed to be new. As an example, the use of a substance for the treatment of migraines could be covered by a patent even if this substance were to be known and used in the field of mechanics; and the same substance could be covered by another patent for its use in the treatment of cancer.

All the above provisions are identical to the provisions of the European Patent Convention except those concerning the patentability of human body elements.

The duration of protection conferred by patents is set out in Article L611-2 of the Intellectual Property Code, which provides for 20 years of protection after the date of filing of the application. However, Regulation (EC) 469/2009 concerning supplementary protection certificates (SPCs) for medicinal products (hereinafter the SPC Regulation) grants supplementary protection to medicinal products in order to compensate for the time between the patent application and the time when the product is granted a marketing authorisation for it to be put on the market.

Under Article 3 of the SPC Regulation, a certificate is granted if:

- the product is protected by a basic patent in force;
- a valid marketing authorisation has been granted;
- it is the first authorisation for the product; and
- the product has not already been subject to a certificate.

An SPC affords its holder an additional five years of protection, for a maximum of 15 years of effective protection as from the date of the grant of the first marketing authorisation. Regulation (EC) 1901/2006 on medicinal products for paediatric use provides, under certain conditions, for a six-month extension of the SPC duration.

In order to obtain a patent covering France, the inventor or his successor in interest needs to file a patent application with the French Patent Office (hereinafter the INPI (*Institut de la Propriété Industrielle*)) or a European patent application designating France with the European Patent Office. The inventor must be mentioned to any patent office where an application is filed, regardless of the ownership rules, although the inventor is entitled to decline such a mention. The application must contain:

- a request for the grant of a patent;
- a description of the invention, including drawings where appropriate;
- one or more patent claims;
- an abstract of the primary information provided; and
- in the case of priority claims, a copy of the prior patent applications.[3]

Patent holders must also pay the registration taxes and annuities. In France, the duration for filing and examination of the patent application is generally around 27 months from the date of filing. The geographical scope of a French patent is all French territory. European patents designating France confer the same rights as would be conferred by a French patent granted in France.

3 Article L612-3 of the Intellectual Property Code.

Patent protection entitles a patent holder to prohibit third parties from using, manufacturing, offering for sale, marketing, selling, importing, exporting, trans-shipping, or using and/or possessing or holding for the aforementioned purposes, any products, compounds, devices, processes or subject matter of the patent.[4] It also entitles the patent holder to prohibit the delivery or the offer to deliver the means of implementing the patented invention. As a consequence, unless authorised by the patent holder, such acts constitute patent infringement. However, patent holders cannot prohibit acts carried out for private use, or for non-commercial or experimental purposes. This includes the preparation of medicines by a pharmacy, and the acts required in the preparation for obtaining full market approval or a visa for medicine advertising.

1.2 Who owns the patent?

The general rule of ownership is set out in Article L611-6 of the Intellectual Property Code, which provides that the right to a French patent belongs to the inventor or his successor in interest.

Multiple individuals and/or entities may share ownership of patents.[5] Article L613-29 of the Intellectual Property Code defines the rules of ownership, where each owner is entitled to:

- exploit the patent for its own profit and grant non-exclusive licences subject to prior notice and equitable compensation to co-owners who do not exploit the invention;
- initiate legal action for infringement, again subject to prior notice to the co-owners; and
- assign or securitise its share in the patent,[6] subject to the following conditions:
    - the co-owners have a three-month right of pre-emption as from the notice of the assignment draft;
    - if the co-owners fail to reach an agreement, the Paris court of first instance has jurisdiction to determine the assignment price; and
    - the parties have a month as from the notice of the judgment or, in the case of an appeal, of the appeal decision in order to renounce the sale or the purchase of the share without prejudice to the damages that may be due, and the renouncing party shall bear the costs.

However, the rules set out by Article L613-29 of the Intellectual Property Code apply unless otherwise agreed by the co-owners. Hence, the co-owners may deviate from these provisions at any time in a co-ownership agreement.

Specific rules apply to inventions made by employees. Indeed, Article L611-7 of the Intellectual Property Code distinguishes different rules of ownership depending on the classification of the employee inventions (the process for classifying

---

4 Articles L613-3 and following of the Intellectual Property Code.
5 Articles L613-29 and following of the Intellectual Property Code.
6 See Paris Court of Appeals, 1st Chamber, September 23 2009, RG 2008/11403, PIBD, 909, III, p1.

employee inventions being set forth in Article R611-1 and following of the same code). Importantly, Article L611-7 of the code does not apply to inventions made by interns or corporate executives, which thus belong to their inventors.

Inventions made by employees can fall within the three following categories, which determine the rule of ownership:

- *Inventions made in the course of a mission:* any inventions made by an employee as part as the performance of an employment contract providing for an inventive mission, or in the framework of studies or research expressly entrusted to the employee by the employer, belong to the employer;[7]
- *Inventions made outside any inventive mission but attributable:* any inventions created by an employee in the course of the performance of his functions, or in the field of the activities of his employer, or as a result of the knowledge or use of specific methods or techniques owned by the employer or data procured by the employer, belong to the employee. However, the employer may exercise its right of attribution of ownership or benefit of all or part of the rights attached to patents protecting such inventions;[8]
- *Inventions made outside any inventive mission and non-attributable:* any other inventions belong to the employee, without any right for the employer. However, employment agreements can include a provision granting the employer a preferential right if the employee decides to assign the invention or to grant a licence for its use.

Article L611-7 of the Intellectual Property Code provides that, for inventions made in the course of a mission and inventions made outside any inventive mission but attributable, a specific written deed of assignment is required and that employees must (under certain conditions) be granted definitive compensation.

When it comes to employment contracts (or company agreements), employers may introduce specific policies regarding employees' inventions. However, as Article L611-7 of the Intellectual Property Code has public policy implications, any policy imposed by an employer must not be less favourable to an employee than the legal provisions in force. Thus, an employer policy may add details – for example, the calculation of the specific remuneration – but it cannot limit employees' rights.[9]

These rules are also applicable to inventions created by civil servants. However, the Intellectual Property Code has additional provisions relating to the classification of the invention and payment of additional remuneration.[10]

1.3 Inventor compensation

As a general rule, parties involved in a transaction resulting in a patentable invention are free to determine the amount and the method for calculating invention compensation, which can take the form of either a lump sum or a royalty. However, if the inventor is an employee, the specific regime set forth in Article L611-7 of the

7 See Paris Court of Appeals, Pole 5, 2nd Chamber, March 28 2014, 13/11578, *Fimurex v Le Duc*.
8 See Paris Court of Appeals, Pole 5, 1st Chamber, April 11 2012, 10/12286.
9 Supreme Court, Commercial Chamber, February 12 2013, 12-12898.
10 Article R611-14-1 of the Intellectual Property Code.

Intellectual Property Code will apply. Because these provisions invoke French public order, employers cannot claim that employees have already been compensated for their inventions by way of their salary.

Also according to Article L611-7, the nature of the inventor compensation will depend on the nature of the invention, as follows:

- For inventions made in the course of a mission, the inventing employee must be granted an 'additional remuneration', on top of his salary, which is determined according to collective bargaining agreements, company agreements or individual employment agreements. Since the above-mentioned legal provisions are of public order, no agreement (whether collective, company or individual) can provide for less favourable remuneration. In that respect, French courts have overruled certain agreements when they are less favourable than the applicable legal provisions – for instance, a collective bargaining agreement providing that the additional remuneration will only be granted if the invention is exploited by the employer within a certain time period after the patent is filed.[11] In order to evaluate the additional remuneration, employers must take into account the economic incentive for employers, the scientific relevance of the invention, the technical difficulties overcome and the level of employee input. There is no requirement that the additional remuneration determination be made in comparison with the inventor's salary.[12] When no contractual provision (in a collective bargaining agreement, a company agreement or an individual employment contract) defines the rules applicable to the additional remuneration owed to an inventing employee, the courts will calculate the amount to be awarded to the employee on the basis of evidence provided by the parties, applying the above criteria. If an employee and the employer disagree on the amount of additional remuneration, the Conciliation Commission for employees' inventions or the Paris Court of First Instance will determine the amount to be awarded.
- For inventions made outside any inventive mission but attributable, the inventing employee must be granted the 'fair price' that is memorialised in an agreement between the employee and the employer. If the employee and the employer fail to reach an agreement, the Conciliation Commission for employees' inventions or the Paris Court of First Instance may determine the amount to be awarded, taking into account any elements which may be submitted by either party, with a view to assessing the fair price taking account of both the initial contributions of each, and the industrial and commercial interest of the invention.
- For inventions made outside any inventive mission and non-attributable, the amount will be determined freely in the assignment or licence agreement entered into between the inventing employee and the employer.

11 See Supreme Court, Commercial Chamber, February 18 2004, 02-17421.

12 See Supreme Court, Commercial Chamber, November 21 2000, 98-11900, Bull civ IV, 179.

If the inventor is a civil servant, the specific regime set forth in Article R611-14-1 of the Intellectual Property Code will apply. For inventions made either in the course of the performance of their functions (which constitutes an inventive mission) or of studies or research expressly entrusted to them, civil servants must be granted a profit-sharing bonus calculated on the profits generated by the exploitation of the invention by public entities that benefit from the invention and on a patent premium incentive. While the profit-sharing bonus takes the form of a royalty, the patent premium incentive is a lump sum that is determined yearly by a joint order of the ministers of budget, public service and research.

1.4 Know-how and trade secrets

There is no legal definition of 'know-how' under French law. Courts have defined it as a package of non-patented practical information which:

- results from experience and testing;
- is secret and is kept secret by material and legal means (ie, not generally known or easily accessible);
- is substantial for the holder (eg, significant and useful); and
- can be identified (ie, described in a sufficiently comprehensive manner).[13]

Know-how, which includes trade secrets, can be thought of as a competitive advantage which is not protected by any registered right. Protection exists, however, and subsists for as long as the know-how is secret: there is no legal expiration date. In France, the unlawful acquisition, disclosure and use of a trade secret may give rise to a civil action based on Article 1382 of the Civil Code.[14] This article, which includes the general principle of civil liability, is applicable to acts of unfair competition and free-riding, which includes the disclosure and/or misappropriation of trade secrets.[15] However, French Courts strictly construe the secrecy requirement, and a failure to maintain secrecy can result in a loss of protection.[16]

While know-how is protected under the principles of general tort law, the narrow category of 'manufacturing secrets' is separately protected by the combined application of Article L621-1 of the Intellectual Property Code and Article 1227-1 of French Labour Code, which qualify the misappropriation of trade secrets as a criminal offence. For many years, some members of the French Parliament have unsuccessfully argued that the protection found in these articles should be extended to all trade secrets. While, to date, this argument has been unsuccessful, the situation may change with the proposed EU directive on the protection of undisclosed know-how and business information (trade secrets) against their unlawful acquisition, use

13 See Supreme Court, Criminal Chamber, September 19 2006, 05-85360.

14 According to Article 1382 of the French Civil Code, the claimant must normally establish (i) a wrongful act (ii) damage and (iii) a causal link, in order to have the right to compensation. Regarding acts of unfair competition, however, the courts tend to infer the prejudice from the wrongful act, without necessarily examining the causal link (Supreme Court, Commercial Chamber, February 27 1996, 94-16885, D1997).

15 See notably Supreme Court, Social Chamber, May 12 1959, Ann. Propr. Ind., 1962, 3, 52; Douai Court of Appeals, July 1 1996, PIBD, 1997, 627.

16 See Paris Court of Appeals, April 12 2013, 12/21643: the Court ordered the assignor of certain know-how to pay €200,000 to the assignee because the assignor had disclosed the know-how in question to one of its subsidiaries after the know-how assignment.

and disclosure, which the European Commission submitted to the European Parliament on November 28 2013. Indeed, Article 1 of the proposed EU directive defines 'trade secrets' as any information which is secret, has commercial value because it is secret, and has been subject to reasonable steps by the person lawfully in control to maintain its secrecy; this definition is broader than the 'manufacturing secrets' category.

Additionally, certain acts relating to the disclosure and misappropriation of know-how can qualify as a criminal offence pursuant to the French Criminal Code, notably under the general provisions on theft,[17] receiving,[18] corruption,[19] fraud, breach of trust,[20] espionage[21] and breach of professional secrecy.[22]

1.5 Employees, consultants and the use of non-disclosure agreements

Protection of know-how is often the subject of confidentiality and non-disclosure agreements signed with third parties which need access to the confidential information. Such agreements are particularly common during contractual negotiations. Obligations of confidentiality and non-competition may also appear in general agreements, such as employment contracts, consultancy agreements, licence agreements, joint venture agreements, outsourcing agreements or research agreements.

No particular laws apply to the drafting of non-disclosure agreements, but such agreements should be drafted carefully to guarantee effective protection. In particular, it is very important to define and precisely identify the information and documents which are considered confidential, as well as the purpose for any disclosure and the permitted uses.

It is worth noting that employment contracts give rise to an obligation of confidentiality such that employees must keep confidential any of the company's know-how learned in the course of performing their contractual duties, even if the contract does not specifically so state.[23]

---

17 Article 311-1 of the French Criminal Code. Theft is punishable by three years' imprisonment and a fine of up to €150,000 for individuals or up to €750,000 for companies.

18 Article 321-1 of the French Criminal Code. Receiving is punishable by five years' imprisonment and a fine of up to €375,000 for individuals or up to €1,875,000 for companies.

19 Article 433-1 of the French Criminal Code. Corruption is punishable by 10 years' imprisonment and a fine of up to €150,000 euros for individuals and up to €1,750,000 for companies.

20 In France, for instance, Article 314-1 of the French Criminal Code provides that a breach of trust is committed when a person, to the prejudice of another, misappropriates funds, valuables or any property that was handed over to him and that he accepted subject to the condition of returning, redelivering or using it in a specified way. This provision could apply to the breach of trade secrets. Breach of trust is punishable by three years' imprisonment and a fine of up to €375,000 for individuals and up to €1,875,000 for companies.

21 Act 68-678 of July 26 1968. Unless otherwise specified by any other applicable law, espionage is punishable by six months' imprisonment and/or a fine up to €18,000 for individuals and companies.

22 Article 226-13 of the French Criminal Code. Breach of professional secrecy is punishable by one year's imprisonment and a fine of up to €15,000 for individuals and €75,000 for companies.

23 See Versailles Court of Appeals, 11th Ch, August 25 2009, 07-1849, *Expert v SAS British American Tobacco*.

2. Context and features of patent transactions

2.1 The principal dealings in a patent

Patent assignments and licences must comply with the general rules for the validity of agreements, as provided by Article 1108 of the French Civil Code. Thus consent, ability to contract, defined object (ie, a valid patent right) and lawful cause should be covered. An inventor's moral rights – which include the right to claim inventorship and to enjoy the benefit thereof – cannot be assigned or licensed. In addition to the general rules applicable to contracts, there are specific rules applicable to patent assignments and licence agreements, as set out next.

(a) Patent assignments

Pursuant to Article L613-8 of the Intellectual Property Code, a French patent or patent application may be assigned in whole or in part by the patent holder. It is important to note that the priority right and any know-how related to the patent are not *de facto* included in the scope of an assignment agreement. Therefore, parties must agree to include them if they wish to do so.[24]

First, the general rules of the French Civil Code related to assignment will apply with equal force to patent assignments.[25] As a consequence, the assignor must deliver the patent and, if expressly agreed by the parties, any related priority right and know-how; and the assignor must also guarantee the full enjoyment of the patent rights by the assignee. Once the patent has been assigned, the assignor must cease the exploitation of the assigned patent and may not disrupt the assignee's exploitation thereof in any way. The remedy in case of violation of the assignor's obligations is either a contract rescission or the payment of damages. Damages may also be awarded to an assignee where the assignor knowingly assigns a defective patent; however, it is common practice for patent assignors to limit or exclude any warranty as far as the patent validity is concerned, and to guarantee only the patent's material existence. Even so, pursuant to Article 1628 of the French Civil Code, clauses excluding the guarantee of the patent assignor from its own disruptive acts are contrary to French public policy and therefore not enforceable.

Secondly, the assignee's main obligation is the payment of the assignment price, and assignment agreements usually include provisions governing contract rescission in a case of non-payment.[26] Additionally, where the price takes the form of a royalty, assignees are often required to exploit the assigned patent and to maintain it by paying maintenance fees. Thus, the assignee must make its best efforts to implement the invention and manufacture and sell the products or other results of the invention. Unless otherwise agreed, the assignee is free to determine its commercial practice. If the assignee fails to exploit the patent, however, the assignor may claim for contract rescission and as a consequence recover its rights to the patent.

Finally, pursuant to Articles L613-8 and L613-9 of the Intellectual Property Code,

---

24 See Paris Court of Appeals, 4th Chamber, January 1991, PIBD 1991, 502, III, p360.
25 See Articles 1604 and following of the French Civil Code.
26 See Paris Court of First Instance, Third Chamber, sect 1, 2004/06467, PIDB 2008, 867, III, p63.

assignment of French patent applications and patents shall be executed in writing; otherwise they are null and void. In addition, assignment deeds must be recorded at the INPI in order to render them enforceable against third parties.

(b) *Patent licences*

Pursuant to Article L613-8 of the Intellectual Property Code, patent applications or patents may be licensed in whole or in part. Licences can be exclusive or non-exclusive for a definite or indefinite duration.

In granting a patent license, the key considerations for licensors are:

- the conditions of effective transmission of the licensed invention and the obligations surrounding any technical assistance;[27]
- the scope of the licence (eg, limitation to a certain technical field);
- the obligation to pay maintenance fees for the duration of the licence;
- the rules applicable to ownership and communication of any improvement made by each party;
- the scope of a 'latent defect' guarantee (where, for instance, it is impossible to achieve the expected result);
- the financial consequences of patent nullity[28] or patent infringement; and
- the authorisation or prohibition of sub-licences.

A licensee's obligations would include personal, effective and serious exploitation of the licensed patent,[29] timely payment of royalties, and compliance with the licensor's quality standards. Licensees would also need to consider non-competition provision during the contract term, and prohibition of any contract assignment.[30]

The price for a licence can be a lump sum or a royalty, or a combination of the two. Parties are free to specify a minimum guaranteed income for licensors and to determine the payment frequency as well as the royalty basis (eg, turnover or profits).

A licensor may request termination if its licensee does not exploit the licensed patent or pay the royalties due. Similarly, a licensee may request termination as a consequence of patent nullity or when the technical implementation of the patent invention is impossible.[31] Damages may be awarded to compensate for the loss of revenue and expenses accrued by licensees to implement the invention or to manufacture the products.

Additionally, under specific conditions Article L613-11 and following of the Intellectual Property Code provides for compulsory licensing for:

- lack of exploitation or insufficient exploitation;
- dependent patents; and
- reasons of public health, national economy and defence.

27 Such a clause should define and precisely identify the scope and conditions of the licensor's technical assistance. See Supreme Court, Commercial Chamber, January 10 2010, 09-11201.
28 This applies specifically in relation to royalties already paid by licensees.
29 This implies the prohibition of sub-licences unless otherwise agreed by the parties.
30 Licensors may also include a termination clause to cover the case where a licensee undergoes a merger or control change.
31 Depending on the nature and gravity of the default, French courts may order a royalties reduction.

As to public health, compulsory licensing is applicable in two sets of circumstances. First, compulsory licensing applies to any patent related to a medicine, a medical device, an *in vitro* diagnostic medical device and any related therapeutic product, as well as their production or manufacturing processes, and any products required for their production. But, secondly, the foregoing applies only where the patent holder deals in the sale of medicines, medical devices or therapeutic products in insufficient quantities or at abnormally high prices, or where it exploits the patent under conditions contrary to public health or under practices definitively determined to be anti-competitive.[32]

Finally, as required for patent assignments, patent licences must be executed in writing. They should also be duly recorded at the INPI to make them enforceable, because although there is no longer an obligation to do so – at least, where they are non-exclusive – it is always advisable to have them recorded in order to avoid any challenge as to their enforceability against third parties.

2.2 Patent rights in corporate transactions

(a) Mergers and acquisitions

Under French law, because a sale of shares or a share purchase agreement only modifies the capital shareholding, assets (including patents) remain the company's property. However, if a transaction consists of a sale of a business, the ownership of the assets, including patents and know-how, will be transferred to the purchaser. In the course of such a sale, the parties must record the transfer of all patents at the INPI to render them enforceable against third parties.

While there is neither a legal requirement nor any other form of obligation to perform due diligence in the context of a merger or acquisition, when it comes to patent due diligence it is advisable to verify: the status of the patent rights (ie, that annuities have been duly paid to keep them in force, and the mechanisms for renewal), the status of ownership, any licences and any other rights granted on the patents (eg, a pledge), their geographical scope, and the existence of any claim or litigation relating to them. Similarly, an analysis of the freedom to operate is recommended in order to assess whether any third-party right could affect the proper exploitation of the patent portfolio. More broadly, the value and transferability of the core business as well as the potential risks must be assessed. In the life sciences sector, companies should thus carefully scrutinise the regulatory approvals required for placing medicinal products on the market, the associated marketing and advertising practices and any available information on pipeline products and clinical trials, product recalls, inspections and complaints. To protect against risks, sellers should make warranties and representations, including that all patent rights are validly registered, that they are not aware of any threats to their validity, and that all necessary licences and permits exist at the time of signing (and the same should be repeated at the time of closing).

Aside from patent due diligence, it is important to ensure that employees and managers with the necessary know-how, expertise and experience are transferred

32 See Article L613-16 of the Intellectual Property Code.

from the seller to the purchaser. To that end, purchasers should identify who holds valuable know-how and determine what the terms and conditions of their employment contracts are. Moreover, it is advisable to investigate who has access to any know-how or trade secrets, and the maintenance of such access mechanisms. The existence of transfer or licence-out agreements and any misappropriation or misuse of know-how and corresponding action taken will also be important to assess. Purchase agreements should clearly state that the seller must destroy duplicate copies of any trade secrets, or that the seller will maintain adequate provisions to safeguard the company's acquired interest in relation to licences held.

When corporate transactions involve a change in control, it is necessary to audit contracts with third parties related to patent licences and core business transactions[33] in order to ascertain their transferability and evaluate the risks associated with potential termination.

(b) Financing

Pursuant to Directive 2003/71,[34] Regulation 486/2012[35] and Article L412-1 of the French Monetary Code,[36] an issuer must include, in the documentation associated with its public offering, summary information regarding the extent to which it is dependent on patents or licences and a description of any research and development (R&D) policies, including the amount spent on sponsored R&D activities. Additionally, according to Article 3 of Instruction 2005-11 issued by the French Securities and Exchange Commission (the AMF), an application filed with a prospectus draft or base document must contain a statement related to patent and trademark rights and quality certifications. Despite these requirements, there is no rule specifically applicable to due diligence prior to an offering.

(c) Joint ventures

When setting up a joint venture, freedom to contract prevails between the parties. The legal regimen of either co-ownership[37] or assignment to a common subsidiary (incorporated by the parties in view of owning and exploiting various rights) may come into play. Usually, both parties contribute the necessary patents rights to the joint venture and remain the sole owners of the contributed prior patent rights unless otherwise specified. As a consequence, the ability to enforce patent rights will be held either by the co-owners of the patent(s) in question or the common subsidiary. As an incentive to encourage such common collaboration, parties may grant each other free, non-exclusive cross-licences.

Public entities participating in joint ventures usually request the full ownership

33 Such contracts are likely to cover research and developments agreements, manufacturing agreements or distribution agreements.

34 EU Directive 2003/71 of November 4 2003 on the prospectus to be published when securities are offered to the public or admitted to trading.

35 EU Regulation 486/2012 of March 30 2012 related to the format and the content of the prospectus, the base prospectus, the summary and the final terms and as regards the disclosure requirements.

36 See Directive 2003/71, Articles 6.4 and 11 of annex I and Article 6.4 of annex X; see also Regulation 486/2012, Article 5.4. of annex XXIII.

37 See above in section 1.2.

– or at least co-ownership – of the patent rights, because it is usually a condition to the receipt of public aid.

In France, joint ventures are generally incorporated as *sociétés par actions simplifiées* (SAS) due to the high flexibility of the statutes and governance requirements associated with this type of entity.[38]

Parties generally anticipate the consequences of the termination of a joint venture. Although there is little precedent on this issue, French courts have given some guidelines as to the consequences of termination if such consequences are not specified in their initial agreement. When the parties to a joint venture do not incorporate a common subsidiary, the courts usually order co-ownership of patents upon dissolution, since the parties had initially decided to collaborate. In contrast, when a joint venture takes the form of a common subsidiary, the courts are likely to rely on the exit provisions under the common corporate law mechanisms (such as any 'shotgun' clause, allowing a shareholder to offer a specific share price to other shareholders, which are then bound to accept or to buy at such a price).

(d) Divestment

'Divestment' is the partial or complete sale by a company of activities or branches, including industrial property rights. Under French law, the accessory follows its principal unless otherwise provided for by the parties. As a consequence, when a patentee decides to dispose of his patent and sell it to a third party, all the rights (except the priority right, if not specified) and licences will be transferred to the new owner. Indeed, Article L613-8 of the Intellectual Property Code provides that licences granted prior to the sale of a patent are not affected by the transfer of such patent.

Regarding warranties and indemnities, the key legal considerations are identical to the ones identified for patent assignments and acquisitions.

(e) Consortium agreements

Freedom to contract prevails between the parties to consortium agreements. During negotiations between parties to a consortium agreement, key issues are confidentiality, each party's contribution (ie, reliability, skills, key persons and technical background), conditions of availability of each party's technical expertise, and the definition of specific requests or needs. While drafting consortium agreements, parties should define:

- the scientific programme planned (ie, the agenda, the parties' contributions including finance, and milestones);
- ownership of patents and know-how;[39]
- planned exploitation and revenue sharing;
- conditions for scientific publication or communication;[40]
- consortium governance; and
- exit clauses.

---

38 See Article L227-1 of the French Commercial Code.

39 Co-ownership may be determined depending on the inventors or technical/scientific field of the results.

40 Generally, publication or communication by one party will involve the prior and complete approval of the other party, will mention the name of the research entity and will keep secret any know-how.

Furthermore, specific rules (as set forth in Articles L413-1 and following of the French Research Code) apply to the participation of civil servants in research activities operated by private companies or consortiums.

2.3 Particular business transactions

(a) *Material transfer agreements*[41]

In France, material transfer agreements are usually executed between a public entity such as a university[42] and a private company. Except for public procurement contracts, freedom to contract prevails between the parties.

Specific considerations relating to material transfer agreements are:

- confidentiality;
- restriction on parting with possession;
- the definition of the permitted purposes and uses;
- the assignment price;
- the ownership of any new intellectual property that the material recipient may generate; and
- non-disclosure.

While transferring toxic and infectious material, the material holder must ensure that the recipient is aware of bio-safety issues, owns the required equipment to manipulate the said material and will assume all associated risks.

(b) *Commercial research and development agreements*

Collaborative arrangements: Due to the uncertainties surrounding results from research, the parties' obligations are usually mainly 'best endeavours' obligations. Key clauses in any agreement are likely to cover:

- the definition of the applicable scientific and technical fields;
- the scientific programme planned (covering agenda, parties' contributions, milestones);
- reporting mechanisms;
- how costs will be shared (including staff and materials);
- ownership of the results;
- conditions of results exploitation (which might be based on a co-ownership agreement or a licence, with revenue sharing);
- confidentiality and non-disclosure issues; and
- the conditions for disclosure of the research results in scientific publications.

Particular issues arise in connection with Regulation 1217/2010[43] which prohibits

---

41 A material transfer agreement governs the transfer of biological materials between entities with a view to conducting biological research or technical evaluation before assignment or licence. The type of biological material involved may include compounds, cell lines, vectors, proteins, viruses or genetic material.

42 For instance INSERM, Institut Pasteur, Assistance Publique des Hôpitaux de Paris.

43 Regulation 1217/2010 of December 14 2010 on the application of Article 101(3) of the Treaty on the Functioning of the European Union to certain categories of research and development agreements.

anti-competitive collusion in R&D agreements. It is therefore prohibited to waive any challenge to intellectual property rights. Nor is it acceptable to prohibit third-party licences or to prohibit competition with the other party through R&D activities in a non-related domain or in the same domain after the termination of the collaboration.

Non-collaborative arrangements: The structure and issues associated with non-collaborative arrangements are similar to those associated with collaborative arrangements. Key considerations for the research principal are the definition of the research or experimentation scope, obligations on researchers to achieve a particular outcome, and the consequences of failure. However, regarding the obligation to achieve a particular outcome, case law mitigates the effect of such provisions by invoking the uncertainty surrounding scientific researches. Regarding ownership of the experimental results, they usually belong to the principal, while the research entity remains owner of the know-how and knowledge used in order to obtain the said results.

(c) ***Contract manufacturing agreements***

Articles L5124-1 and following of the Public Health Code strictly regulate the manufacture of medicinal and therapeutic products[44] in France. Companies wishing to manufacture such products must comply with the regulation applicable to pharmaceutical establishments and have prior authorisation from the ANSM (France's medicinal products protection agency), as well as a pharmacist (or pharmacists) as a shareholder or in executive management. Pursuant to Article R5124-2 of the Public Health Code, 'manufacturing' also includes the purchase of the source products, quality control, batch release and storage. Manufacturers must also comply with the goods practices issued by the ANSM as updated in March 2014.[45] Specifically, manufacturers must maintain written and updated records of the specifications, formulas, manufacturing and conditioning instructions, processes, reports and records related to each manufacturing operation. They must also hold specific documentation for each batch and for experimental products.

Manufacturing agreements must be executed in writing. Before entering into such an agreement, parties must verify the manufacturer's skills, experience, technologies, manufacturing processes and premises. Key clauses concern:

- the transmission of any relevant information related to the product to be manufactured and its confidentiality;
- compliance with the legal standards and with all necessary marketing authorisations;
- purchasing of the source materials and conditioning products;
- quality control and the conditions under which batches are released; and
- product liability.

---

44 See also Articles R5124-1 and following of the Public Health Code.

45 "Manufacturing good practices", ANSM, Ministry of Heath Official Bulletin2014/1*bis*.

(d) *Product distribution agreements*

Distribution of medicinal and therapeutic products is strictly regulated by Articles L5124-1 and following of the Public Health Code. Under French law, only pharmaceutical laboratories, certain wholesalers and depositaries, and pharmacists are legally entitled to purchase and sell medicines and medical devices.

Wholesalers are the main intermediary in medicinal distribution, and are allowed to purchase for resale, store and deliver non-experimental medicines to other business entities (thus, they are not allowed to sell medicines directly to patients). They are subject to specific requirements, notably regarding the content of their inventory, delivery lead time and the geographical scope of their activity.[46] Alternatively, laboratories can distribute their products directly to pharmacists or pharmacists' groups. Since 2009, certain pharmacists' groups (*centrales d'achat pharmaceutiques*) have been allowed – on behalf of themselves, other pharmacists or groupings known as *structures de regroupement à l'achat* – to purchase and store non-reimbursable medicines. The last group may also purchase and store non-reimbursable medicines and any other product that pharmacists can sell, but only if the pharmaceutical establishment is a member of the group. Only pharmacists are allowed to offer for sale and sell medicines to the public.

Pharmaceutical laboratories and wholesalers can also enter into contracts with depositaries, which may not own the products of the entity on behalf of which they act, nor purchase the products for resale.[47] Depositaries may be representatives, commission agents[48] or commercial agents. In addition to the Public Health Code requirements, Articles 1984 and following of the French Civil Code govern parties' agreements. When depositaries are in charge of negotiating, entering into contracts on behalf of laboratories or prospecting for new clients, they become commercial agents,[49] which then implicates them in public order issues. Specifically, principals are legally required to give written and sufficient prior notice to commercial agents before terminating or deciding not to renew contracts, to respect the duration of prior notice provided by Article L134-11 of the French Commercial Code and to compensate commercial agents in the event of termination or non-renewal of the contracts.[50]

Distribution agreements are subject to the general rules applicable to commercial agreements relating to the transparency and framework of commercial negotiations[51] and to the prohibition of anti-competitive practices. Distribution agreements must be executed in a framework agreement by March 1 of each year, and must specify tariffs, discounts, payment terms diverging from their standard terms and conditions, and when the distributor will render commercial services to its suppliers.[52]

---

46 See Article R.5124-59 of the Public Health Code.

47 Article R5124-2 of the Public Health Code.

48 They are commission agents when they act in person without acting on behalf of laboratories – Article L132-1 of the French Commercial Code.

49 See Articles L134-1 and following of the French Commercial Code.

50 French Courts usually award compensation equivalent to between 18 and 24 months of commissions.

51 See Articles L441-6 and L441-7 of the French Commercial Code.

52 Practices that do not respect Article L441-7 of the French Commercial constitute criminal offences punishable by a fine of up to €75,000.

Additionally, Article L138-9 of the French Social Security Code provides rules applicable to discounts and commercial benefits granted by suppliers of pharmaceutical products to their clients.

(e) *Co-promotion agreements*

Co-promotion agreements consist of the joint marketing of pharmaceutical products by at least two laboratories. Parties may decide to market a product under a single trademark (co-promotion or co-branding) or two trademarks (co-marketing). Such agreements aim at accelerating product implementation in the market and increasing the commercial impact. Here, freedom to contract also prevails between the parties.

The main considerations will be the trademark issues (ownership and use), pricing policy and geographical scope. However, owing to competition issues (including pricing agreements and market partitioning) and to France's strict regulation of the promotion and advertising of pharmaceutical products, co-promotion agreements have become less frequent.

(f) *Clinical trial agreements*

France implemented EU Directive 2001/20 related to the implementation of good clinical practice in the conduct of clinical trials on medicinal products for human use through Article L1121-1 and Articles R1121-1 and following of the Public Health Code. French standards lay down few binding requirements, but those specifically set out cover:

- prior and positive notice from France's Ethics Committee (the *Comité de Protection des Personnes,* CPP) and from the ANSM;
- written agreements;
- obligations on the sponsor to subscribe to appropriate insurance cover;
- prior and informed consent of the patients participating in the trial; and
- free participation and free provision of the trial products.

Before entering into a clinical trial agreement, a protocol must be established, mainly aiming at defining the therapeutic area, the investigational medicine, requirements pertaining to trial subjects (numbers, selection or exclusion criteria, consent), details about the concrete conduct of clinical trials, financial considerations and confidentiality. Confidential data usually includes the protocol, the case report form on each trial subject, and the investigator's brochure.

In addition to the parties' respective obligations and responsibilities, the term, the fees and confidentiality matters, parties usually include clauses relating to the ownership of the results, the conditions of the disclosure of the trial results in scientific publications, the archiving period for each trial subject, and also the possibility for the laboratory to audit the other party (including processes and premises) or the opportunity for the investigator to outsource the clinical trials.

New rules on clinical trials will apply when Regulation 536/2014 of April 16 2014 on clinical trials on medicinal products for human use, repealing EU Directive 2001/20, become applicable (which will be no earlier than May 28 2016).

Clinical trial agreements involve substantial data processing, and data transfers to countries or private/public entities. Such activity is subject to Articles 53 and following of Act 78-17 dated January 6 1978 on Data Files, Data Processing and Individual Liberties,[53] as well as methodology standard MR-001 which provides specific requirements for the processing of personal data for the purpose of medical research.

3. Issues associated with patent transactions

3.1 Regulatory law and compliance

As a member state of the European Union, France must comply with Directive 2001/83 on the Community code relating to medicinal products (hereinafter 'the Community code'), which has been implemented in the Public Health Code. France must also adhere to Regulation 726/2004 laying down Community procedures for the authorisation and supervision of medicinal products for human and veterinary use and establishing the European Medicines Agency, and to the SPC Regulation.

(a) Marketing authorisations

Because of its key impact on human health, the pharmaceuticals sector is highly regulated. Consequently, a medicinal product may only be placed on the market if it complies with statutory requirements, the purpose of which is to verify its quality, safety and efficacy. A medicinal product may only be placed on the market if a marketing authorisation has been granted either by the ANSM or by the European Commission. A marketing authorisation is granted for a period of five years and may be renewed thereafter indefinitely.

The European Commission grants marketing authorisations for medicinal products referred to in Annex to the Regulation 726/2004 (mandatory scope), such as medicinal products containing a new active substance for treatment of cancer, diabetes and viral diseases, medicinal products developed by means of recombinant DNA technology or monoclonal antibody methods, or medicinal products designated as orphan medicinal products pursuant to Regulation 141/2000. The Commission may grant marketing authorisations for other medicinal products (optional scope), such as medicinal products constituting a significant therapeutic, scientific or technical innovation. When a medicinal product falls outside the scope of Regulation 726/2004, the ANSM is solely competent to grant its marketing authorisation, through a national procedure, a decentralized procedure or a 'mutual recognition' procedure.

(b) Data exclusivity

Companies may apply for a marketing authorisation through the abridged marketing authorisation procedure. Under this procedure, applicants are not required to provide the results of pre-clinical tests and clinical trials as part of their marketing authorisation application, provided that they can demonstrate that the product

53 French data protection legislation is frequently updated in order to implement EU regulations.

under consideration is a generic product, that is, a medicinal product which has the same qualitative and quantitative composition in active substances and the same pharmaceutical form as a reference medicinal product, and whose bioequivalence with the reference medicinal product has been demonstrated by appropriate bioavailability studies.

This abbreviated application may only be submitted eight years after the initial marketing authorisation for the original medicinal product was granted (the so-called 'data exclusivity period'). In addition, the generic product can only be placed on the market 10 years after the initial marketing authorisation (the so-called 'market exclusivity period'). If a new therapeutic indication with a significant clinical benefit has been approved for the reference product during the first eight years following the initial marketing authorisation, the reference product benefits from an additional one year of marketing exclusivity (the so-called '8+2+1' formula). The same regime applies for marketing authorisations granted at the EU level,[54] as well as for marketing authorisations granted on the basis of French national procedures.[55]

The initial marketing authorisation – together with all its variations and extensions as well as any additional dosage strengths, pharmaceutical forms, administration routes or presentations authorised in relation to the medicinal product covered by the initial marketing authorisation – constitutes a global marketing authorisation. This means, for the purposes of data and marketing authorisations, that variations and extensions of the first marketing authorisation as well as new marketing authorisations deriving from the initial marketing authorisations, do not extend the data and marketing exclusivity period of the reference medicinal product (unless they fall under an exception – for example, for an orphan drug under Regulation 141/2000).

In accordance with Article 10(6) of the Community Code, French legislation allows an exception to data exclusivity protection in relation to "studies and tests required to obtain a marketing authorization for a medicinal product and the action needed to achieve them and obtain the authorization" (the so-called 'Bolar' exemption). France has, however, broadened the scope of the exemption provided by Article 10(6) of the Community Code to include studies and trials carried out by innovative companies. Thus, under French law the exemption covers studies and trials in support of self-standing applications for medicinal products, and applications for new combinations of authorised active substances.

(c) *Patent extensions*

Conditions for the grant of an SPC have raised many questions and have been the subject of numerous court cases, which have shaped the interpretation of Regulation 469/2009. Even so, a number of questions relating to patent extensions still remain unanswered today.

---

54 Article 14(11) of Regulation 726/2004.
55 Article L5121-10-1 of the Public Health Code.
56 See, Paris Court of First Instance, February 1 2006, PIBD, 829, III, p317; Paris Court of First Instance, January 7 2009, PIBD 2009, 891, III, p181.

(d) ***Role and importance of selected regulatory aspects of transactions***

Table 1 sets out significant points in relation to selected regulatory aspects of transactions, including their associated risks and how those risks can be mitigated.

Table 1: Selected regulatory aspects of transactions

Issue	Deal structure to which this issue applies	Possible implications /risks	Possible risk mitigation measures
Existence of licences and authorisations for all products and business activities	Share deal, possibly asset deal and sale of business, depending on whether licences are transferrable	Risk of violation of law and the associated sanctions	• Regulatory due diligence • Representation by the seller that all necessary licences and permits exist at the time of signing (and the warranty should be repeated at the time of closing) • Important to check licences for manufacture, distribution and marketing • Important to check licences granted to the target, and product-specific licences
Pending authorisations for a specific product	Share deal, asset deal, sale of business	Negative decision by the regulator – ie, the authorisation not being granted	• Representation by the seller that the application has been duly, timely and correctly filed • Indemnity from the seller in the case of non-approval

3.2 Commonly disputed matters

Pursuant to Article L615-17 of the Intellectual Property Code, the Paris Court of First Instance has exclusive jurisdiction over any actions involving patent rights, including infringement and non-infringement declaratory actions, as well as nullity actions. Questions of infringement and validity are examined by the same court, almost always within the same proceedings. The Paris Court of First Instance also has jurisdiction over cases related to employee and civil servant inventions and to any additional payable remuneration.[56]

However, commercial courts may have jurisdiction to rule on disputes arising in patent transactions where the disputes only concern the performance of contractual obligations.[57] If, however, commercial courts have to assess a patent's validity, they will have to refer the case to the Paris Court of First Instance, which will have exclusive jurisdiction on this issue.[58]

3.3 Competition rules

Antitrust issues have a growing impact in patent and regulatory proceedings for life sciences companies.

First, unlawful agreements (known as 'concerted practices') and abuses of a dominant position are prohibited at the EU level by Articles 101 and 102 of the Treaty on the Functioning of the European Union, as well as under French law through Articles L420-1 and following of the French Commercial Code. In France, the pharmaceuticals sector is subject to particular scrutiny by the French Competition Authority (FCA). The FCA takes the view that it can examine any action by a pharmaceutical laboratory directed to health professionals that is aimed at compromising generic substitution, in particular where the laboratory in question holds a dominant position in the relevant market. Practices such as denigration of competitors' products, predatory pricing, quantitative discounts and tie-up practices may constitute abuse of a dominant position.[59]

Secondly, patent settlement and licensing agreements may fall under Article 101 of the Treaty on the Functioning of the European Union and Article L420-1 of the French Commercial Code, which prohibits agreements which restrict competition. Official guidance on the compatibility of such agreements with the EU regulation is provided by the 'transfer of technology' block exemption given in Regulation 772/2004 and the accompanying guidelines.

Thirdly, certain clauses are subject to a higher level of scrutiny, including agreements entailing a reverse payment by the originator in return for postponing the entry of the generic medicines onto the market ('pay for delay'). To date, the FCA has not ruled on whether such payments are acceptable, but is expected to follow the position taken by the European Commission.

3.4 Securitisation of patent rights

Patents holders can securitise their patent assets (including patent applications) for the payment of any debt, including third-party debt – although the existence of insolvency proceedings will greatly reduce a patent holder's capability to grant such a security to a creditor. Applying Articles L613-8 and R613-55 of the Intellectual

---

57 See Supreme Court, Commercial Chamber, June 7 2011, 10-19030: the commercial court has jurisdiction to pronounce on the alleged breach of the prohibition to disclose know-how attached to a patent licence.

58 See Reims Court of Appeals, Civil Chamber, January 17 2011, 10/02420: the Court of Appeals upheld the decision of a commercial court which declined its jurisdiction because, in order to pronounce on contractual issues, it had to assess the scope of the rights opposed by the claimant in relation to the patents and subject matter of the accused contract.

59 See FCA: decision 03-D-35, July 24 2003, *Novartis Pharma*; decision 07-MC-06, December 11 2007, *Schering-Plough*; decision 07-D-09, March 14 2007, *GSK*; decision 13-D-11, May 14 2013, *Sanofi-Aventis France*; decision 13-D-21, December 18 2013, *Schering-Plough*.

Property Code together, the person or entity granting the security interest must be the patent holder, and deeds that grant, assign and/or waive a patent security must be executed in writing and recorded with both the local tax authorities[60] and the INPI in order to render them enforceable against third parties.[61]

Pursuant to Articles 2337 and 2355 of the French Civil Code, a patent security confers on the beneficiary the right of priority and the right to follow property into the hands of any assignee. Therefore, if a patent is the subject of a security and is transferred, the existing security interest will remain in force and at the same hold priority ranking.

3.5 The effect of insolvency

The effects of insolvency will differ depending on which party to an agreement is insolvent.

If a licensor is insolvent, he may be unable to pay the maintenance fees of the licensed patent. In bankruptcy proceedings, the legal administrator or the liquidator appointed by the court will be in charge of paying such fees. However, if the maintenance fees on a patent are not paid, the patent is automatically revoked and ceases to exist. Licensors or bankruptcy administrators have been known to request reinstatement after the fact, but French courts usually uphold the revocation, finding that an insolvency or a bankruptcy proceeding affecting a patent holder does not constitute a legitimate excuse for failure to pay maintenance fees as required by Article L612-16 of the Intellectual Property Code. If this is the case, any licensing agreements pertaining to the revoked patent also cease to exist.

However, where a patent remains in force despite bankruptcy or insolvency proceedings, French law also provides for the continuation of associated contractual obligations. For example, if, during the course of a bankruptcy proceeding, the bankruptcy authorities decide to assign the patent in question, Article L613-8 of the Intellectual Property Code provides that this act does not alter the licensee's rights and the licence agreement will remain into force until its termination. Additionally, bankruptcy authorities may require that a licensor continue its ongoing contracts or impose the compulsory transfer of a licence to a third party.[62]

As for licensees in insolvency or bankruptcy proceedings, as a general rule they remain obligated to perform their contractual duties, specifically to pay the royalties due. Importantly, Article L622-21 of the French Commercial Code prohibits the automatic termination of licence agreements based on a licensee's insolvency or bankruptcy. This prohibition has public policy implications and therefore applies to – and even overrides – licence agreements which specifically provide otherwise, even if governed by foreign laws.

60 See Article 680 of the French Tax Code and the note in BOI-ENR-DG-20-30-30-20 of September 12 2012.

61 Recordation at the French patent office can be executed at the request of (i) one of the parties to the deed or (ii) the patent holder. To that end, the French patent office requires that the executed security agreement be submitted together with, where appropriate, its translation into French.

62 See Colmar Court of Appeals, 1st Chamber, June 13 1990, RG 457/90.

3.6 Taxation

All expenses related to patent registration or maintenance constitute, at the patent holder's discretion, deductible expenses of depreciable intangible assets. Under French law, the assignment of patents, technical know-how or marketing authorisations are treated as depreciable intangible assets. With respect to royalties earned from licensing agreements, the general rule is that they are a deductible expense for the licensee and taxable income for the licensor.[63] However, the administrative Supreme Court has found that patent or know-how licences may, under certain circumstances, be classified as intangible assets held by the licensee.[64]

3.7 Hot topics

There are currently four hot topics worthy of mention here.

First, the 'transfer of technology' block exemption under Regulation 772/2004 and related guidelines expired on April 30 2014. The exemption has been replaced via EU Regulation 316/2014 of March 21 2014, which entered into force on May 1 2014. The main features of the new rules relate to 'patent pools', no longer automatically exempted clauses, and guidance on settlement agreements. Patent transactions will have to be carefully reviewed in light of this new set of rules.

Secondly, the tension behind clinical data transparency is of major importance for life sciences companies. While it was revising Directive 2001/20 on clinical trials and the EMA policy on access to documents, the biggest challenge for the European Commission and the EMA was to strike the right balance between transparency on the one hand and protection of commercially confidential information on the other, because such information might be used by competitors, which in turn could discourage R&D efforts. With the adoption of EU Regulation 536/2014 of April 16 2014 on clinical trials on medicinal products for human use, repealing Directive 2001/20, it remains to be seen how the regulation will be applied by member states, although the regulation will not become applicable before May 28 2016 at the earliest.

Thirdly, pharmaceutical research is no longer only about discovering new molecules, and the focus of research is tending to shift towards new and improved uses of known molecules. Yet 'incremental innovation', as it is known, is not adequately rewarded by the patent system in France. This is illustrated by recent decisions which adopted a narrow construction of 'second medical use' claims, excluding dosage regimens from patentability in contradiction of a decision of the European Patent Office's Enlarged Board of Appeal and other national rulings in Europe.[65]

Finally, another highly debated area in the field of pharmaceutical patents concerns the conditions of the grant, scope and duration of SPCs. Regulation 469/2009 has been the subject of numerous court cases but many questions still remain unanswered today.

---

63 Tax doctrine BOI-BIC-CHG-40-20-10, §10, of September 12 2012.

64 See State Council, August 21 1996, 154488, SA *Sife*; State Council, October 16 2009, 308494, *Pfizer Holding*. See also Tax doctrine BOI-BIC-CHG-20-10-10, §70, of November 12 2012.

65 See Paris Court of First Instance, November 9 2010, *Teva v Merck*; Paris Court of Appeals, March 12 2014, *Eli Lilly v Teva.*

Germany

Nicole Battistini-Kohler
Manja Epping
Michael Gaul
Christof Höhne
Matthias Hülsewig
Anja Lunze
Steffen Wirsing
Taylor Wessing

1. Patents, SPCs and know-how: foundations for transactions

1.1 Patents and supplementary protection certificates

(a) Overview

Patents are intellectual property rights that can protect so-called 'technical' inventions – that is, product and process inventions. In contrast with utility models, patents are an examined protective right, meaning that an examiner must assess the requirements for patentability before a patent is granted. A patent has a duration of up to 20 years, calculated from the application date onwards, and confers exclusive rights on its proprietor. In pharmaceutical cases, the exclusive rights may be extended for a product covered by the patent for a maximum of a further five years by way of a supplementary protection certificate.

To obtain patent protection in Germany, an application must be filed either with the German Patent and Trademark Office (the DPMA) or with the European Patent Office (the EPO). If a patent is granted by the DPMA, the patent will have a prefix 'DE' before its publication number. If the proprietor seeks protection in more countries than just Germany, he can file his application with the EPO: prosecution and grant of the patent is handled centrally for all European Patent Convention member states in which the proprietor seeks protection. Once granted, the patent becomes a bundle of national patents in the selected countries. These bundled patents have an 'EP' prefix before their publication number.

(b) Requirements for obtaining a patent

An invention may be eligible for patent protection if the claimed subject matter is not expressly excluded from patent protection by Articles 52 and 54 and Rule 23 of the European Patent Convention 2000, Sections 1, 1a, 2 and 2a of the German Patent Law or EU Directive 98/44/EC – subjects excluded cover, for instance, scientific theories, methods of treatment, mathematical methods, and violations of public order. This can be an issue in the life sciences sector, as highlighted by the recent decision of the European Court of Justice in the *Brüstle* case.[1] In this case, it was

1 ECJ, Case C-34/10 *Oliver Brüstle v Greenpeace*.

found not to be permissible to patent a process that involved the removal of stem cells from a human embryo where the effect was that the embryo would be destroyed.

If the subject matter is deemed patentable, the following prerequisites must be fulfilled to obtain patent protection:

- *Novelty:* Under Article 54 of the European Patent Convention and/or Section 3 of the German Patent Law, the invention must be novel. This requires that the invention does not form part of the state of the art. The novelty criterion is absolute – any written or oral disclosure to the public worldwide can be novelty destroying. The patent system is a first-to-file system; there is no grace period in Germany for prior publications of the invention. The date for assessing novelty is either the application date or, if applicable, the date of a prior application by the same party, to which priority may be claimed (the priority date).
- *Inventive step:* Under Article 56 of the European Patent Convention and/or Section 4 of the German Patent Law, an invention is considered as involving an inventive step, having regard to the state of the art, if it is not obvious to a person skilled in the art. To ascertain inventive step, it is determined who the person skilled in the art is. A 'person skilled in the art' is a legal concept – accordingly, it is not a physical person and may even be a team of scientists. This is often the case in the life sciences, since drugs are commonly developed by a team of chemists, doctors and formulation experts. Customarily, an invention will be formulated by a 'problem–solution approach' – that is, the problems of the state of the art are identified over which the claimed invention provides a solution. An invention will customarily be found to involve an inventive step if, in finding the solution, the inventor had to traverse prejudices or overcome (technical) burdens.
- *Supporting disclosure:* An invention as claimed must be disclosed in such a fashion that an average person skilled in the art is enabled to carry it out. Whether a disclosure is enabled is assessed by:
  - determination of who the average person skilled in the art is (see previous bullet point);
  - the specialist knowledge of the average person skilled in the art;
  - the common general knowledge of the average person skilled in the art; and
  - the amount of work the average person skilled in the art can reasonably be burdened with to achieve the aim of the invention.
- *Industrial applicability:* An invention must be industrially applicable to qualify for patent protection. 'Industrial applicability' is assumed when an invention can be manufactured or used in any kind of industry, including agriculture. The aim of this requirement is to avoid the patentability of general, abstract theories. The criterion for industrial applicability is viewed rather liberally and is satisfied if at least the possibility of an industrial application exists.

(c) *Structure of a patent*

A patent (and a published patent application) typically has three elements:

- *A description:* This aims to explain the invention in detail, customarily by identifying the state of the art, the problems with the state of the art, and the way the invention solves these problems.
- *Worked examples, figures or diagrams:* These provide details of ways in which the invention can be worked.
- *The claims:* These are precisely worded paragraphs at the end of the patent which identify exactly what is claimed (ie, what is protected by the patent). There are independent claims and claims (known as dependent claims) referring back to independent claims. When assessing the scope of protection, it is necessary to assess the meaning of the wording used in the claims. In this regard it has to be noted that a patent is considered to be its own lexicon of terms used.[2] Accordingly, an interpretation of the claim wording should be carried out by:
    - looking at the literal context of the term in the claim wording;
    - consulting the description; or
    - consulting the worked examples, figures or diagrams.

It is thus not, for instance, allowed to resort to literature (eg, dictionaries) outside the patent in order to interpret the claim wording. This becomes clear in an example: if the claim wording requires that a certain claim feature of a device is 'round', this does not automatically mean that it has to be round. If the description of the patent defines 'round' as square, this term would have to be interpreted as requiring a square configuration.

Although most patents will conform either to claims protecting a product or a process, or both, there are forms of claims particularly found in life sciences patents that are not easily categorised as either. Such claims are formulated as follows:

- 'X for the purpose of Y' – this is construed in German case law as meaning that the claim protects a product X, which is suitable for the purpose Y.[3] In the context of life sciences, such claims are typically seen in the form of a pharmaceutical compound for use in therapy. The scope of protection is in this case limited to the specific customisation of X for the purpose of Y.
- 'Second medical use' claims – the medical use claim form causes a potential difficulty when attempting to claim any subsequently discovered use of pharmaceutical X for the treatment of a different disease. This is because claims to methods of treatment are prohibited. In order to avoid the appearance of such a claim, the European Patent Convention has sanctioned the use of a claim in the form 'X for use in treating Y' for claims dating from December 13 2007. Prior to this, the so-called 'Swiss' claim in the form of 'X in the manufacture of a medicament for treating Y' had been used – and such claims predating the European Patent Convention 2000 are still in force.

2 BGH GRUR 1999, 909 – *Spannschraube.*
3 BGH GRUR 1992, 305, 307 – *Heliumeinspeisung;* BGH GRUR 1983, 733 – *Hydropyridin.*

(d) *Rights granted under a patent*

Once granted, a patent confers upon its proprietor the following exclusive rights (depending on the patent category):

- *Product patent:* making, disposing of, offering to dispose of, using or importing the product claimed or keeping it for disposal or otherwise; and
- *Process patent:* using the process claimed or offering it for use in Germany without permission from the patent holder; and the disposal of, offer to dispose of, use or import of any product that is obtained directly by means of the process claimed, or the keeping of any such product whether for disposal or otherwise.

The above-mentioned exclusive rights may not be carried out by third parties directly (direct infringement), or indirectly (indirect infringement).

Indirect infringement of a product claim occurs, for instance, if a third party manufactures, disposes of, sells, offers or imports into Germany a part of a patented product, as long as this part qualifies as an essential element of the invention or relates to an essential element of the invention. Indirect infringement requires a so-called 'double domestic relevance' – for example, an offer will only be considered infringing if the offer takes place in Germany and the product is to be sold in Germany.

Indirect infringement of a process claim may occur, for instance, if only certain acts of a patented process are carried out in Germany. A particular requirement in this regard is that significant relevant effects of the invention are obtained in Germany.[4]

Besides direct and indirect infringement, infringement by the doctrine of equivalents must be considered. In this regard, the claim wording is not considered the absolute boundary of protection; rather, obvious variants that a person skilled in the art would take into account are also covered.[5] The following prerequisites have to be fulfilled for infringement by the doctrine of equivalents:

- The product of the potential infringer achieves the same technical effect as the invention claimed in the asserted patent.
- An average person skilled in the art at the priority date would have been able to devise the product of the potential infringer when reading the patent description, without requiring inventive activity.
- An average person skilled in the art at the priority date would consider the product of the potential infringer and its configuration as comparable to the claimed technical solution of the asserted patent.

It has to be noted that although a patent confers exclusive rights on the proprietor, this does not imply that the proprietor is automatically allowed to make use of the patented teaching without being capable of infringing the rights of third parties – just having a patent does not provide automatic clearance to make use of the patented teaching.

---

4 OLG Düsseldorf, BeckRS 2010, 12415.

5 The application of the doctrine of equivalents has been limited in recent jurisprudence of the German Federal Supreme Court: BGH GRUR 2011, 701 – *Okklusionsvorrichtung*; BGH GRUR 2012, 45 – *Diglycidverbindung*.

(e) *Enforcing a patent*

Since only the patent proprietor (or his exclusive licensee) is allowed to make exclusive use of the claimed teaching, he can prevent third parties from infringing his patent and any exclusive rights conferred by the patent.

In Germany, patent infringement cases are tried before specialised patent chambers at the competent regional court. In total, there are 12 regional courts having competence to rule on patent infringement matters in Germany. Most of these patent chambers almost exclusively handle patent matters and therefore have experienced judges. The most experienced – based on the number of filings per year – are the patent chambers of Düsseldorf, Mannheim and Munich.[6]

Germany has a bifurcated system. Accordingly, invalidity as such is not a defence in patent infringement proceedings. Rather, infringement and validity are decided on separate tracks, with the infringement court having the option to stay the infringement proceedings in accordance with Section 148 of the German Civil Procedure Code pending the outcome of a parallel validity attack (ie, the opposition against the patent in suit filed within the opposition period or (only after the expiration of the opposition period) a nullity action in the Federal Patent Court against the patent in suit). This is particularly favourable for the patentee, as the validity track before the EPO (opposition) or German Federal Patent Court (nullity action) customarily moves slower than the infringement track. Infringement litigation (proceedings on the merits) generally takes around 10–14 months from filing to a first-instance decision, whereas the validity track at the Federal Patent Court takes about 20–24 months to a first-instance decision. Accordingly, there may be an 'injunction gap', in which the infringement court has awarded the complaint in accordance with the claims – which can be preliminarily enforced – before validity is considered (and the patent potentially revoked).

Usually, a plaintiff will assert four claim categories in a patent infringement complaint:

- a cease-and-desist claim;
- a claim for accounting;
- a claim for removal/recall and destruction of the infringing device, and
- a claim to declare the defendant liable for compensatory damages.

If successful, a plaintiff can provisionally enforce the individually awarded claims of the judgment (customarily against posting a security bond for damages, in case the judgment is subsequently reversed on appeal). In Germany, an injunction is a regular consequence of law for (direct) patent infringement.

It has to be noted that in Germany damages are only awarded for actual loss or damage suffered (no punitive damages). The amount of damages to be paid is subject to further, separate proceedings – damages may be calculated according to infringer's profit, lost profit or reasonable royalty. In practice, a calculation of infringer's profit and reasonable royalty are most popular; a calculation by lost profit is only

6 Kühnen/Claessen, GRUR 2013, 592: in 2011, 475 patent infringement cases were filed in Düsseldorf (about 40% of all German cases), 265 in Mannheim (about 25%) and 174 in Munich (about 15%).

applicable in limited cases because the plaintiff has to demonstrate that all customers would have bought only his product, which in practice only works in a monopoly market (which may be the case with blockbuster drugs). A calculation of lost profits also, of course, forces the plaintiff to disclose his profit calculation.

In the life sciences sector, and in particular in pharmaceutical cases in which a generics company is the infringer, the plaintiff often tries to obtain a preliminary injunction (which only covers an injunction and in some cases also accounting). To obtain a preliminary injunction in Germany, these prerequisites must be fulfilled:

- *Basis for an injunction:* A demonstration of patent infringement is necessary. As preliminary injunction proceedings are only summary proceedings, only limited evidence need be made available. Moreover, as the court only has limited time to look at the merits of a case, infringement must be clear-cut.
- *Requirements for a preliminary injunction:*
  - The claimant has to act swiftly, once becoming aware of potential infringement. As a rule of thumb, the claimant has to file a request within four weeks of becoming aware of potential infringement – which demonstrates that he is in need of urgent legal relief.
  - The validity of the patent has to be secured. As preliminary injunction proceedings are summary in nature, the court does not have time to scrutinise validity. It is thus customarily required by the courts that the patent has a 'certificate of validity' going beyond the mere fact that it was granted.[7] Such a certificate may state that the patent has survived opposition/nullity proceedings, or that noteworthy companies are licensees to the patent (and thus believing in its validity).
  - A preliminary injunction must be required by the claimant – that is, he has to demonstrate that he would suffer considerable harm and that the case cannot be deferred to proceedings on the merits.

Preliminary injunctions can be granted on an *ex parte* basis (without hearing the respondent) within a couple of days – even within a few hours if necessary – or on an *inter partes* basis (with an oral hearing) within a couple of weeks. To mitigate the risk of an *ex parte* injunction, a potential respondent may file protective letters with the competent court, in which he provides his non-infringement and/or invalidity arguments. These arguments can then be considered when a request for a preliminary injunction is filed. As a rule of thumb, it only makes sense to file protective letters if strong non-infringement and/or invalidity arguments are available. If these are not available, a protective letter can have the reverse effect and increase the risk of an *ex parte* injunction.

(f) ***Supplementary protection certificates***

A system of extending the term of protection for pharmaceutical products after patent expiry, by so-called supplementary protection certificates (SPCs), was introduced in 1993 by Council Regulation (EEC) 1768/92 (now EU Regulation

7 OLG Düsseldorf, InstGE 12, 114 – *Harnkatheterset.*

469/2009, hereinafter the SPC Regulation). The purpose of SPCs is to remedy the common problem that patent term for many pharmaceutical products can be significantly eroded by the time taken for a marketing authorisation to be granted – limiting exclusivity in the market and thereby commercial return. An SPC can alleviate this problem by providing up to five years of further monopoly protection for specific 'products' that were covered by the expired patent.

In order to obtain an SPC, there are three key conditions that must first be satisfied:[8]

- The product must be protected by a basic patent in force.[9] The purpose of this provision is to reflect the intention of the SPC legislation to prolong monopoly protection for products that are already patented.
- A valid authorisation to place the product on the market as a medicinal product must have been granted.[10] This must be the first authorisation to place the product on the market as a medicinal product.
- The product must not already have been the subject of an SPC.[11] The purpose of this provision is to prevent further SPCs for modifications of the same product.

It should be emphasised that an SPC does not prolong the term of a patent. Hence, many other potential products that fall within the scope of protection of the basic patent will not be covered by an SPC.

1.2 Who owns the patent?

The inventor can apply for patent protection and subsequently owns the granted patent. Where the inventor is working for a company and the invention qualifies as an 'employee invention' which the employer wants to claim, the employer may file the application and subsequently become the owner of the patent. The same applies if the inventor has sold his invention to a third party. An invention qualifies as an 'employee invention' if the inventor is employed by an employer while making his invention and the invention was made in the course of his employment.

Inventions unrelated to employment and made, for instance, in the inventor's spare time at home, do not customarily qualify as employee inventions. If an employee invention is made, the inventor has to provide his employer with a notification. Where the employer does not object within a period of four months starting from the date of receipt of the notification, it is assumed that he wants to claim the invention.[12] This, in turn, gives rise to a claim for compensation on behalf of the employee. Where the employer objects within the four-month period, the employee is free to use the invention as his own.

8 Article 3 of the SPC Regulation.
9 Article 3(a) of the SPC Regulation.
10 Article 3(b) of the SPC Regulation.
11 Article 3(c) of the SPC Regulation.
12 Section 6 German Employee Invention Law. The situation is inverse for inventions notified to the employer by the employee before the 1 October 2009: the invention belonged to the employee if it had not been claimed by the employer within the 4 months period. This has been a frequent source of litigation and loss of ownership of patents in the past.

Independent contractors (for instance a consultant) are not considered employees. Accordingly, if the person hiring the independent contractor wants to own inventions made during the time of his work, a separate agreement is required.

If the invention has several inventors, these form (absent a specific contract) a community of part-owners in accordance with Section 741 and following of the German Civil Code.[13] This has the consequence that they are jointly entitled to use the invention. Each inventor may sell his share without the consent of the other inventors; an (exclusive) licence may, however, only be granted with the consent of the other part-owners.[14]

Disputes arising as to ownership of a patent application or patent are brought in a so-called 'vindication action' before competent civil courts.

1.3 Inventor compensation

An employee inventor is compensated once the patent has been granted and has an economic value for the employer. The most common method for calculating employee inventor compensation is yearly royalties until the patent lapses; alternatively, lump sum payments can be agreed upon.

When calculating yearly royalties, several factors are taken into account: the net sales of the product (which net sales were generated in a year?), the contribution ratio of the inventor (was he the sole inventor?), a common royalty amount (what royalties are customarily paid?) and the personal share (did the company motivate the invention, or did the inventor come up with the invention on his own?).

1.4 Know-how and trade secrets

Know-how can be protected from unlawful dissemination by the German Unfair Competition Law, but only while it is secret information. Unlike patent protection, there is no specific threshold that secret know-how has to meet (it does not have to be inventive etc); furthermore, a patent lapses 20 years after the application date, while know-how protection may be available longer, as long as it is kept secret. Another key difference is that a patent application and a subsequent patent are published and thus made available to the general public, while know-how has to be kept secret. In this connection, a patent provides straightforward exclusive rights to its proprietor for a certain time, whereas it may be difficult in practice to avoid the public dissemination of secret know-how by employees and consultants. This may in particular apply when an employee (for instance, from the research and development department) does not leave the company on good terms.

It is often a strategic decision as to how to protect a technical invention. This in particular applies for certain production processes: if these are patented, they are disclosed to the public with the patent application/subsequent patent and, accordingly, a third party would factually know how to use it. With some process claims, it may be difficult to ascertain infringement, in particular if it cannot be

---

13 R Schulte and T Kühnen, *Patentgesetz: Kommentar* (in English: *Commentary on Patent Law*), Heymann, 8th edn, 2008, section 6 at no 21.

14 B. Krasser, *Patentrecht (Patent law)*, CH Beck, 6th edn, 2009, section 19(V)(b).

inferred from the final product how this was produced. Accordingly, on the one side there would be exclusive rights afforded by a patent, but on the other hand it would be factually difficult to enforce this patent; furthermore, the patent would lapse after some time. As a result, know-how protection may in some cases be a better option for protection than a patent.

1.5 Employees, consultants and protecting know-how

As indicated in section 1.4 above, know-how can only be protected while it is secret. This, in turn, requires non-disclosure agreements (generally clauses in employment contracts) with employees or consultants working for a company, so as to mitigate a risk of dissemination to the public. If, then, an employee or consultant disseminates secret know-how to the public, this will be a breach of a non-disclosure agreement. Even so, the action will destroy know-how protection for the information by its having become public knowledge. As stated earlier, it is therefore a challenge to maintain know-how protection.

It is also difficult to maintain a full non-disclosure obligation with employees once they have left the company. Under German law this may, depending on the exact circumstances, be considered anticompetitive for the former employee, giving rise to a claim for compensation for maintaining secrecy after having left employment.

2. Context and features of patent transactions

2.1 The principal dealings in a patent

(a) *Patent assignments*

The assignment of a patent under German law can be conducted by way of a simple contract that does not require written form, although it is recommended. The only exception is the assignment of a European patent application that requires written form according to Article 72 of the European Patent Convention. Although there is no legal requirement under German law to conduct a change of recordal in the register of the DPMA in order to render the transfer to be legally effective, such a change is recommended.[15] The assignment of a patent confers all ownership rights to the assignee. German law does not foresee the necessity of a consideration for the transfer to be effective.

Since under German law a change of recordal in the register is not required for the transfer to be effective, the assignee will want to include in the assignment agreement a warranty from the patentee that he is the sole owner of the respective patent and that the patent is unencumbered with third-party rights.

(b) *Patent licences*

A licence agreement is a contractual relationship under the German Civil Code

15 Pursuant to Section 30(3) sentence 2 of the German Patent Act, a patentee is not entitled to sue for patent infringement against a third party if he is not registered as the patentee in the DPMA patent register. Moreover, an unregistered patentee is not entitled to other benefits regarding the patent register.

between the patentee, who becomes the licensor, and a third party, the licensee. In contrast to a patent assignment, where the ownership of the patent is shifted to the assignee, a licensor retains patent ownership but grants the licensee the right to use the patented technology in accordance with the provisions of the licence contract.

The licence is the right granted by a licence agreement. According to Section 15(III) of the German Patent Act, this right has 'succession protection', whereby the assignment of the patent or the grant of another licence to use the patent does not affect licences previously granted to other parties. Therefore a licence is occasionally called a "quasi right *in rem*" by German legal authors.

There are various reasons for a patentee to license his technology. Besides the reason of generating revenue, other motives of the licensor can be bottlenecks in production or distribution, cost savings by outsourcing, or the lack of a market, research and development knowledge or specialisation in particular industries or technologies. By obtaining a licence, the licensee can exploit new technology without having the risk of research and development, or he can enhance his product portfolio for other reasons.

Licence agreements are also made to end infringement disputes or to avoid suchlike by cross-licensing disputed patents between companies. Sometimes a licence agreement is used for concealed distribution of profits, by assigning the patent to an affiliate and afterwards paying high royalties for the licence granted to the assignor.

There are a number of aspects of licence agreements that need to be considered when negotiations for such an agreement are being undertaken. The most important of these are described in the remainder of this section.

Types of licence: There are various types of patent licence, namely an exclusive licence, a sole licence and a non-exclusive licence. Each is described further as follows:

- *Exclusive patent licence:* By granting an exclusive licence, the licensor waives his right to exploit the patented technology himself or to license the patent to third parties. The licensor still owns the patent but only the licensee is allowed to exploit and utilise the technology under the provisions of the agreed terms. This exclusivity is typically mirrored in the level of royalties payable.

  The licensee is entitled to enforce the patent in court proceedings in his own name. Furthermore, the licensee can grant sub-licences to third parties unless this right has been excluded or limited in the licence agreement. The grant of an exclusive licence can be entered in the DPMA patent register to publicise the extensive rights of the licensee (see Section 30(IV) of the German Patent Act).

  It is the prevailing view that, under German law, an exclusive licence obliges the licensee to make use of the licensed technology. It is advisable to define this obligation in more detail in the licence agreement. In addition, licence agreements often provide for minimum royalties to guarantee a certain amount of royalty income.

  By granting exclusive 'field of use' licences that are limited to certain purposes or applications, the licensor can grant more than one exclusive licence – that is, one for each definable purpose or application.

- *Sole patent licence:* Within a sole licence, the licensor grants to a single licensee the right to use the patent and commits himself not to license the patent to another party. In contrast to an exclusive licence, the licensor retains the right to use the licensed technology himself. In practice it is advisable to define in the licence agreement the scope of rights retained by the licensor.
- *Non-exclusive patent licence:* This type of licence grants the licensee the right to use the patent, but the licensor retains the right to use the licensed patent himself and to grant further non-exclusive licences to third parties.

Unless otherwise stated in the licence agreement, neither a non-exclusive licence nor a sole licence gives the licensee the right to grant sub-licences or conduct court proceedings to enforce the patent against infringers.

Licence scope: Parties to a licence agreement are free to determine the territory applicable to the licence agreement. Under Article 73 of the European Patent Convention, a European patent application may be licensed in whole or in part for the whole or part of the territories of the designated contracting states. For a granted European patent, it should be specified in the agreement for which national parts of the patent the licence is granted. Precise provisions setting forth the applicable territory should be part of each licence agreement to avoid unintended overlaps. In this regard it has to be considered that it comes to exhaustion in the European Union and European Economic Area once a product has been marketed in a member state in accordance with the licence, even if it was territorially limited.

The scope of a licence can be limited by 'field of use' licences. Because a patent confers on the patentee the exclusive right of making, offering, putting on the market or using, importing or possessing for these purposes,[16] the scope of the licence can be limited to one or several of these fields.

In case the licensee is using the licensed technology for research and development during the term of the licence, it should be clarified what happens to improvements or the technology made by the licensee (ie, whether they are covered by the scope of the existing licence or whether an additional one is necessary).

Licence duration: The duration of a licence is a topic to cover in the agreement to prevent any problems in the future. Therefore a licence should not end before the patent protection does, unless there is an explicitly stated reason to. In the absence of any such provision, the licence and therefore the duty to pay royalties both end with the expiry of patent protection. If royalties are paid after the expiry – for example, because the parties agreed on lower royalties in exchange for a longer duty to pay – this must be expressly stated in the agreement.[17]

16 Section 9(1) of the German Patent Act.

17 Guidelines on the application of Article 101 of the Treaty on the Functioning of the European Union to technology transfer agreements, 2014/C 89/03, Section 187.

Royalties: Royalties are the reward for the right to use the patented technology and, depending on the specific situation, the following provisions or a combination of them are conventional to calculate them:

- Lump sum payments are paid at the date of commencement of the licence term and cover the whole term. It should be clarified in the agreement whether the lump sum is partly refundable in the event of early termination.
- Royalties per unit produced are calculated by a fixed amount per piece or a certain percentage of the sales price. Whereas a fixed amount provides a certain security for the licensor, he can benefit from any rise in the sales price if a percentage is set – but he will likewise get less if the price drops.
- Royalties based on a percentage of the turnover of the licensee are common in the life sciences industry, based on net sales, which are defined as the gross amount invoiced for the licensed products less sales returns and allowances, customs and taxes as well as freight and insurance costs. In order to prevent conflicts, the licence agreement needs to contain precise definitions and provisions regarding the method of calculating the turnover and the permitted deductions from gross sales.
- Annual minimum royalties are often combined with royalties per piece and royalties based on a percentage of turnover to secure a basic amount of royalties for the licensor, especially when an exclusive licence is granted.

In complex contracts, the level of royalties depends on other contractual considerations. Conceivable are front-loaded deals with high upfront payments, early-milestone payments, and low or even no periodical royalties; alternatively, royalties could comprise back-loaded deals with no payments at the beginning, late-milestone payments and higher royalties during the period of agreement.

Ownership of improvements: The allocation of improvements (eg, new inventions regarding a licensed patent) is a hot topic in cases where the licensee is assigned to perform researches on behalf of the licensor. Since the licensor is willing to obtain the right on all improvements, the licence agreement has to describe precisely who will own the improvements.

An obligation to assign rights on improvements to the licensor is problematic in respect of European competition rules and may be considered a restriction or distortion of competition (see also section 3.4 below).

Furthermore, it is important to be aware of the German Employee Inventions Act. Under this act, the licensee has to claim inventions reported to him by his employees in order to obtain ownership of the improvements and pay mandatory compensation in return.

Warranties and indemnities: The licensor's obligation to compensate for damage to the licensee caused by breach of contract is statutory in German law and stipulated in section 280 of the German Civil Code. Thus, the concept of warranties and indemnities is unusual within the German jurisdiction.

A distinction must be drawn between contractual provisions that are subject to the statutory liability mentioned above which is – unless limited – a fault-based liability for negligence and intent, and independent warranties according to Section 311 of the German Civil Code, which establish a liability without fault.

It is common for the licensor to guarantee, via an independent warranty, that he has the ownership of the licensed patent and that he is not aware of any intellectual property right of a third party that may be infringed in the course of using the patent as intended.

In most cases, the licensor will not give an independent warranty that the patent is and will remain valid, since it can never be fully excluded that prior art is identified on the ground of which the patentability of the respective patent can be successfully challenged. The parties therefore usually limit the scope of the warranty and oblige the patentee only to warrant that "to the best of his knowledge" the licensed patent is valid. However, it is common for the patentee to warrant that:

- he is the sole owner of the patent and the patent is unencumbered by third-party rights;
- he is not aware of reasons for nullity and cancellation, actions for annulment or third-party offences against the licensed patent;
- any appropriate rights to the invention underlying the patent are validly claimed from employees under the Employee Inventions Act; and
- all renewal fees have been paid in time.

Termination: In general, a licence is granted for a certain period of time and expires automatically, but the parties often push to obtain rights to terminate the agreement before that time. Whereas the licensor would like to have some leverage in case the licensee is in default with the royalties, the licensee wishes to cancel the licence and stop paying royalties as soon as the patented technology becomes unnecessary or his company gets into financial difficulties.

Owing to its nature of being a continuing contractual relationship, a licence agreement can always be terminated for exceptional reasons under German law. According to Section 314 of the German Civil Code, each party may terminate an agreement for a compelling reason without a notice period if the terminating party, taking into account all the circumstances of the specific case and weighing the interests of both parties, cannot reasonably be expected to continue the contractual relationship until the agreed end or until the expiry of a notice period. Such a termination is possible within a reasonable period after obtaining knowledge of the reason for termination; a period of two months is in most cases considered to be reasonable.

2.2 Patent rights in corporate transactions

This section highlights several characteristics within German law and outlines certain aspects of corporate transactions within the life sciences sector. At the same time it specifies appropriate intersections with the tax, finance, employment and intellectual property rights regimes.

(a) *Mergers and acquisitions*

Share deals and asset deals are available as possibilities for transactional dealings under German law. They have in common that, unlike a joint venture, they lead to a clean cut and usually generate the immediate flow of the principal consideration for the shares or assets sold – ie, the purchase price. These two merger and acquisition (M&A) transaction models can be differentiated easily, since in a share deal the purchaser acquires the entirety of the legal entity – thus inclusive of all assets, claims, liabilities, contracts and employees – whereas in an asset deal every single contract and every asset has to be transferred individually (so-called 'singular succession' under German law).

Within the life sciences sector, asset deals represent the majority of cases. This transaction model, besides leaving the principal risk of insolvency of the selling entity with its owner, furnishes the buyer with the advantage that he is not only protected against hidden liabilities but is also capable of choosing between interesting and less interesting assets. Moreover, in most of the cases the acquired assets spread across more than just one entity. As a consequence, a share deal is only possible if all assets of the target business are transferred to one single legal entity.

Generally, asset deals are dependent on the resolution of the following three key issues:

- A principle of German property law demands that all assets, contracts and rights to be transferred must be exactly defined. That requires a usually time-consuming compilation of detailed annexes to the contract. Furthermore, where there is a complete assumption of contracts and takeover of liabilities, third-party creditors necessarily must declare their consent.
- From an employment law perspective, it is necessary to consider the automatic transfer of employees to the purchaser's company, under the applicable rules for the takeover of business undertakings, if the purchaser acquires the whole business unit and the employees do not disagree with the transfer of employment relationships. Should they object, they will remain with the selling entity, leaving the seller with the possible need to lay off the retained employees and possibly to establish a reconciliation of interest and a negotiated social plan. In any case, it needs to be verified whether any issues under the German Works Constitution Act are triggered.
- Asset deals in an international context – and cross-border deals are the standard in life sciences transactions – must deal with determination of the applicable law. Generally, the parties are free to choose the applicable law. By derogation from this principle, the following subject matters are wholly or partly governed by local laws: intellectual property law, employment law, regulatory law, tax law and antitrust law. In practice, a short review of governing laws should be made at the beginning of an asset deal in order to ensure that the right local counsels are involved in a timely manner to review local legal issues.

Besides these key issues, life sciences transactions involve a number of regulatory issues. While pharmaceutical products require a marketing authorisation, medical devices need a European 'CE' mark, based on the appropriate CE certificates, before they can be placed in the EU market. Attention should be paid to the fact that marketing authorisations for EU countries can only be obtained by entities seated or with a branch established within the territory of the European Union. This is quite similar to the topic of wholesale dealers' licences. Those licences can also only be granted to entities seated or having a branch within the territory of the European Union. From a purchaser's perspective, said regulatory requirements might give rise to the set-up of a subsidiary/branch within EU territory. Further, within the scope of such transactions the necessity of requesting new CE certificates might appear. To make sure that the target business can be continued or started by the purchaser as soon as possible, timing is of the essence.

Frequently, developments in the life sciences sector become increasingly government-funded. If assets are sold, this can lead to the possible obligation that granted benefits have to be repaid. It is therefore advisable to negotiate with the authorities to not demand repayment of the grants before transactions are performed.

(b) Financing

In the life sciences sector, the two primary reasons for an initial public offering (IPO) of shares are, first, the procurement of shareholders' capital due to the high costs for ongoing research and development and, secondly, providing an exit route for founders and institutional investors in order to realise some or all of their previous investment.

However, there are many risks that need to be addressed before and during the transaction. From an investor perspective, in the life sciences sector the main risk that exists is where the company's value mostly depends on intangible assets such as patents and know-how. These values cannot be clearly established. For the company, the IPO triggers the obligation to prepare a prospectus describing the company for potential shareholders. If mistakes have been made in preparation of the prospectus, this leads to an extensive liability under German law (so-called 'prospectus liability').

(c) Joint ventures

Joint ventures take a different approach to M&A transactions. While the latter lead to a clean cut in the relationship between the seller and the buyer, joint ventures are set up for intermediate to long-term ongoing collaboration. Usually, joint ventures are set up to jointly develop and market new products. Key issues to be resolved for both parties are the scope of licences, the allocation of IP rights, the allocation of responsibilities, escalation mechanisms, exclusivity and non-compete provisions, technology transfer upon termination, and the control rights.

If the collaboration is not meant to be as tight as in a joint venture, but is nevertheless arranged for a certain duration, parties might conclude a mere licence agreement. In such a case, IP rights remain with the licensor and the licensee's rights to use the intellectual property are limited according to the scope of the licence. The advantage for the licensor is a steady stream of income from the licence fees, while the licensee's advantage is to not be directly concerned with business risks to the licensed intellectual property.

(d) ***Divestment***

Due to the current rapid development of the technological environment in the life sciences markets, a continuing need exists for companies to divest unprofitable business areas. The easiest way to divest a certain business or technology is to assign the patents, know-how and other assets to an acquirer, but only where the assets are organised in a simple fashion within one business unit and not spread across various entities. In most cases, it will be required to sell and assign a variety of rights, together with other assets and including the transfer of employees engaged in working with these rights and assets.

When a straightforward asset deal is not feasible – for example, for tax or organisational reasons – a divestment could also be structured through a carve-out of the target business to an existing special purpose vehicle or through a newly-to-be-established legal entity, which will then be sold and transferred by way of a share deal. For a carve-out, German law mainly offers the following options:

- spin-off by way of transferring the target assets to one or more other entities held by the transferring entity (possibly together with other shareholders);
- spin-off by way of transferring the target assets to a newly-to-be-established legal entity held by the transferring entity;
- split-off by way of transferring the target assets to one or more existing entities held by the shareholders of the transferring entity; and
- split-off by way of transferring the target assets to a newly-to-be-established legal entity to be held by the shareholders of the transferring entity.

Under all of these options, the prerequisites set forth by the German Transformation Act, the German Works Constitution Act and German tax law have in particular to be observed.

(e) ***Due diligence***

The fundamental aim of due diligence is to provide a realistic overview of the risk profile of the target business. Within the typical areas of legal, financial, commercial and technical/operational due diligence, legal due diligence usually encompasses an analysis of the most important contracts, the situation regarding intellectual property rights and employment matters, as well as regulatory affairs. In many cases, the due diligence is preceded by initial evaluations, business discussions and – in particular in the life sciences sector – non-disclosure and confidentiality agreements before a potential acquirer is given extensive insight into the target business.

The further aims of due diligence are to work out the preferred deal structure, to draft appropriate representations and warranties, and to determine whether other contractual agreements besides the sale-and-purchase contract need to be concluded. It further needs to be clarified which entities are to be involved as contracting parties and whether specific arrangements are required for reorganisation or even restructuring measures in connection with the acquisition of the target business. In the life sciences sector, a particular focus area for due diligence is patents and other IP rights, marketing authorisations, CE certificates and other issues of regulatory compliance.

As in every M&A transaction, in addition to the usually complex analysis of the relevant legal issues, practical circumstances impose a challenge on the collection and co-ordinated evaluation of all external input received by law firms, patent attorneys, regulatory advisors and tax and finance advisors. For this purpose, specific persons need to be determined for each party to the transaction to ensure a centralised flow of communications.

2.3 Particular business transactions

(a) *Material transfer agreements*

A material transfer agreement (MTA) is a contract ruling the transfer of research materials from one company to another for research and development (R&D) purposes. The MTA defines the rights of the provider and the recipient regarding the materials and derivatives. Biological materials such as cell lines or proteins are the most frequently transferred materials, but MTAs may also be used for other types of material, such as chemical compounds for common medicinal products.

Important topics to be covered by an MTA include:

- definition of the material and limitation of use;
- confidentiality obligations;
- allocation of arising intellectual property; and
- publication rights, especially if a university is a contractual party, since university employees have a right to publicise inventions made by them in the course of their teaching and research activities (Section 42(1) of the Employee Inventions Act).

(b) *Commercial research, development and collaboration agreements*

R&D agreements and collaborations range from simple contract research agreements to complex collaboration agreements between companies pursuing a joint R&D project where they combine their patented technologies and grant licences to each other. An example of a more complex agreement would be the use of Party A's patented technology to reformulate a medicinal product manufactured and distributed by Party B.

Often both parties have an interest in the outcome of the activities. Therefore contracts have to define precisely the ownership of patentable technology or know-how arising in the course of the cooperation. Back-licences for a party not getting ownership of the new technology guarantee that both can make use of the technology once the cooperation has ended.

In a typical partnering agreement between two pharmaceutical companies, the envisaged collaboration extends from research and development (eg, the joint clinical development of a compound) to the authorisation of a medicinal product in one or more indications, and the commercialisation of the product until the end of the patent term and beyond. In many cases the co-development is followed by an exclusive right of the licensee to commercialise the product in a defined territory. The important points to cover in this type of agreement include:

- the scope of the licence granted (Should it cover only one specific product or

a research platform as a whole? Does the licensor retain any rights to develop, promote or sell the product?);
- the allocation of new (foreground) intellectual property; and
- the consequences of termination of the agreement (Should the licensor be put in a position to continue developing and/or selling the product in the contractual territory? If yes, a grant-back of licences to new know-how or intellectual property owned by the licensee may be required.) Clauses of this kind are often complex and require diligent drafting.

(c) Contract manufacturing agreements

A pharmaceutical entrepreneur may not have the means or skills to manufacture a medicinal product himself. In this case, he has the option to outsource the manufacturing to a specialised company. According to Section 9 of the Ordinance for the Production of Medicinal Products and Active Substances, based on Chapter 7 of the EU Guidelines for GMP published on basis of Article 47 of Directive 2001/83/EC, a contract manufacturing agreement has to fulfil the following requirements:
- Any manufacturing operation carried out under contract shall be the subject of a written contract;
- The responsibilities of each party shall be clearly defined; and
- There shall be no subcontracting of any work without written authorisation from the contract giver.

Issues to consider in contract manufacturing agreements include the following:
- The duties of the contract giver – for example, the scope of its obligations to examine each batch of the product delivered, and the relevant deadlines for notifying any product defects – should be clearly defined.
- Limitation of liability of the contract acceptor can be followed – if no approval has been obtained previously – by problems with the contract giver's liability insurance in cases where the limitation leads to damages to be paid by the contract giver without the option to take regress.
- Provisions for a process transfer should be part of any agreement if it cannot be entirely ruled out that the contract giver wants to manufacture the product himself (or transfer the manufacturing to a third party) and specific practical processes or additional know-how would be necessary for this purpose. This is particularly important if the product to be manufactured under the contract manufacturing agreement is a biological medicinal product.

Contract manufacturing agreements will almost always contain licences for the use of patented technology owned by the pharmaceutical entrepreneur and therefore provisions referring to licences are equally required.

(d) Product distribution agreements

A distribution agreement is a contract between the manufacturer of the product in question and a distributor to distribute and/or sell the products manufactured within a defined territory. Common types include exclusive distribution by the distributor within the contractual territory, semi-exclusive distribution in parallel with the manufacturer and non-exclusive distribution.

Important topics to be governed include:

- the scope of the agreement (exclusive/non-exclusive) and the territory covered;
- the products which are the subject of the agreement;
- ordering terms, minimum quantities and transfer of ownership;
- delivery terms, pricing provisions and terms of payment;
- allocation of regulatory responsibility; and
- licences for IP rights necessary to enable the distribution (patents, trademarks).

Where medicinal products are the subject of distribution, some additional topics to consider arise. It is important to determine, for example, whether the medicinal product in question has already obtained a marketing authorisation in the contractual territory and – if not – which party is responsible for filing the necessary application.

A distribution authorisation is also needed to wholesale medicinal products in the European Union. To obtain such authorisation, a distributor must have suitable and adequate premises to ensure proper storage and distribution, as well as designated staff – in particular, a qualified person designated as responsible (Article 79 of EU Directive 2001/83/EC).

In addition, distributors must comply with Article 80 of EU Directive 2001/83/EC and – among other things – keep records of any transaction in which medicinal products are received, dispatched or brokered, and immediately inform the competent authority of medicinal products they receive or are offered which they identify as falsified or suspect to be falsified.

(e) Co-promotion agreements

Co-promotion (also called 'co-advertising') is a form of collaboration through which a marketing authorisation holder, which generally produces the medicinal product in question, agrees with another pharmaceutical company that its marketing team will also promote the medicinal product – for example, by visiting healthcare professionals.

Co-promotion should not be confused with co-distribution, a concept unique to Germany whereby a marketing authorisation holder grants another pharmaceutical company the right to make use of its marketing authorisation to distribute the relevant medicinal product in parallel with the marketing authorisation holder under its own name and on its own account.

Co-marketing should also be distinguished from co-promotion. The term 'co-marketing' covers collaborations where a pharmaceutical company assigns a so-called 'duplicate' of a marketing authorisation for an authorised medicinal product to another pharmaceutical company, to enable that company to sell and distribute the medicinal product in parallel as a marketing authorisation holder under its own name, trademarks and responsibility.

In contrast, the co-promoting pharmaceutical company is involved in neither the production nor the distribution of the medicinal product, since its marketing team only promotes the medicinal product and refers to the marketing authorisation holder when it comes to the completion of purchase agreements with customers.

A co-promotion agreement should contain detailed provisions in relation to the promotional activities, such as:

- the target group and territory of the promotion;
- the size and qualification of the team conducting the promotion; and
- the regulatory provisions to comply with during the promotion.

With regard to the fee paid to the promoting partner, payment of a contractual lump sum would require considerable trust and knowledge regarding the effectiveness of the promotional activities. Therefore, more frequently a share of turnover is agreed, to reduce the risk for the marketing authorisation holder and incentivise the co-promoting partner through a performance-related fee.

(f) Clinical trial agreements

A clinical trial agreement is an agreement that covers clinical research in human healthy volunteers or patients. Clinical trials are necessary to prove efficacy and safety of medicinal products, usually funded by a pharmaceutical company (the sponsor) and conducted by a physician (investigator) and his employer, a public or private hospital or an academic institution (the clinical trial site). Often, clinical trials are administrated by a contract research organisation acting on behalf of the sponsor.

The division of responsibility for liabilities arising due to injury caused to the trial subjects, as well as compliance with the regulatory requirements, are two of the main points to consider for clinical trial agreements. Improvements and new intellectual property rights regarding the medicinal product are less likely to arise from clinical trial agreements than from any of the agreements mentioned before. Nevertheless, provisions governing the allocation of arising IP rights and improvements are usually included.

Topics that have to be considered in this regard include the Employee Inventions Act, especially when the trials are carried out in a university clinic. According to Section 42(2) of that act, an employee of a university is free not to report an invention to his employer, which means that ownership of the invention may not transfer to the employer (see section 2.1(b) above). Furthermore, under the same provision, a university employee has the right to publicise inventions made by him in the course of his teaching and research activities after two months at the latest, which may jeopardise a patent application because it cannot usually be filed within such a short time frame.

The legal intention of the provisions in Section 42 of the Employee Inventions Act is to protect academic freedom, a constitutional right of university employees. Therefore an employee cannot in advance waive these rights in relation to his employer (the university) but, according to the prevailing views of legal authors, may do so in relation to the sponsor of the clinical trial. As a consequence, clinical trial agreements are mostly made tripartite among the sponsor, the clinic and the investigator heading the trials. Furthermore, investigators should sign separate agreements with the sponsor to waive the mentioned rights against the sponsor.

Another emerging issue is compensation for inventions assigned to the sponsor. While it was common to consider compensation for inventions to be covered by the consideration payable for the clinical trial, nowadays clinics increasingly demand additional payments for each invention made during a trial. In any event, inventions made and assigned by university clinics, which are in general public entities, must be rewarded, whether by a properly calculated consideration payable for the clinical trial or by additional payments. Otherwise, the assignment of inventions could be considered as unauthorised indirect state aid granted to the sponsor and could therefore be prohibited under Article 107(1) of the Treaty on the Functioning of the European Union.[18]

3. Issues associated with patent transactions

3.1 Regulatory law and compliance

Patent transactions in the life sciences sector usually involve issues of regulatory law. This section explains some of the most important terms, regulatory laws and product regimes that are commonly part of patent transactions and agreements in the sector.

(a) Regulatory bodies in Germany

In Germany there are a large number of different authorities which are responsible for certain parts of pharmaceutical regulation, supervision and pricing. Two in particular should be mentioned:

- The Federal Institute for Drugs and Medical Devices (*Bundesinstitut für Arzneimittel und Medizinprodukte*; BfArM) is the regulatory body responsible for medicinal products (for human and veterinary use).
- The Paul Ehrlich Institute (PEI) is the competent regulatory authority for serums, vaccines, blood preparations, bone marrow preparations, tissue preparations, tissues, allergens, advanced therapy medicinal products, xenogenic medicinal products and blood components manufactured using genetic engineering.

Both the BfArM and the PEI are executive agencies of the Federal Ministry of Health. In relation to marketing authorisations for medicinal products, the PEI (Vaccines and Biomedicines) or BfArM (all other medicinal products) is responsible for applications made through the national procedure and will act as the competent authority where Germany is designated as the reference member state in the mutual recognition procedure or the decentralised procedure.

It should also be noted that German local authorities are in charge of supervising the manufacture, testing, storage, distribution, marketing and sale of pharmaceuticals in Germany.

18 Framework for state aid for research and development and innovation, 2014/C 198/01, section 2.2

(b) *Clinical trials*

Before starting a clinical trial, pharmacological and toxicological tests must be carried out (eg, via animal tests) and the foreseeable risks and inconveniences have to be weighed against the anticipated benefit for the individual trial subject and other present and future patients in a risk–benefit evaluation.

Further formal requirements to start a trial are a favourable opinion by an independent local ethics committee and the approval by the national competent authority (BfArM or PEI). Additionally, any individual trial subject (or if a minor, the legal representative) has to be informed by a physician about the nature, significance, risks and implications of the clinical trial and has to give written consent. Even so, an insurance policy is mandatory that provides coverage in the event that a person dies or his or her body or health is injured during the course of the clinical trial.

(c) *Marketing authorisation*

In order to obtain an authorisation to place a medicinal product onto the market, an application must be made to the BfArM or PEI (Sections 21 or 77 of the Medicinal Products Act, as applicable) or to the European Medicines Agency (EMA) if the centralised procedure applies (Article 4(1) of Regulation EC 726/2004). An application must contain specific information, be presented in accordance with the requirements set out in Annex I of Directive 2001/83/EC and follow the guidance published by the EMA.

Marketing authorisation will be refused if the applicant has not properly or sufficiently demonstrated the quality, efficacy and safety of the medicinal product. Formal mistakes or errors in the procedure of the authorisation can also lead to a refusal – for example, if documents provided by the applicant are incorrect or the labelling or package leaflet is not in accordance with the law.

In general, a marketing authorisation is valid for five years. Once renewed after five years, the authorisation is valid for an unlimited period.

(d) *Data exclusivity*

Data exclusivity is a regulatory protection against generic competition. A product authorised within the European Union for the first time enjoys a period during which preclinical and clinical trial data may not be referenced in an application for a marketing authorisation of another company.

For products authorised via the centralised procedure after November 20 2005, and via any other procedure after October 30 2005, the rules determining exclusivity follow the so-called '8+2+1' approach:

- During the first eight years, data exclusivity applies. After eight years, a generic company can file an application for market authorisation based on the preclinical and clinical trial data submitted for the originator product, but it still cannot place the generic product on the market.
- After 10 years from grant of the marketing authorisation, the generic company can market the generic, unless the innovator product qualifies for a further one year of exclusivity.

- One additional year is granted if, during the first eight years, an authorisation for a new therapeutic indication is obtained, which is held to bring a significant clinical benefit.

(e) *Orphan drugs*

'Orphan' drugs are medicinal products for rare diseases (also called orphan diseases). To qualify for 'orphan' designation, a medicinal product must meet the following criteria set forth in Regulation (EC) 141/2000:

- The product must be intended for a life-threatening or chronically debilitating condition affecting not more than 5 in 10,000 persons in the European Union, or for a life-threatening, seriously debilitating or serious and chronic condition and it is unlikely that the marketing product would generate sufficient return to justify the necessary investment; and
- no satisfactory method of diagnosis, prevention or treatment of the condition concerned has so far been authorised.

Benefits of an orphan designation are the possibility of obtaining assistance with the development of the medicine, reduced fees (including fees for pre-authorisation activities such as protocol assistance, application for marketing authorisation, and inspections) and a 10-year period of market exclusivity following authorisation, during which no directly competing or similar product can be placed on the market.

(f) *Medical devices*

Medical devices are those devices that – in contrast with medicinal products that achieve their principal intended action in or on the human body by pharmacological, immunological or metabolic means – typically act by physical means, including mechanical action, physical barrier, or replacement of or support to organs or body functions.

Rules relating to the safety and performance of medical devices were harmonised in the European Union in the 1990s. The core legal framework consists of three directives, which are all currently under review and intended to be replaced by two EU regulations (on medical devices and in vitro diagnostics, respectively). Medical devices may only be placed in the EU market if they meet the requirements laid down in the directives, as assessed in specially regulated conformity assessment processes – in most cases (except for low-risk products) with the involvement of a private or public entity called the 'notified body'.

If conformity with the applicable directive has been confirmed, the product can be marked with a 'CE' mark.

(g) *Advanced therapy medicines*

An 'advanced therapy medicinal product' (ATMP) is a medicinal product which is:

- a gene therapy medicinal product;
- a somatic cell therapy medicinal product; or
- a tissue engineered product.

Similar to the provisions for orphan drugs, special incentives for small and medium-sized enterprises (SMEs) were implemented whereby fees payable to the EMA for scientific advice or the application for marketing authorisation are reduced.

ATMPs often offer groundbreaking opportunities for the treatment of disease and injury but have to fulfil stricter post-authorisation requirements (eg, risk management and traceability) since the safety and efficacy of such novel drugs cannot be determined finally during the authorisation procedure.

(h) Biosimilars

'Biosimilars' are medicinal products developed to be similar to an existing biological medicine, a medicine that contains one or more active substances made by or derived from a biological source (eg, erythropoietin, growth hormones, insulin, anti-TNF and monoclonal antibodies). Biological medicines, including biosimilars, are generally characterised by large and complex active substances and the variability in molecules of the same active substance.

Like the reference product, biosimilars have a degree of natural variability (eg, glycosylation). They are not the same as generics, which have simpler chemical structures and are considered to be identical to their reference medicines.

As for all medicinal products manufactured by using biotechnological processes, biosimilars must obtain marketing authorisation via the centralised procedure. Directive 2004/27/EC introduced an abridged pathway to obtain marketing authorisation. As information on the reference medicine is already available, the amount of information on safety and efficacy needed to recommend a biosimilar for authorisation is usually less extensive than the amount needed to authorise an original biological medicine. Details regarding the necessary non-clinical and clinical studies, as well as comparability exercises, are set out in various guidelines issued by the EMA.

3.2 Commonly disputed matters

(a) Termination without notice

Frequently, parties enter into disputes about the right to terminate a licence agreement for due cause. According to German law (Section 314 of the German Civil Code), each party may terminate an agreement for the performance of a continuing obligation for due cause without a notice period. 'Due cause' exists if the terminating party, taking into account all the circumstances of the specific case and weighing the interests of both parties, cannot reasonably be expected to continue the contractual relationship until the agreed end or until the expiry of a notice period.

What to consider as due cause is not defined in the German Civil Code and has to be assessed on a case-by-case basis. In order to avoid court proceedings or arbitration with an uncertain outcome, it is advisable to make corresponding provisions in an agreement. The parties should think about potential future scenarios and agree on certain reasons to be considered as causes for termination. If a party is given the right to terminate the agreement because the economic benefit achieved with the agreement is not as high as expected, the relevant economic parameters should be clearly defined.

(b) *Claims of the licensor in case of an exclusive licence*

During recent years, Germany's Federal Court of Justice (the BGH) handed down three decisions regarding the issue of whether a patent owner, who has granted an exclusive licence and is therefore not allowed to make use of the patented technology, can conduct court proceedings to enforce the patent against infringers and, in particular, claim damages based on patent infringement. The court stated that, even in the case of an exclusive licence, the licensor can still enforce the patent against infringers independently of the licensee because the licensor's position as the patentee remains with him and therefore he does not have to tolerate the unauthorised use of the patented technology.[19]

The licensor can also claim damages in his own name if he has ongoing financial interests in the licence – for example, if he participates in the use by the licensee of the licensed patent via royalties based on the turnover generated with the patented technology.[20]

However, unless the licensee has assigned his claims to the licensor, the licensor can only claim his own share of damages because, under the German Civil Code, licensor and licensee are not treated as joint and several creditors. Nevertheless, the parties are free to claim damages in joint proceedings, which has the advantage that no party has to set out its exact loss or damage ratio.

To avoid parallel or double proceedings, the parties should agree in advance on provisions for joint proceedings or the assignment of claims to either party in the event of patent infringement.

(c) *Patent applications*

Licensing a patent application is often a source of dispute, since a number of issues need to be considered during the drafting of the corresponding agreement.

The parties have to agree whether the licence shall only be valid if the patent is granted, or if royalties are also due for the unprotected technological know-how. It should also be clarified whether the applicant guarantees the registration of the patent with the licence agreement and therefore has to pay damages if it is denied.

A further topic to cover in the agreement is the question as to how to proceed if claims of the patent are amended during the application process, and the effect this has in relation to the scope of the licence and the agreed royalties.

(d) *Definitions*

It can be observed generally that disputes often arise from imprecisely drafted agreements. Financial contract terms particularly, such as the definition of milestones or the method of calculating turnover (net sales) or the amount of minimum royalties, require some attention before an agreement is signed.

3.3 European and national competition rules

Article 101(1) of the Treaty on the Functioning of the European Union (hereinafter

19 BGH, May 20 2008: X ZR 180/05 – *Tintenpatrone*.
20 BGH, April 5 2011: X ZR 86/10 – *Cinch-Stecker*.

the Treaty) prohibits agreements that have as their object or effect the restriction or distortion of competition within the EU internal market. Article 101(3) of the Treaty provides that this prohibition may be inapplicable in certain circumstances where the overall benefits of an agreement outweigh its anticompetitive effect. On the basis of this provision, the European Commission has issued several regulations, accompanied by comprehensive guidelines, for different types of agreements (the so-called 'block exemptions'). The block exemption regulations serve as a safe harbour for those agreements that meet the requirements of the applicable regulation but do not preclude individual assessment under Article 101(3) of the Treaty of an agreement that does not meet the requirements.

National competition law follows the same rules. Section 2(1) of the Restraints on Competition Act reiterates the standards set by Article 101(3) of the Treaty, and Section 2(2) of the Restraints on Competition Act explicitly refers to the block exemption regulations issued by the European Commission which apply by reference even in cases where competition between EU member states is not affected.

(a) The technology transfer block exemption

The relevant block exemption for many patent and know-how licence agreements in the life sciences sector is EU Regulation No 316/2014 on the application of Article 101(3) of the Treaty on the Functioning of the European Union to categories of technology transfer agreements (technology transfer block exemption, TTBE) which entered into force on May 1 2014. A technology transfer agreement is considered to be an agreement under which one party (or each party) grants to the other a licence to use its particular technology to produce contracted products. In order to qualify for an exemption under the TTBE rules:

- the agreement must contain the grant of at least one licence;
- the contractual parties must not have a combined market share of more than 20% for the relevant product, service or technology (30% each, if non-competitors); and
- no so-called 'hardcore' restrictions must be part of the licence agreement.

'Hardcore' restrictions here include, in the case of agreements between undertakings that are not competitors, provisions that restrict (with a few exceptions) a party's ability to determine its price when selling products to third parties, or the territory into which the licensee may passively sell the contractual products. If the agreement is concluded between competitors, hardcore restrictions also include the limitation of output; and stricter limits on price fixing, as well as on the allocation of markets and customers, apply.

Excluded restrictions are clauses that are not classified as hardcore restrictions and are not part of the safe harbour. Such excluded restrictions include:

- obligations of the licensee to assign to the licensor, or grant an exclusive licence to the licensor in respect of, the licensee's improvements or new applications of the licensed technology;
- obligations on the licensee not to challenge the validity of the licensor's intellectual property rights; and

- in case of non-exclusive licence agreements, the contractual right of the licensor to terminate the agreement in the event that the licensee challenges the validity of licensed IP rights.[21]

3.4 Securitisation of patent rights

Life sciences companies often face the problem that self-created IP rights are valuable but difficult to monetise. As an alternative to licensing out the patented technology, it has become more and more common for companies to securitise their patents in order to realise a part of the lifetime value of the patent at once. Thereby companies can increase their own capital and reduce the use of debt capital.

A typical form of securitisation is the assignment of the patent or the right to exploit the patent to a legal entity (commonly a 'patent exploitation fund') that issues shares to investors to collect the money needed. In return, the patentee receives a lump sum, based on the estimated royalty income or the value of the patented technology, and mostly a certain percentage of the revenues generated by the assigned right. The valuation of a patent is complex and costs between €5,000 and €40,000, depending on the scope of the patent.

Thereafter, the assignee licenses the patent to third parties and – if the technology is still needed to continue the former operations – also licenses it back to the life sciences company, to refinance the paid lump sum and make a profit.

For small and medium-sized life sciences companies, this kind of securitisation brings the advantage that specialised patent exploitation funds are usually more commonly used in monetising intellectual property rights and thereby generate more revenue than the company itself would have been able to.

3.5 The effect of insolvency

Under Germany's Insolvency Act, insolvency occurs when a debtor is unable to meet its mature obligations to pay. This is presumed to be the case if the debtor has stopped payments (Section 17 of the act). Over-indebtedness is also a reason to open insolvency proceedings for a legal person and exists if the debtor's assets no longer cover its existing obligations to pay – unless it is highly likely, considering the circumstances, that the enterprise will continue to exist (Section 19 of the act).

The insolvency itself has no impact on the validity of a patent agreement and any related licence. Whether the agreement and the licence remain in force depends on the party being insolvent and the decisions made by the insolvency administrator. It must be noted that, according to the prevailing view among legal authors, any contractual right to terminate a contract in a case of insolvency of a party is invalid because it would circumvent the intention of Germany's Insolvency Act.

Despite the significant economic importance of technology licensing, German insolvency law does not contain any specific provisions on the impact of insolvency on licence agreements for either the licensor or the licensee. Two attempts made by the Federal Department of Justice in 2008 and 2012 to initiate special legislation to protect a licensee against the consequences of insolvency of its licensor failed before

21 See Paragraph (125) of the Guidelines to the TTBE.

the legislative process had even started. As a result, the legal situation remains unclear and its development largely depends on the interpretation of the general legal provisions by the courts.

(a) An insolvent licensor

If the licensor is insolvent, the insolvency administrator can choose between performance and non-performance of the licence agreement, if the mutual obligations were not, or not completely, performed by the parties to the licence agreement at the date when the insolvency proceedings were opened (Section 103 of the Insolvency Act).

The wording "not, or not completely, performed" has raised a number of unresolved questions and problems in the context of licence agreements. In cases with a defined licence term and periodical royalty payments, the contract is considered to be not performed completely. If the licence granted is unlimited in time and the royalties were paid at once, it is possible to consider the licence agreement as fully performed, although even this is not undisputed and, according to prevailing case law, depends on the individual circumstances.

If the insolvency administrator chooses non-performance, the licence agreement technically ends and the licence falls back to the licensor. The licensee can claim for damages for non-performance of the contract, and register his claim to be included in the insolvency schedule with the insolvency administrator.

(b) An insolvent licensee

In general, the insolvency administrator has an interest to continue using the licence if the licensed technology is necessary to keep the insolvent business running.

However, according to the prevailing views of legal authors, the licensor may terminate the licence agreement, and thereby regain its licence, if the licensee does not pay the royalties due during insolvency. Claims for unpaid royalties due prior to the insolvency cannot justify a termination by the licensor and can only be registered to be included in the insolvency schedule with the insolvency administrator.

As mentioned in section 3.5(a) above, the insolvency administrator can choose between performance and non-performance of the licence agreement with the same consequences set out above if he chooses non-performance.

3.6 Taxation

(a) General

Royalty payments received by a German individual, company or corporation are subject to German income tax. Unlike in many other European countries, such as Belgium and Luxembourg, there is no special IP tax regime applicable to royalty payments in Germany (eg, no special tax allowances or reduced tax rates).

Where a corporation or individual has unlimited income tax liability in Germany, the worldwide income of such corporation or individual will be taxed in Germany.

The corporate income tax rate for corporations amounts to 15.825%, including solidarity surcharge (flat tax). In the case of a partnership or limited partnership, income tax will be levied on the profits at the level of the respective partner. The individual tax rate depends on the total income of the partner and varies between 14% and 45%. In addition to (corporate) income tax, the profits may be subject to trade tax. The trade tax rate depends upon the applicable municipal collection rate. Typically, the trade tax rate amounts to 14% to 16%.

(b) *Withholding tax*

Where a licence is granted by a foreign licensor to a German licensee, the royalty payments may be taxed at source (ie, in Germany). The withholding tax rate amounts to 15.825%, including solidarity surcharge. However, if according to an applicable double tax treaty no or reduced withholding tax may be levied, the licensor may apply for an exemption from or reduction in the withholding tax. The relevant application forms are available online at www.bzst.bund.de. Whether an exemption may be granted should be analysed in each individual case. However, in the case of intra-group licence agreements between European entities, a withholding tax exemption will usually be granted.

(c) *Value added tax*

The value added tax (VAT) rate is currently 19%. A reduced VAT rate of 7% applies if the licence refers to the grant, assignment or exercise of rights that are subject to the Copyright Act (eg, software licences).

Where a licence is granted from a foreign licensor to a German licensee, the transaction will be subject to German VAT. In such case the royalty payment should be charged net of VAT, since the licensee, not the licensor, must remit the VAT to the German tax office (reverse charge). The licensor receives the royalty payment net of VAT and must issue an invoice without VAT. Furthermore, the licensor is required to refer explicitly to the licensee's VAT remittance obligation in the respective invoice.

3.7 Hot topics

Two hot topics currently exist with regard to patent law and administration in Germany: one on sub-licences and the other on insolvency protection. Each is briefly described further below.

(a) *Sub-licences*

A sub-licence is a licence granted by a licensee. Where an exclusive patent licence has been granted, it is always assumed – unless otherwise stated – that the licensee is allowed to grant sub licences. The reverse applies in the case of non-exclusive licences: the licensee is never permitted to grant sub-licences, unless otherwise stated in the licence agreement or permitted afterwards by the licensor.

One highly debated question has been the validity of a sub-licence upon termination of the head licence. The Federal Court of Germany has recently issued two decisions which have preserved the validity of a sub-licence upon termination of the head licence.[22]

---

22 BGH, June 19 2012: I ZR 24/11 – *Take Five*; BGH, June 19 2012: I ZR 70/10, I ZR 24/11 – *M2Trade*

The court stated that each sub-licensee had made investments based on its sub-licence and considered that, because the sub-licensee could not influence or foresee early termination of the head licence, he should therefore have the right to continue using the intellectual property rights concerned under the terms of the sub-licence. In exchange, the head licensor is entitled to claim the assignment of the claim from the head licensee for the payments stipulated in the sub-licence agreement. The head licensor thereby becomes entitled to claim the sub-licence royalties from the sub-licensee directly. The sub-licence, however, continues between the head licensee and the sub-licensee, so the head licensee will have to continue to perform the sub-licence without receiving payment from the sub-licensee.

Both cases involved in these judgments concerned the licensing of copyright, so that the 10th division of the BGH, which is responsible for patent law, was not directly involved in the decisions. Nonetheless, it has confirmed that it stands in agreement.

Some consequences flow from these judgments. If the head licensor has no interest in sub-licences that may survive termination of the head licence, he has the option not to permit sub-licences at all. However, in many cases, especially when the licence is granted to a company with subsidiaries, this is not an option for the licensee. A better option is to include a provision in the head licence agreement stating that all sub-licences shall end simultaneously with the head licence. The agreement has to ensure that this provision is not only contractual but a 'quasi right *in rem*'. Therefore it should be stipulated that the licensee is only permitted to grant sub-licences under a resolutive condition stating that the sub-licence ends simultaneously with the head licence and all licensed rights fall back to the head licensor.[23] In this case the sub-licensee can only obtain a sub-licence from the licensee with the resolutive condition present, because the licensee cannot grant to the sub-licensee a right that he does not possess and a *bona fide* purchase of rights is not possible under German law.

(b) Insolvency protection for patents

Patents are often the main assets of life sciences companies. For this reason, companies often want to secure their patents and concentrate their relevant patent rights in a specific legal entity in order to minimise the overall risk that arises from daily business operations, costly research and development, the liabilities of the company or any other business risk.

In this regard, a life sciences company can found a new affiliated company that has the sole purpose of administering the company's IP rights, and it can assign all of its patents to this company. Afterwards, the new company grants back a licence under the assigned patent to the life sciences company and former owner of the patent. To avoid the risk of insolvency, the life sciences company needs to ensure that the new company has the financial means to fulfil its duties, even if these duties are minor and the risk is low. Therefore sufficient royalties must be agreed and paid to the new company.

23 C Klawitter, *GRUR-Prax* 2012, p425.

In cases where the life sciences company has one major licensee and both wish to protect the patent against insolvency, it is possible to assign the patent to a new company the shares of which are held by the licensor and the licensee. In addition, the parties have to stipulate in the company agreement the retirement from the new company of a partner in a case of insolvency of that partner. Provided that the licensee paid an adequate price for the shares in the new company (based on the value of the patent), the patent is, in a case of insolvency, not a part of the insolvency estate and the insolvency administrator has only a claim for compensation against the new company for the retirement of the insolvent partner.

The authors are grateful to Peter Hellich for his assistance in the preparation of this chapter.

India

Disket Angmo
Fiona Desouza
Jose Madan
Khaitan & Co

1. Patents, SPCs and know-how: foundations for transactions

1.1 Patents and supplementary protection certificates

Patents and know-how provide strong foundations for transactions in India related to science and technology including the burgeoning field of life sciences. The Patents Act 1970, governing patents in India, has gone through several major amendments in the past to be in line with the patent provisions obtaining in major developed jurisdictions outside India. In general, that act provides patent protection to innovative products and processes in the various fields of science and technology, including life sciences, although certain developments – such as methods of treatment of human beings and animals, methods of agriculture and, under certain circumstances, new forms of known substances – are still to be brought under the umbrella of patentable inventions.

Patent jurisprudence is still developing in India. There are no major statutory provisions under Indian patent law specifically for the life sciences sector, nor are there specific patent case laws in this sector to provide guidance for the Indian courts. In addition to the limited general patent case law available in India, the courts may consider leading case law in major foreign jurisdictions to decide cases of patent-related disputes in the life sciences sector in India in future.

The general legal provisions and case law discussed in the rest of this chapter are applicable to all sectors of science and technology in India, including the life sciences sector.

(a) *Rights of patentees*

A patent entitles the patentee to the exclusive right to prevent third parties, who do not have his consent, from using a patented process or from making, using, offering for sale, selling or importing into India for those purposes a patented product, or a product obtained directly from a patented process.

(b) *Patent term and renewal*

An Indian patent affords protection to an invention across India for a maximum term of 20 years, with the term calculated from the filing date of the application in India. Renewal fees are paid annually.

Supplementary protection and patent term extensions are not available in India.

(c) ***Patentability criteria***

Patentability criteria for obtaining a patent in India are novelty, inventive step and industrial applicability. Considerations for novelty and inventive step are those applicable in other major jurisdictions. Industrial applicability means that the invention must be capable of being used in industry.

(d) ***Opposition and revocation***

A patent application can be opposed before grant, and a patent can be opposed within a year of publication of grant or revoked at any time thereafter, on various grounds provided in the Patents Act.

(e) ***Compulsory licences***

Under India's Patents Act any person, including a current licensee of a patent holder, who has an interest in a patented invention can file an application with the Controller General of Patents, Designs and Trademarks (the Controller) for the granting of a compulsory licence under a patent. Such an application can be filed after three years from the date of issuance of the patent on any of the following grounds:

- The reasonable requirements of the public with respect to the patented invention have not been satisfied; or
- The patented invention is not available to the public at a reasonably affordable price; or
- The patented invention is not worked in the territory of India.

The term 'worked in the territory of India' has been interpreted by the courts to mean local working, namely commercially manufacturing a patented product or using a patented process in India.

In March 2012, the Controller granted a compulsory licence to Natco for Bayer's patented anti-cancer drug sorafenib tosylate (Nexavar®).[1] This is India's first and only compulsory licence to date for a pharmaceutical product since the amendments in the Patents Act were introduced in 2005 permitting patentability of pharmaceutical products. The Controller found that all three grounds for the grant of a compulsory licence were satisfied in this case. Natco was granted a licence to manufacture and sell sorafenib tosylate at a price of INR8,800 (about US$140) for one month's treatment, as against Bayer's product, which had cost patients INR280,000 (about US$4,500).

(f) ***Infringement***

Since what constitutes an act of infringement is not defined in the Patents Act, any violation of the rights of patentees mentioned above may be construed as an act of infringement. The burden of proof is usually on the plaintiff alleging infringement. In some cases, the burden of proof shifts to the defendant, as in the case of a patent for a process to manufacture a new product or where an identical product is being made by the alleged infringer but the patentee is unable to determine the process actually used.

---

1 *Natco Pharma Ltd v Bayer Corp* [Compulsory Licence Application 1 of 2011, decided on March 9 2012].

(g) ***Ability to deal***

Subject to the conditions prescribed in the Patents Act relating to joint ownership of patents, and subject also to any rights vested in any other person of which notice is entered in the register of patents, the person registered as grantee or proprietor of a patent has the power to assign, grant licences under or otherwise deal with the patent, and to give effectual receipts for any consideration for any such assignment, licence or dealing, provided that any equities in respect of the patent may be enforced like those of any other movable property.

1.2 Who owns the patent?

The rules of law applicable to ownership and devolution of movable property generally apply in relation to patents. Patents are included in the definition of moveable property, according to the Supreme Court of India.[2]

Joint owners of a patent hold equal undivided shares in the patent, and each owner can work the invention to his own benefit without giving an account to the other joint owners. However, a joint owner cannot assign his share in a patent or grant a licence under the patent without the other joint owners' consent.

India has been contemplating enacting a statute similar to the US Bayh–Dole Act. A 2007 bill, titled Public Funded Research and Development (Protection, Utilisation and Regulation of Intellectual Property) Bill, aims to regulate publicly funded research and patenting activities associated with such research.

A person who develops an invention is the inventor and natural owner of the development. If the inventor was employed at the time of his developing the invention, and if the invention arose directly out of and/or in the course of his employment, then the employer of the inventor is entitled to obtain the patent in the employer's own name. The employer must provide proof of right to file the patent application in India, which can be in the form of an employment agreement or an assignment or execution of a declaration prescribed under the Patent Rules 2003 in India.

Section 69 of the Patents Act envisages that if an assignee or a licensee fails to apply for registration of a document under which an assignment of a patent has been made to him, he will be precluded from relying on the document as proof of his title to the use of the patent.

Although judicial pronouncements in the area of patent licences are few, the law in this regard is well founded in India. Some applicable features are as follows:

- An owner of a patent can grant a licence, but a joint owner of a patent cannot grant a licence without the consent of the other joint owner(s).
- A licence or creation of interest in a patent is not valid unless the same is in the form of an agreement in writing.
- In order for an agreement to be enforceable, it should be stamped with the requisite stamp duty or value.
- The terms and conditions of an agreement should be explicit.
- An application for registration of an agreement should be filed with the Indian Patent Office either by the licensor or by the licensee.

2 *Sunrise Associates v Govt of NCT of Delhi* [Civil Appeal 4552 of 1998].

- A document registered at the Patent Office will be effective from the date of its execution and not from the date of registration.

The terms of licence are kept confidential by the Controller, if requested by the patentee or licensee.

1.3 Inventor compensation

There are no statutory provisions regarding inventor compensation, nor is there case law in this regard to provide guidance to the Indian courts. However, several companies in India have started incentivising innovation by rewarding inventors for their developments.

1.4 Know-how and trade secrets

Unlike the statutory patent protection provided for inventions, there is currently no special statute protecting confidential information, know-how and trade secrets in India. Such information is protected by common law and contract law.

Duty to maintain confidentiality subsists – at times even in the absence of an express agreement – based on the principles of equity and natural justice. Proof of proprietorship over the confidential information or know-how may be established by maintaining proof of creation of a trade secret, either by mailing the information to oneself and retaining a postmarked and sealed envelope or by depositing a dated copy of the information with a third party.

1.5 Employees, consultants and protecting know-how

Parties to a contract dealing in non-patented inventions can create safeguards by including confidentiality and non-compete provisions which survive beyond the term of the contract.

As regards enforcing non-compete and confidentiality provisions which remain in force after the term of employment contracts, Indian courts have held that reasonable restrictions may be enforced, especially when the same relate to the protection of trade secrets.

There are three essential requirements for breach of confidence to exist:

- The information must have the necessary quality of confidence.
- It must have been imparted in circumstances importing an obligation of confidence.
- There must be unauthorised use of the information to the detriment of the party who communicated it.

Remedies available for breach of confidence depend mainly on the contractual provisions. The contract may provide for resolution of disputes through arbitration or litigation. Remedies granted by the courts or arbitrators may either be specific relief or an award of damages. The remedies available to the owner of trade secrets are to obtain an injunction preventing the offending licensee from further disclosing the trade secret, requiring return of all confidential and proprietary information, and compensating the owner for any losses suffered through disclosure of the trade secret.

The Indian courts have upheld a restrictive clause in a technology transfer agreement, imposing negative covenants on a licensee not to disclose or use the information received under the agreement for any purpose other than that agreed in the said agreement. Further, in a case where there was an absence of a contract, the Delhi High Court awarded an injunction based on the principles of equity.[3] The court held that, even in the absence of an express confidentiality clause in a contract, confidentiality is implied and the defendant is therefore liable for its breach of the confidentiality obligations.

2. Context and features of patent transactions

2.1 The principal dealings in a patent

(a) Patent assignments

Assignment of a patent in India confers on an assignee all rights in the patent, as would be conferred on an inventor and owner of a patent. Thus if the intention is to completely acquire a patent, it should be assigned. Assignment should necessarily be made in writing for it to be valid in the eyes of law – for instance, assignment by virtue of a deed of assignment is valid and binding. Such assignment by virtue of the deed of assignment must also be recorded in the Register of Patents, maintained by the Patent Office, for enforcement purposes. Assignment can take place for a granted patent as well as a patent application.

Before assignment of a patent, an assignee must conduct proper due diligence to ensure that the patents or patent applications being assigned actually exist and that the assignor has proper authority to assign them. Additionally, the assignee must also ensure that the patents being assigned are free from encumbrances and that the assignment does not in any way violate the legal rights of any third party. Appropriate warranties and indemnities should be obtained from the assignor for all of the foregoing.

(b) Patent licences

Another common way of acquiring rights in a patent is by way of licence. An owner of a patent can grant to anyone a licence to use and exploit that patent. For a licensor, a patent licence proves useful for better commercial exploitation of the patented invention and beneficial because it does not involve giving up all rights in the patent; for the licensee, a licence is beneficial by virtue of the ability to exploit a patented invention. There are several forms of licences that can be granted. Some common ones are those known as express, statutory, implied, exclusive and non-exclusive.

A licence must be set down in writing to be valid. A licence agreement must always specify the exact nature and extent of rights granted. The categories and details of patents that are licensed should be clearly mentioned; future improvements and updates to the licensed product should also be covered. Whether the licensee has a right to sub-licence and the terms and conditions if such a right is

---

3 *John Richard Brady v Chemical Process Equipments P Ltd* [AIR 1987 Delhi 372].

granted should be agreed upon to avoid issues in the future. The territorial extent of the licence should also be clearly determined. According to the patent laws of India, exclusive grant-back provisions in licence agreements for patents are construed to be restrictive conditions and held to be invalid, so licence agreements should not contain such provisions.

The determination of the rates of royalty is a contractual affair; they should therefore be decided upfront. Royalty rates are mainly based on sales, profits or the number of units sold. The assignment or licensing of a patent between a resident and a non-resident entity in India is governed by the Foreign Exchange Management Act 1999 and the rules framed thereunder, which regulate and manage foreign exchange in India; so royalty payments made in transactions between a resident and a non-resident of India will be subject to the provisions of that act and its associated rules.

Registration of a patent licensing contract in the Register of Patents in India is mandatory for a licensee. However, a licensor may also register the licensing contract.[4]

As regards the doctrine of *caveat emptor* in patent transactions, it may be noted that India's Sale of Goods Act 1930 provides for implied representations on the part of the seller that it has the right to sell the goods, and an implied warranty that the buyer shall enjoy quiet possession thereof, and that the goods shall be free from any charge or encumbrance, unless the circumstances of the contract are such as to show a different intention. Moveable property falls within the definition of 'goods', so these provisions should apply to patent transactions as well.

In spite of the above provisions, proper due diligence is advisable for all patent and know-how transactions. Therefore, before obtaining a licence to a patent, a licensee should conduct proper due diligence to ensure that the rights being licensed actually exist and that the licensor has proper authority to license those rights. Additionally, the licensee must ensure that the rights being licensed have not been given to any third party and the licence does not in any way violate the legal rights of any third party. Appropriate warranties and indemnities should be obtained from the licensor for all of the above.

It may be noted that in the absence of an express or implied warranty regarding the validity of a patent, the invalidity of the patent cannot be pleaded by an assignee in an action for the purchase price.[5] If an assignor had merely professed that he had made an invention and had obtained a patent for the same, and there was no express or implied warranty regarding the validity of the patent, it cannot be assumed in the absence of fraud that the assignor meant to assert that the patent was indefeasible. The assignee, having known what the invention was and having had means of ascertaining its value, contracted for the patent acting on his own judgement, and hence the assignee is bound to pay the amount agreed to.[6] However, where a patentee has asserted and guaranteed the validity of the patent in an assignment, the assignee can, in a suit arising from the assignment agreement, set up the invalidity of the patent.[7]

---

4 *Sunrise Associates* (see note 2 above).
5 *Hall v Conder* (1857) 26 LJCP 138.
6 *Hall v Conder* (1857) 2 CBNS 22, 53.
7 *Nadel v Martin* (1906) 23 RPC 41 at 53 (HL), followed in *Henderson v Shields* (1907) 24 RPC 108.

2.2 Patent rights in corporate transactions

(a) Mergers and acquisitions

Acquisition by business sale can happen either by 'slump sale' or asset purchase. A 'slump sale' essentially means the transfer of one or more undertakings as a result of the sale for a lump sum consideration without values being assigned to the individual assets and liabilities in such sale. An asset sale is a transaction where certain assets are cherry-picked and transferred, and individual values are assigned to each asset; there is no transfer of liability. Acquisition can also take place by way of a share sale, which involves acquisition of equity shares of a company. In such a case, all liabilities get transferred. Therefore, in the case of a slump sale or a share sale, it is specifically important to have proper warranties and indemnity provisions to protect the purchaser against liabilities that get passed on.

Patent due diligence is an important aspect of any merger or acquisition, when the transaction involves a transfer of patents. When acquiring an entity that owns patents, the prospective acquirer should ensure that the patents are held in the name of the target entity being acquired and not in the name of an affiliate or any individual such as a director of the target. If the patent is not in the name of the target, an appropriate transfer of the patent should be effected before consummation of the transaction.

It is best to check the online records of each registration for any undisclosed documents. Details of oppositions filed with the Controller are published in the online records of India's Patent Office. However, where patent revocation proceedings are filed at the Intellectual Property Appellate Board (IPAB), the details are not published in the online records of the Patent Office, nor in the online records of the IPAB as such proceedings are kept confidential. Therefore, inspection of the physical file at the Patent Office should be considered.

Pending litigation details need to be ascertained from the entity being acquired. Such details may not be available from online resources and so all documents and details need to be provided by the target entity. Requisite warranties and indemnities need to be taken into account but, more importantly, the role played by the patent in such litigation needs to be ascertained in order to determine whether the transaction or the business's future prospects will be affected by an adverse outcome to the litigation – and if so, how severely.

Transfers of patent rights in transactions usually happen by means of assignments or other agreements and due recordal of the same with the Controller.

(b) Financing

When a company decides to go public and float its shares on a public exchange, there are several actions it needs to undertake with regard to patents owned by it. There are several stock exchanges in India. However, the most popular are the National Stock Exchange (NSE) and the Bombay Stock Exchange (BSE). There is an exchange platform for small and medium-sized enterprises (SMEs), both on the NSE and BSE. The NSE platform for SMEs is named Emerge and the BSE platform is called BSE SME Exchange.

When a company owns patents, before going public it needs to conduct an audit of all of those patents. This will involve getting the patents valued, ascertaining their fair value. Extensive disclosures need to be made regarding the business of a company that is going public, and this includes all patents and patent applications, which will have to be disclosed in the flotation prospectus.

(c) ***Joint ventures***

When entering into a joint venture, either party/partner may have patents in its name. Instead of pooling the patents and having the patents registered in the name of the joint venture company, it is better to have the partners as proprietors and provide a licence to the joint venture for exploitation of the patents. In this way, once the joint venture is terminated there will be no problem related to transferring the patents from the joint venture back to the partners.

Enforcement of patents can be the responsibility of the joint venture company and hence, vicariously, of the officers/representatives of the joint venture company.

(d) ***Divestment***

During divestment of a business, it is important to ensure that the patents, amongst other intellectual property that is part of the business being divested, are clear of all claims, restrictions etc that may hinder a transfer. It is also important to conduct a thorough analyse and decide which patents to completely divest with the business to ensure that, once the business is divested, a situation doesn't arise where a patent that has been divested proves to be essential for the remaining business. If such a situation is pre-empted, proper provisions need be made so that the remaining business still has the right to exploit such patents (for example, a licence agreement entered into at the time of divestment).

Another issue that needs consideration in relation to patents and know-how or trade secrets is the need to ensure employee confidentiality. Agreements in restraint of trade are generally considered invalid under Indian law. However, clauses which prohibit an employee from disclosing commercial and trade secrets have been held to be not in restraint of trade and therefore held to be valid by the Indian courts.[8] Therefore, such provisions must be clearly laid down before divesting a business: all information that is confidential needs to be set out clearly, and the extent and duration of such confidentiality should also be clearly stated in writing.

(e) ***Consortium agreements and commercial research and development agreements***

When parties get together to work on a research project, there can be several combinations of agreements. Parties can own the patents jointly or own them in proportion to the funds contributed, or one party can own them and license them to the other party, etc. Whatever the agreement, it is very important to set it out in writing before any research or other work under the agreement begins, to avoid bad blood and cost later on. It should be agreed in writing as to whom the inventions created by workers/employees of the research project will be assigned and who will exploit the patents and know-how and when.

Confidentiality is a very important aspect of such arrangements where there are

8 *Embee Software Private Ltd v Samir Kumar Shaw* AIR 2012 Cal 141.

several parties involved, and with diverse interests. It is extremely important to enter into clear and stringent confidentiality and non-disclosure agreements among the parties to such arrangements, as well as all employees and workers.

Appropriate warranties and indemnities need to be obtained in relation to use of patents belonging to the parties to a consortium agreement for the purposes of a research project. The warranties and indemnities also need to state that no third-party rights are infringed and that there exist no restrictions to use of the relevant patents. This is essential to protect the project from claims from third parties.

Under India's Biological Diversity Act 2002, appropriate permissions from the National Biodiversity Authority or State Biodiversity Board, as the case may be, should to be obtained for patenting, exporting or otherwise dealing with biological material from India.

2.3 Particular business transactions

(a) Material transfer agreements

Specific guidelines have been issued by the Ministry of Health and Family Welfare, under which a committee has been constituted to consider and evaluate applications for the import or export of human biological material, or infectious biological material or human biological waste material for commercial purposes. The Indian Council of Medical Research (ICMR) prescribes a format for such material transfer agreements and processes such applications.

(b) General: clarity of specification

In all types of transactions involving patents, it is essential to know exactly which patents will be covered, mention them in writing with absolute clarity, specify the extent of their use or avoidance of use, and lay out the rights and obligations of all parties to the transaction well in advance.

When a party to a transaction is allowed to use certain patents to perform its obligations under an agreement, it is important not only to lay out the rights and obligations of such party and the extent of use of the patents, as mentioned in the preceding paragraph, but also to take into consideration other intellectual property rights that may be involved, such as brands, trademarks, copyrights and designs. This is because use of a patent should not lead to misuse of other intellectual property. For instance, in a product distribution agreement it is essential that the distributor affixes only such marks on the product as may be permitted by the proprietor or other party to the agreement.

(c) Clinical trial agreements

Clinical trials in India are governed by the Drugs & Cosmetics Act 1940 and Schedule Y of the associated Drugs & Cosmetics Rules 1945. Clinical trials, whether for clinical investigation or any clinical experiment by any institution, have to be conducted with the written permission of the Drug Controller General of India. The *Indian Good Clinical Practice Guidelines* also lay down principles and guidelines for conducting clinical research in India.

The sponsor of a clinical trial should enter into a formal and legal agreement or contract with the researcher(s) or their institution(s).[9] The responsibilities and liabilities of the sponsor, researchers and research institutions must be clearly agreed and stated in writing in the agreement before the start of the research project.

3. Issues associated with patent transactions

3.1 Regulatory law and compliance

The Drugs & Cosmetics Act 1940, together with the rules framed thereunder, is the main statutory basis governing the regulation of pharmaceuticals in India. At present, regulatory laws in India for pharmaceuticals are not linked to patent laws. However, several branded multinational companies have been seeking to link the regulatory approval of a drug product to its patent status – a practice that is referred to as 'patent linkage' in industry parlance.

In 2010, the Supreme Court of India dismissed Bayer's plea for patent linkage. Bayer had appealed to the Supreme Court against the Delhi High Court's dismissal[10] of Bayer's writ petition against the Union of India, the Drug Controller General of India (DCGI) and Cipla Ltd, wherein Bayer sought an order that the DCGI should consider the patent status of its drug, sorafenib tosylate, marketed as Nexavar, before granting marketing approval to any generic versions of the drug. Bayer contended that the DCGI was aware of the patent status of its drug; hence the granting of marketing approval would amount to a violation of Bayer's patent rights. Bayer also argued that if the DCGI granted approval, Cipla's drug would infringe Bayer's patent. The writ petition had tried to forge a link between patents and the drug regulatory approval process – a link that does not exist under Indian law. Patent linkage is seen as something that is against public health interests in India as it results in a delay in the entry of cheap, generic medicines into the market.

The so-called 'Bolar' exemption is, in principle, provided under India's Patents Act. It is possible for any drug manufacturer/developer/originator to use the technology of a patented drug to generate data and to demonstrate bioequivalence that would assist in the regulatory or marketing approval of a generic product, while the patent is still in force.

Legal provisions for patent extensions and data exclusivity were under consideration in the past; however, these did not materialise in law.

New plant varieties are protected under separate legislation, namely the Protection of Plant Varieties and Farmers' Rights Act 2001. Genetically modified seeds may be protected under this legislation. The Genetic Engineering Approval Committee is the Indian regulatory authority dealing with approvals of genetically modified organisms.

3.2 Commonly disputed matters

As yet, patent disputes in India have mainly been related to patent infringement and

---

9 Paragraph 3.1.2 of the *Good Clinical Practice Guidelines*.
10 *Bayer Corp v Union of India* [LPA 443/2009].

patent invalidation proceedings. Suits relating to the declaration of a court as to non-infringement, for any relief in a case of groundless threats of infringement proceedings, or for infringement of a patent cannot be instituted in a court lower than a district court. Where a counterclaim for revocation of the patent is made by the defendant in a suit filed in the district court, the suit along with the counterclaim is transferred to the respective High Court for decision. There are no special courts in India for hearing patent-related matters; both patent matters and contract law matters are heard by civil courts in India.

3.3 Competition rules

Under the Competition Act 2002, certain transactions meeting specified financial thresholds are required to be notified to the Competition Commission of India (CCI) for its approval. The notified transaction cannot be completed until the CCI gives its approval.

The financial threshold tests employed to determine whether the CCI must be notified involve an analysis of the worldwide and Indian assets and turnover of both the directly transacting parties: the acquiring group and the target. If any of the threshold tests provided under the Competition Act are satisfied, then the transaction must be notified to the CCI unless a formal or informal exemption applies.

Section 140 of the Patents Act contains provisions for the avoidance of certain restrictive conditions in contracts relating to sale, lease or licensing of patents. For instance, no licensee can be estopped from challenging the validity of a patent.[11] Also, conditions requiring exclusive grant-back and coercive package licensing are regarded as unlawful. Coercive package licensing refers to the practice of mandatorily granting to the licensee licenses under other patents which may or may not be essential to the use of the patent actually being licensed. This practice is considered unlawful in India, since it would compel the licensee to accept licenses under patents that the licensee may or may not desire.

Furthermore, conditions requiring the acquisition of any article other than the patented article from the seller, lessor or licensor, or any restriction on such acquisition from any other person, are also regarded as unlawful.

3.4 Securitisation of patent rights

In a normal financial transaction, securitisation involves creating a security interest in the asset that is securitised, in favour of the lender; such interest is in any future receivables from the asset. When patent rights are securitised, the procedure is the same – that is, the rights in the patents will be transferred to a special purpose vehicle for a lump-sum amount which is given to the borrower. This special purpose vehicle then issues security receipts to investors, which are backed by the patent rights. The returns on such rights will be based on whether the patent is worked on by the borrower and a good value is realised, or whether the patent is licensed and good money (royalty etc) is realised. Therefore, the returns on the investment will depend upon the working of the patent.

---

11 *Enercon (India) Ltd. v Aloys Wobben* [Order 18 of 2013].

In India, any entity that acquires an interest in a patent is required to record such interest with the Patent Office. This is important so that third parties are aware of such interests and no *mala fide* transactions are consummated in a securitisation process such as to hinder the rights provided to investors.

3.5 The effect of insolvency

In India, insolvency proceedings depend on the nature of the entity. There are voluntary as well as involuntary proceedings. However, in either case, the assets of the entity are subject to being disposed of. Licence agreements generally do and should contain provisions that lay down the effect of one party's insolvency. Most commonly, the agreement terminates on insolvency of one contracting party.

3.6 Taxation

(a) *Patent transfer by way of slump sale*

If patents are transferred as part of a slump sale transaction, then gains arising out of such a transaction will be subject to tax as capital gains computed as set out in the next paragraph.

In the case of a slump sale, the excess of sale consideration over the net worth of the undertaking is considered a capital gain in the hands of the transferor. The net worth is defined as the aggregate value of total assets of the undertaking transferred as reduced by the value of liabilities of such undertaking as they appear in the books of account. If the business is owned and held by the transferor for more than 36 months, then gains are taxable as long-term capital gains and currently subject to tax at the rate of 20% (plus a surcharge[12] of 5% and an education cess[13] of 3%); if the business is not owned for more than 36 months, the capital gain is taxed as a short-term capital gain and subject to tax at the rate of 30% (plus a surcharge of 5% and an education cess of 3%).

(b) *Patent transfer by way of itemised sale*

If patents are transferred individually and not as a part of a larger transaction, once the capital gain is determined it will be taxed as a short-term capital gain.

(c) *Tax implications for the payment of royalties under a licence arrangement*

Indian tax laws involve withholding tax when royalties are paid by an Indian entity under a licence arrangement. Therefore, such taxes must be kept in mind while entering into any licence agreements.

3.7 Hot topics

Two 'hot topics' are currently worthy of mention: that of compulsory licensing and that of so-called 'evergreening'. Each is discussed further below.

---

12 The surcharge is applicable at the rate of 5% if the total income of the company exceeds INR10 million.

13 An education cess at the rate of 3% is applicable on the aggregate amount of income tax and surcharge.

(a) *Compulsory licensing*

Compulsory licences are not common in India. Since the commencement of the product patent regime in 2005, the Controller has granted only one compulsory licence – to Natco in 2012 for Bayer's patented anti-cancer drug sorafenib tosylate (Nexavar).[14]

The Controller found that though the Patents Act requires any one of the grounds for grant of a compulsory licence to be satisfied, in this case all three grounds had been satisfied, namely:

- *The reasonable requirements of the public had not been met:* According to Natco, about 23,000 patients require sorafenib tosylate in India annually. In contrast, Bayer submitted that only 8,842 patients in India require the patented drug annually. The Controller observed that the drug could not have been made available to more than 200 patients in 2011 since Bayer had sold only 593 boxes of the drug in that year.
- *The patented drug was not available to the public at a reasonably affordable price:* Bayer was charging INR280,000 (about US$4,500) for one month's treatment with Nexavar, which is unaffordable for a majority of the patient population in India.
- *The patented invention had not been worked in India:* Although Bayer had manufacturing facilities for oncology drugs in India, even after four years from the date of grant of the patent Bayer had still not worked the invention in India, nor had it voluntarily granted a licence to anyone else, including Natco, to work the invention in India.

Some significant points of the Controller's order are summarised below:

- Natco was ordered to pay Bayer a royalty of 6% of the net sales of sorafenib tosylate under a compulsory licence. (The IPAB later increased this to 7% when Bayer appealed against the Controller's order, but the IPAB upheld the Controller's order as regards the grant of the compulsory licence itself.)
- The price of sorafenib tosylate sold by Natco was not permitted by the Controller to exceed INR8,800 (about US$140) for a pack of 120 tablets (one month's treatment). Natco was required to furnish details of sales to both Bayer and the Controller on a quarterly basis and also to make payment to Bayer on a quarterly basis.
- The compulsory licence granted was non-exclusive and non-assignable.
- Natco was not permitted either to import sorafenib tosylate or to outsource production, but was required to manufacture sorafenib tosylate at its own manufacturing facility.
- The compulsory licence did not include any right to represent publicly or privately that Natco's product is the same as Bayer's or that Bayer is in any way associated with Natco. Also, the physical appearance, trade name and packaging of Natco's product had to be visibly distinct from Bayer's product.

---

14 *Natco Pharma Ltd v Bayer Corp* (see note 2 above).

Thus, it is clear that compulsory licences are not commonly or easily granted in India as the grounds for applying for a compulsory licence have to be strictly satisfied, the application is required to go through a rigorous procedure for determining whether a compulsory licence should be granted, and even when the licence is granted it is subject to stringent conditions.

Bayer appealed to the IPAB against the Controller's order, but since the order was upheld by the IPAB, Bayer appealed before the High Court of Bombay which, on 15 July 2014, upheld the IPAB's order. Bayer now has the option of appealing to the Supreme Court of India against the order of the High Court of Bombay. The issue is expected to be a long-drawn-out battle fought in the milieu of a rapidly progressing intellectual property environment in India, where the final judgment could go either way.

(b) Evergreening

Section 3(d) of the Patents Act is intended to be an anti-'evergreening' provision. According to Section 3(d), a new form of a known substance is not patentable in India unless the new form shows "enhancement of the known efficacy" of that substance. The Supreme Court in its recent judgement[15] held that, in relation to chemicals and especially pharmaceuticals, if the product for which patent protection is claimed is a new form of a known substance with known efficacy, then the subject product must pass Section 2(1)(j) of the Patents Act, relating to patentability criteria, and Section 2(1)(ja) relating to inventive step; additionally, it must clear the test of enhanced efficacy as provided in Section 3(d) as read with its explanation. In its judgement, the Supreme Court denied Novartis a patent in India for its beta crystalline form of imatinib mesylate (Glivec®).

Novartis had argued that the beta crystalline form of imatinib mesylate had, among other things, improved bioavailability, lower hygroscopicity, improved thermodynamic stability and improved flow properties. However, the Supreme Court did not construe these physico-chemical properties as an enhancement of the known efficacy of the known closest prior-art substance. Also, the Supreme Court remarked on the absence of comparative *in vivo* animal data to show improved therapeutic efficacy. With regard to the interpretation of the term 'enhancement of the known efficacy', the Supreme Court reasoned that "not all advantageous or beneficial properties are relevant, but only such properties that directly relate to efficacy, which in [the] case of medicine ... is its therapeutic efficacy". This decision made it very clear that, in order to obtain a patent for a new form of a known substance in India, the applicants must show in their patent specification sufficient proof that the new form of a base compound or a known substance is therapeutically more efficacious than the base compound or the known substance. Experimental data to establish that the new form has a therapeutic advantage will in future be essential in determining whether the new form will qualify for a patent in India in relation to Section 3(d) of the Patents Act.

---

15 *Novartis AG v Union of India* [Civil Appeal 2706-2716 of 2013].

16 *In Re Indian Patent Application No 106/DELNP/2008*, decision dated December 26 2012.

In a separate decision of the Controller in 2012,[16] prior to the above Supreme Court judgment, the Controller interestingly accepted *in vivo* absorption data which demonstrated that the claimed polymorph of rifaximin had a 100-times-reduced toxicity compared with the known form of rifaximin disclosed in the prior art. The Controller accepted this data submitted during examination on an affidavit sworn by one of the inventors, judged it to satisfy the Patents Act's Section 3(d) requirements, and granted a patent for the application.

There are still further clarifications required regarding the interpretation of Section 3(d) of the Patents Act. For example, the courts have not yet touched upon whether *in vitro* cell culture data would be acceptable for proving therapeutic efficacy. Also, at this point there are no set guidelines or thresholds stipulated for what exactly would constitute improved therapeutic efficacy (such as what number of animals or patients should be tested with the new form and the known substance, what will constitute a statistically significant improvement, etc). So applicants may provide data that would be generally acceptable to the scientific community for establishing improved therapeutic efficacy of a drug. Also, if a smaller amount of a new form is able to produce the same therapeutic effect as a larger amount of an older form, whether this difference in potency would be regarded as an improvement in therapeutic efficacy is unclear. Judicial clarifications on these issues in future will greatly benefit stakeholders and are, hence, keenly awaited in all pharmaceuticals, chemicals and life sciences cases involving Section 3(d) of the Patents Act.

Israel

Uri Fruchtman
Liad Whatstein
Dr Shlomo Cohen & Co

1. Patents, SPCs and know-how: foundations for transactions

1.1 Patents and supplementary protection certificates

(a) *Nature and term*

Patents are the key form of intellectual property relevant to the life sciences in Israel. The basic patent term is 20 years from the date of filing in Israel or entry into the national phase for Patent Cooperation Treaty (PCT) applications, although pharmaceutical patents may be eligible for an extension of the term. Essentially, Israeli intellectual property laws closely follow the substantive framework in common law countries and provide adequate protection to innovators.

One exception is with respect to the peculiar and rather restrictive system of Patent Term Extension (PTE), which was designed with the aim that the Israeli PTE will be the shortest among corresponding term extensions or supplementary protection certificates (SPCs) in other pharmaceutical markets. Accordingly, while subject matter eligibility is to a great extent similar to the US PTE system, the term of extension is determined based on multiple linkages to extensions granted in many other countries and is also subject to multiple caps. Amendment 11 of Israel's Patents Act (enacted January 2014) retained the PTE linkage system but reduced the number of reference countries to include only France, Germany, Italy, Spain, the United Kingdom and the United States.

(b) *Validity and scope*

The substantive law of validity under Israeli legislation is similar to the principles recognised under UK and US law. A patentable invention is "an invention, whether a product or a process in any field of technology, which is new and useful, can be used industrially and involves an inventive step" (Section 3 of the Patents Act). An invention is deemed 'novel' if it was not published, in Israel or abroad, before the priority date. It is generally not legitimate to mosaic together a number of prior-art documents to anticipate the invention. In contrast, when evaluating inventive step, all common general knowledge should be reviewed and 'mosaicing' is permitted to the extent it is not an inventive exercise by itself. The addressee of the prior art is an "ordinary skilled person" who does not exercise inventive faculties. An inventive step may not necessarily be significant, and the simplicity of the invention does not prevent protection.

In determining the scope of protection, Israeli courts interpret claims on the basis of purposive construction of the specification as a whole. The protection afforded to the patentee extends against literal infringement and also to the essence of the invention. The UK doctrine of variants and the US doctrine of equivalents are routinely considered and applied. While it is not resolved to what extent the US doctrine of 'file wrapper estoppel' applies in Israel, statements by the patentee during prosecution amounting to the clear and unmistakable surrender of subject matter will be considered by the courts.

The protection of pharmaceutical and biotechnological inventions under Israeli law and practice is essentially equivalent to the protection provided in the United States and Europe, with some notable exceptions and differences in prosecution. A method of human treatment cannot of itself be the subject matter of patent protection. However, both 'use' and 'Swiss-type' claims are eligible for patent protection; veterinary methods are also acceptable.

Selection inventions were for many years considered difficult to prosecute in Israel, but in 2011 the Supreme Court of Israel held that the validity of selection inventions should be evaluated on the basis of 'ordinary' novelty/inventive-step rules. Natural products are considered patentable, provided only that the claim does not cover the product in its natural environment. It is yet to be seen whether the US Supreme Court decision in *AMP v Myriad Genetics* 569 US 12-398 (2013), which held that isolated naturally occurring DNA is not patent eligible, will have any impact on local practice. Whether reach-through claims are acceptable has also not yet been considered in Israel.

(c) Examination and publication

Israel is a strict examining country. The Israeli Patent Office conducts a prior-art search and examines the application on various other grounds – for example, unity of invention, or sufficient support for the claimed invention. In addition, the Patent Office relies on the examination and search results of parallel applications in the leading examination jurisdictions of Europe and the United States.

Before examination, an applicant is required to file with the Patent Office an information disclosure statement concerning prior-art references. Since the enactment of Amendment 10 of the Patents Act in July 2012, third parties are allowed to supplement that information within a period of two months after the applicant submits its disclosure statement to the Patent Office. Unlike the recent change to the US Patents law under the America Invents Act of 2011, Amendment 10 of Israel's Patents Act does not enable third parties to submit comments on the relevance of the prior-art references.

An applicant can base the Israeli application on a corresponding granted patent or allowed application in one of the 'recognised countries' in the Israeli Patent Office (ie, Australia, Austria, Canada, Denmark, Germany, Japan, Norway, the Russian Federation, Sweden, the United Kingdom, the United States and the European Patent Office). When invoking this modified means of examination, the claims of the Israeli application must be rendered identical to the granted claims of the foreign patent on which it is based. In such a case, the application shall be deemed to have complied with the requirements of novelty, inventive step, enablement and support. Caution should be exercised before invoking this procedural route, though, as it could potentially present some serious pitfalls to the patentee. Among others, examination in Israel is stayed if the corresponding patent is opposed. According to at least one court case, the presumption of validity (in infringement proceedings) might also be jeopardised.[1]

For many years, Israeli patent applications were kept confidential until they successfully passed examination and were accepted by the Patent Office. Amendment 10 of the Patents Act revoked the dependency between allowance and publication. Under the revised law, non-PCT applications are published after 18 months from the earliest priority date. PCT applications are published no later than 45 days from entry into the Israeli national phase. Upon publication, the file becomes open to public inspection.

(d) *Opposition and revocation*

Following acceptance of the application, there is a three-month non-extendable opposition period. If an opposition is filed, the patent will not be granted until the opposition proceedings are resolved. In addition, following grant and up to the end of the patent term, any interested party may petition the Patent Office to revoke the patent.

The pre-grant opposition system still existing under Israeli law has been severely criticised and can be abused to block patent grants. Although opposition proceedings are conducted at the Patent Office, they are equivalent in nature and procedure to litigation in court. Agents also have a right of audience at the Patent Office but, in view of the contentious nature of opposition proceedings, it is important that they be conducted by patent litigation counsel with litigation skills and experience.

(e) *Infringement and exceptions*

Essentially, any commercial use or any use by a commercial entity constitutes infringement (production, use, offer for sale or sale, or importation for one of the foregoing – as well as export, which is not specifically mentioned in the Patents Act). However, experimental use in order to improve the invention or to develop a new invention is exempt. Also exempt are experimental acts for the purpose of obtaining a marketing approval in Israel or in another country whose laws contain a 'Bolar'-type defence.

1.2 Who owns the patent?

The Patents Act provides, as a basic principle, that the inventor is the rightful owner of a patent. The exception to the rule is employee inventions. In the absence of an agreement to the contrary, any invention of an employee that has been developed as a result of the employment and during the period of employment will be the employer's property. In determining whether an invention is an employee invention, courts will apply a flexible set of rules and will analyse the relationship

1 MCA 14993/07, *Decell Inc. v I.T.I.S Traffic Service Ltd. Tak-Dis* 2008(2) 11160.

between the invention and the employee's duties. Most importantly, if an employee holds a senior position in the organisation, he is deemed to be under special fiduciary duty to further the interests of the employer. For such employees, the scope of inventions belonging to the employer will be broader and it will be easier to establish the causal connection between the invention and the employment.

1.3 Inventor compensation

In the absence of an agreement between the employer and the employee regarding compensation in consideration for an invention, a special Compensation Tribunal determines whether an employee is entitled to compensation and the scope of such compensation. The tribunal considers the nature of the employee's position, the relationship between the invention and the employee's duties, the employee's initiative in developing the invention and the potential to commercially exploit the invention. Until recently it was generally assumed that an express waiver of royalties by the employee is legal and binding. However, in *Bayer v Plurality Ltd*[2] the Supreme Court commented (without reaching a final decision) that it might not be permissible to waive a right to employee compensation. Notwithstanding, in *Barazani v Iscar Ltd*[3] the Compensation Tribunal concluded otherwise, and further held that in certain circumstances a general waiver of rights is sufficient to waive royalties. The tribunal's decision has been appealed to the Supreme Court and it remains to be seen whether the decision will stand. Although proceedings before the Compensation Tribunal have only rarely been invoked in the past, in recent years there has been a modest but notable increase in such proceedings.

According to regulations published by the Ministry of Finance in 2011, physicians and researchers in the public health system are now entitled to royalties of 35% of net revenues from lawful commercialisation of their employee inventions. In parallel, Israeli regulators have invested substantial resources in recent years to enforce the rights of the State of Israel in service inventions whenever they are smuggled into the private market without the state's consent.

1.4 Know-how and trade secrets

In addition to strong patent protection (which nevertheless discloses a patented technology), most pharmaceutical and biotechnological developments rely on undisclosed information and know-how. The Commercial Wrongs Act provides trade secret protection for information that:

- is not easily discoverable;
- provides a competitive business advantage; and
- has been kept secret by reasonable means.

Both business information and technological information are eligible for protection. A trade secret is created when all the requirements for its creation are fulfilled for the first time, and protection expires when the information first becomes

2 LCA 3564/12, *Bayer v Plurality Ltd Tak-Sup* 2012 (3) 4552.
3 Tribunal's decision of May 4 2014 in *Barazani v Iscar Ltd* (unpublished).

publicly disclosed. Trade secrets are protected only against misappropriation and do not confer any monopoly. Reverse engineering is allowed under the Commercial Wrongs Act.

As can be seen from the foregoing, trade secrets supplement patents by safeguarding non-patentable technology. In addition, if the technology is patentable but difficult to reverse-engineer, trade secrets may provide an alternative means of protection over patents.

The life of a trade secret may be longer (or shorter) than of a patent. However, trade secrets provide a much weaker protection than patents.

1.5 Employees, consultants and protecting know-how

Under the Commercial Wrongs Act, any person has a duty not to disclose or otherwise use confidential information acquired by virtue of contractual or fiduciary relationships. In contracts of employment, such non-disclosure obligations are deemed to be implied. Generally, these duties will extend beyond the period of employment and as long as the trade secrets remain confidential. Such non-disclosure obligations must not extend to restrictions on what is legally deemed to be the general art, skills or professional experience gained by the employee.

Post-employment non-competition obligations intended to protect trade secrets will rarely ever, if at all, be implied.[4] Explicit non-compete covenants will be enforced only insofar as they are necessary to prevent misappropriation of the employer's trade secrets, and must be limited in time (generally, to not more than six months).. In practice, in order to be valid, contracts of employment should preferably include special remuneration in consideration for such post-employment non-compete covenants and, even then, enforcement will be highly selective if at all.

In principle, the above obligations and analysis can also be applied to consultancy relationships.

2. Context and features of patent transactions

2.1 The principal dealings in a patent

(a) Patent assignments

Assignments take place for a variety of reasons, including a lack of resources to conduct research and development (R&D) or in order to fund R&D for a more promising technology. In practice, however, they are not common in Israel's life science industry, owing (among other things) to a no-assignment policy at universities and medical centres, where most pharmaceutical and biotechnological inventions originate.

According to the Patents Act, invention and patent rights may be assigned pursuant to a written agreement. However, the Supreme Court has held that this requirement to put the agreement in writing is merely probative and not constitutive.[5] The Patents Act further provides that a patent assignment must be

4 CA 6601/96, *AES Systems Inc v Sa'ar* PD 54 (3), 850.
5 CA 635, 520/80, *Rosenberg v Rubinstein* ILR 38 (1) 85; FH 2/84, *Rosenberg v Rubinstein* ILR 38 (3) 689.

recorded in the Patent Register in order to be valid against third parties. Whether this provision is intended to exclude the operation of other laws or general doctrines regarding conflicting contracts or equitable owner/beneficiary protection against creditors has not yet been sufficiently considered in Israel.[6] However, it is assumed that, at least if the matter comes before the Supreme Court, the court would hold that the general laws apply.

A typical patent assignment should follow the plain language of the Patents Act. In order to record an assignment, the assignee should normally file with the Patent Office a short-form deed of assignment: submitting the entire agreement is also feasible, but unwarranted. The entire agreement will often include the assignment of supplementary trade secrets and know-how. When the technology is assigned at an early stage, before any actual development, it is customary to include milestones of development which determine the level of consideration depending on the prospective success and commercialisation of the technology.

Depending on the bargaining power of the parties, the agreement might include various warranties and indemnities. Standard warranties are those that state that, to the assignor's best knowledge, no third-party rights to the patent exist, and that exploiting the patented invention would not infringe any third-party rights. At least according to some District Court case law, the performance of due diligence by the assignee is a prerequisite for invoking the warranties of the assignor.

(b) Patent licences

Licences play an important role in the commercialisation of life science patents in Israel. Although industry is changing, the typical scenario still often involves a pharmaceutical/biotech invention created and patented in the academic sector or the emergent public health sector and licensed (via the institution's technology transfer office (TTO)) to a start-up that develops the technology and takes it through the initial regulatory pathway (ideally up to and including phase II clinical trials). At this juncture the start-up – which now not only holds a licence to the basic technology coupled with in-house proprietary intellectual property (supplementary patents, trade secrets and know-how) but also drained financial resources – will seek a global pharmaceuticals company to carry the product through phase III, the regulatory approval stage and launch. Licensing the technology (including sub-licensing the university patents) is a common means to achieve that goal.

Under the Patents Act, only an exclusive licensee (in parallel with the patentee) may bring patent infringement proceedings or file for patent term extension. Only a licence that excludes the right of the patentee itself to exploit the patented invention is considered exclusive. A sole (co-exclusive) licence will not suffice.

Limited exclusive licenses (eg, by field of use or medical indication) are not

---

6 In one Magistrates Court decision, it was held in effect that a creditor's right of attachment over a debtor's patents trumps the right of a yet unrecorded assignee (see CF 46420-08-10, *IXI Mobile Ltd v Luzzatto & Luzzatto Patent Attorneys* (published in the Nevo database)). However, it is questionable whether this patent decision is in line with the general Supreme Court precedents, starting with Civil Appeal 189/95 *Soldier Treasury Bank v Mazal Aharonov*, which provide that the purchaser of specific property may take priority over the creditors of the seller.

uncommon, but it is doubtful whether they may be deemed exclusive for the purpose of the Patents Act. In order to mitigate this gap, a limited exclusive licence agreement should include a duty of cooperation by the patentee-licensor.

Similarly to a patent assignment, a licence (whether exclusive or not) must be recorded in the Patent Register in order to be valid against third parties. In practice, the Patent Office will not bar a petition for patent term extension filed by an unregistered exclusive licensee if it is taking the necessary steps for recordal. However, a recent decision by the Patent Commissioner suggests that this practice might not apply in cases of unreasonable delay in filing for recordal.

Consideration for a licence deal rarely involves a single lump sum and often provides for a mixture of upfront, milestone and sale-derived payments. Where the licensor is a public institution and the licensee is, typically, a start-up company, the upfront payment is in many instances only intended to demonstrate the licensee's commitment to the project, and milestone payments may also be relatively modest (at least until phase III). Sometimes, an equity share in the start-up licensee is also included. Where the licensor is the start-up company and the licensee is a big pharmaceutical company, the upfront payment may be higher since the licensor is likely to depend on it for its continued survival. However, the big pharmaceutical companies are not keen to pay large upfront sums before the technology is fairly well proven.

University TTOs often insist on a covenant by the licensee to bring the product to market within a prescribed time and according to prescribed milestone deadlines. Licensing university patents for the purpose of blocking competition is generally against the policy of the universities, even if the licensee is willing and able to compensate the university for any loss of future royalties for sales.

One area in particular in which a university's licensee can impact the terms of the transaction involves milestone deadlines. Biopharmaceutical licence agreements are especially at risk of unrealistic milestones since they are still sometimes drafted with a view to traditional drug development timelines, which are more structured and over a relatively short time span. In addition, the regulatory landscape for 'small molecule' pharmaceuticals is also becoming somewhat less structured, especially in phase III trials. University TTOs are generally not in a hurry to terminate licences due to failure to meet milestones, unless they have lost faith in the licensee. However, the price of unrealistic milestones may still be high – there are obvious costs involved in negotiating extensions (with or without monetary penalties) and, more importantly, even a low-risk threat of termination might turn out to be a deal breaker for a prospective 'big pharma' sub-licensee.

The Supreme Court has long held that a post-patent-expiration royalty clause violates basic principles and cannot be enforced. However, carefully drafted de-scaled post-patent-expiration royalty clauses in hybrid licence agreements (covering both patents and trade secrets/know-how) remain valid. Even so, enforcing post-patent-expiration royalties will always be more expensive and less certain. Thus, a licensor in a hybrid licence agreement which involves high-value trade secrets or know-how may be less willing to accept smaller upfront payments on account of relatively large downstream patent and (de-scaled) post-patent royalties. While not generally advisable for either party, a separation of the patent and the trade secrets or know-how components into two independent agreements should sometimes be considered – but depending on the nature, level of specificity and relative value of the licensed trade secrets or know-how and the interests of the parties.

2.2 Patent rights in corporate transactions

(a) Mergers and acquisitions

In a classic merger transaction (ie, one without the need for a court order), a company assumes all the assets and liabilities of the target company, which is liquidated. A separate assignment of the patent rights of the target company is not required.

The patent licence rights of the target company are likely to be absorbed as part of the merger, in the absence of provisions to the contrary in the licence agreement (and since the transfer of rights following a merger is a consequence of the law, these provisions should probably explicitly forbid a merger or any other change in control, and not simply a transfer or assignment of rights). However, the issue has yet to be considered in Israel, and the courts may be attentive, among others, to US federal case law that supports a presumption of non-transferability.

In a reverse triangular merger, in which it is the target company that formally survives the merger, the (indirect) acquisition of the patent licence rights of the target company (which becomes the buyer's subsidiary) will likely not be even remotely susceptible to a presumption of non-transferability. Thus, when considering a triangular merger (for example, when a foreign company wishes to acquire the rights of an Israeli company, and bearing in mind that direct mergers between foreign and Israeli entities are not possible under Israeli company law), a reverse triangular merger should be preferred in this respect over a forward triangular merger. That being said, an explicit and sweeping change-in-control clause in the licence agreement will overcome even a reverse triangular merger structure.

In a share purchase transaction, the buyer does not actually acquire ownership of any company patents, but it may gain *de facto* control of all of them. Traditionally, buyers might have considered a share purchase transaction over a merger in order (among other things) not to assume the liabilities of the acquired company arising out of securing patent rights or commercialising inventions to which third parties may claim rights. While this consideration still holds true, its scope and weight have recently been undermined by the Supreme Court. In the *Levin* case,[7] the Supreme Court held that where patents of the company are based on a misappropriation of trade secrets, its shareholders who sell their shares for profit may be ordered, under certain circumstances, to make restitution to the rightful owner in accordance with the laws of unjust enrichment. On the other hand, although traditionally whether the shareholders of a company did not know that its ownership of patents was questionable had no bearing when the company alone had been sued (being a separate and existing legal entity), recently one district court decision has held that

7 CA 8945/09, *Sofia Levin v Israel Discount Capital Markets & Investments Ltd* Tak-Sup 2011 (2) 1020.

this question may nevertheless be relevant to the remedies and court costs awarded if liability of the company is established.

In an asset purchase transaction, the buyer may acquire the seller's entire patent portfolio or it may cherry-pick patents following a thorough patent-by-patent due diligence process. In either case, the buyer will most likely fail to gain ownership of patents that turn out to be the property of third parties, since according to the vast majority of Israeli case law the 'market overt' exception[8] does not apply to intellectual property rights.

(b) *Financing*

Stock trading is an important route for financing R&D biopharmaceutical companies in Israel. The Tel Aviv Stock Exchange (TASE) is the only stock exchange in Israel and it operates under the supervision of the Israeli Securities Authority (ISA). Biomed companies that meet certain threshold criteria are included in TASE's TA-Biomed Index. TASE is a member of the World Federation of Exchanges as well as the International Options Market Association, and it is considered by the US Securities and Exchange Commission as a designated offshore securities market. In March 2011, the European Securities and Markets Authority paved the way for Israeli companies to list their shares in the European Union, based on their Israeli prospectuses.

Companies whose shares are listed on the major US stock exchanges or the London Stock Exchange's main market may, under certain conditions, list on the TASE as well. These cross-listed companies must submit to the ISA just the prospectus and periodic and immediate reports that they submit to their respective US/UK regulators and are exempt from ISA's disclosure-related rules (including the patent-related disclosure requirements discussed below).

Until recently, there were no specific listing/prospectus/disclosure/transparency rules applicable to patents in Israel. Many companies used to specify in the prospectus and various reports to the TASE basic information regarding their patents and whether they relate to the company's core or non-core technologies. However, some companies used to issue only laconic statements stating their belief that their patents were adequately protecting their technology, and the companies did not report specific patent-related events. In October 2012, the ISA published draft guidelines of reporting requirements for biomed companies that (among other things) dealt with patents. Following a public consultation process, the ISA decided to move for issuing the guidelines as secondary legislation in the form of an amendment to the Securities Regulations (mainly the Securities Regulations (Periodic and Immediate Reports), 1970 and the Securities Regulations (Details, Structure and Form of Prospectus), 1969). In March 2014 the ISA published the draft regulations for further public input. Since industry is for the most part already following the patent-related guidelines, and since the ISA already introduced certain amendments to the

8 Under Section 34 of the Israeli Sale Law, 1968, a good faith purchaser of stolen movable property prevails over the rights of the original owner, if the goods have been acquired from a merchant in the ordinary course of his business and the purchaser has taken possession of the goods.

guidelines following discussions with industry, there is little chance of further change when the final regulations are issued.

According to the draft regulations, biomed companies must disclose in the prospectus and periodical reports to the TASE information relating to their 'material patents'. A 'material patent' is defined as any patent that protects a substance, a process for its manufacture, its use or a drug product containing it (or a process for manufacture of the drug product), or a medical device that requires marketing approval, the expiration or lapse of which might significantly damage the company's business results.

With respect to such patents, the following information must be disclosed (in a tabular format):

- patent number;
- patent description (whether covering a composition of matter, process, formulation, combination, active pharmaceutical ingredient, use, second medial indication or other);
- patent rights (eg, ownership, licence);
- expected expiration date; and
- countries where the patent has been granted.

In addition, with respect to pending material patents at an advanced stage of prosecution (ie, a national or regional phase application as opposed to a provisional application or an international phase PCT application), the following information must be disclosed:

- patent application description;
- expected patent rights;
- priority date;
- application date; and
- countries where the patent has been filed.

In addition to the above standard tabular information, a company must disclose the following information:

- any opposition/difficulty/delay the company is aware of with respect to its pending and registered material patents, including any legal proceedings brought against the company concerning its pending or registered material patents and which according to the company's assessment will negatively impact their grant or the company's rights in the registered patents;
- legal proceedings to which the company is a party concerning infringement of material patents in which the company has rights that – according to the company's assessment – may have significant implications for the company's business activity;
- possible legal as well as financial exposure in the event of a company's material patent application being rejected;
- with respect to any material patent that is pending more than five years, the reasons, to the best of the company's knowledge, for the delay in registering the patent;

- with respect to any material patent the company does not own, the identity of the entity which will own the technology which the company will develop based on the material patent, and any limitations imposed by the owner of the material patent with respect to the specific use the company may make of the material patent, including the period or territory of permitted use;
- with respect to any material patent that has lapsed in the year of report, the implications of the lapse on the company's business activity;
- with respect to any material patent that is expected to lapse in the year after the periodical report date, the company's assessment of expected implications for the company's business activity;
- any drug product which has been recognised as an orphan drug and whose contribution to the business results of the company, if it were to be successfully developed, would be material;
- any drug product, whether successfully developed or not, whose contribution to the business results of the company would be material, and which has gained some product exclusivity (eg, a patent term extension, data exclusivity or paediatric exclusivity – the company must specify the nature of the granted exclusivity); and
- the company's strategy for handling its intellectual property, including the goals that it is seeking to achieve by utilising its intellectual property; the steps it has taken in order to achieve these goals and the state of affairs regarding the achievement of these goals at the time of the periodic report; and details regarding the identity and qualifications of the person on behalf of the company who is in charge of handling and managing its intellectual property, if such a person has been appointed.

In addition to patent-specific information the company must disclose, in its prospectus and periodical reports, detailed information regarding the royalties it pays or receives as a consequence of using its, or third parties', intellectual property.

With respect to disclosure in the framework of immediate reports, the company must specify patent-related information (including, with respect to a newly granted patent, the patent division of rights, if any, and, with respect to all other triggering events, the company's assessment of expected implications for the company's business activity) when one of the following triggering events occur:

- the granting of a material patent;
- the expiry of a material patent in which the company had rights;
- a rejection of a material patent application; or
- the company has been made aware of a lawsuit having been filed that claims a material patent used by the company infringes third-party rights.

Finally, according to the draft guidelines a biomed company may – but is not obligated to – supplement the prospectus and periodical reports with an opinion by a patent attorney with respect to the strength of its patents (including a freedom-to-operate analysis).

Notwithstanding the pending disclosure regulations targeting specifically biomed companies, in recent years the ISA's general policy has been in favour of shortening the disclosure-related reports, with the aim of improving them and providing to the public only material information required for making sound investment decisions. For example, with respect to material agreements and collaboration agreements, the ISA's guidance is that companies should avoid specifying in their reports standard contractual clauses and remedies intended to protect their rights and prevent exposure to risks, such as are found in non-disclosure and intellectual property provisions; it is when the agreement does not include such standard clauses and remedies that the duty of disclosure arises. It has yet to be seen whether the development of the ISA's general policy will lead to adjustments in the disclosure requirements specific to biomed companies.

In another development, in January 2014 the inter-ministerial Committee for the Promotion of Publicly Traded Research and Development Companies published recommendations intended to encourage investments in R&D companies through TASE. The committee recommended granting ISA the power to exempt biomed companies from disclosure requirements for prospectuses and current reports. The stated purpose of the committee's recommendations was to allow the ISA to adjust local disclosure requirements to bring them into line with US and UK practice. In June 2014, the Ministry of Finance published an exposure bill aimed at implementing the committee's recommendations. It is assumed that the bill, if enacted, will have no impact on the disclosure requirements for biomed companies under the pending regulations.

(c) Joint ventures

In joint ventures, unless agreed otherwise each party retains its previous patent rights, and ownership of new technology that was developed as a consequence of the joint venture is allocated to the inventor (or his employer). When several parties to a joint venture have contributed to the development of a new patentable technology (in a manner that qualifies as 'inventing' under the Patents Act), they will jointly own any resulting patent(s).

Under the default rules of the Patents Act, a co-owner gains an equal share in the patent, even if its inventive contribution was less than equal. A co-owner is not entitled to license its share in the patent without the consent of the other co-owner, but a co-owner may:

- reasonably exploit the patent – subject always to payment of adequate royalties or a share of profits to a second co-owner if, as a consequence of the exploitation, the second co-owner is not able to reasonably exploit the patent;
- assign its share in the patent;
- bring patent infringement proceedings; or
- file for patent term extension.

The courts have wide discretionary powers to order a co-owner how to act with respect to a patent.

Generally, joint ownership of patent rights (let alone deferring to the default rules of joint-ownership under the Patents Act) is not recommended and is prone to create conflicts. Future patent entitlements should be explicitly allocated to one of the parties to the joint venture agreement. While various paradigms intended to provide 'efficient' entitlements in biomed joint venture agreements are sometimes considered, often the relative bargaining power of the parties will determine the outcome. Even so, two recommendations can be put forward: first, the party that is to own the joint venture patents should have optional licence rights to the basic patents of the other party so as to provide freedom to operate for improvements in the basic patent; and, secondly, the researchers should be employees of the parties, or have signed proper assignments – otherwise their entitlements (if they are the inventors) will trump the allocation of patent new rights under the joint venture agreement.

(d) Divestment

Patent divestment may arise in situations where the company discontinues exploitation of the technology and tries to recoup its investment by monetisation of the technology. Otherwise, patent divestment (ie, assignment of the patents) can be selected as an alternative to licensing. This is a risky strategy. Whereas in licence relationships the patentee retains its proprietary rights in the patents, when the patents are assigned the assignor generally retains contractual rights for a future royalty stream and possibly additional future milestone payments. If the assignee of the patents becomes insolvent, the patents may be further assigned to third parties within the framework of insolvency or liquidation proceedings, and the assignor's contractual rights for prospective payments will no longer be enforceable. Therefore, generally, unless a significant upfront payment is secured, divestment of the patent rights should be avoided and a licence relationship with appropriate clauses regarding termination and reversion of rights in cases of insufficient exploitation is to be preferred.

If, nevertheless, a decision has been made to divest through assignment, it is recommended to meticulously screen the divested patent portfolio for any generic (platform) patents that may potentially be relevant for other company developments, and to single them out as 'licensing-only' patents.

Generally, each company employee's confidentiality obligations would apply with respect to trade secrets and know-how transferred out of the company, as long as they remains secret. However, if continuing collaboration will be required to help the acquiring entity absorb the divested technology, existing confidentiality agreements of relevant employees should be updated to state that the employees are obligated to maintain confidentiality of information received from the acquiring company as part of the collaboration between the two companies, even if only for the sake of caution.

(e) Consortium agreements

Consortium agreements will generally require a more complex structure than 'ordinary' joint venture agreements, by reason of the multiplicity of the parties involved. There are several basic models of consortium agreements but, of course, each specific agreement will need to be tailored to fit the characteristics of the particular research project and the respective bargaining positions of the parties involved. A typical model which characterises several consortium collaborative efforts is where each member of the consortium owns the rights in the intellectual property that it creates and grants the other parties non-exclusive rights to use such rights either for the purpose of the specific project or for any other purpose. Such model is mandatory in consortiums that operate under the Israeli Magnet programme – the funding and support programme for R&D on generic (platform) technologies by the Chief Scientist at Israel's Ministry of Industry, Trade and Labour. Another possible model is where a 'leading' party is nominated who is solely entitled to exploit the results.

Other than various extreme models (reciprocal unlimited non-exclusive licences to all members of the consortium at one end of the scale, and a leading party with exclusive rights at the other), there is a spectrum of interim models in which the rights are allocated to the parties in different ways and with demarcation lines appropriate to each specific research project and to the commercial interests of the parties involved. In any event, due to their potential complexities, consortium agreements require (inasmuch as their structure is not dictated under government-related programmes) careful drafting and effective dispute resolution mechanisms, in order to prevent scenarios whereby legal threats hinder the exploitation of technology.

2.3 Particular business transactions

(a) *Material transfer agreements*

Material transfer agreements (MTAs) are usually intended to facilitate collaborative research and development based on proprietary technology developed by one of the parties, and to regulate the transfer of tangible research materials between the parties. The agreement defines the rights of the provider and the recipient with respect to the transferred materials and their derivatives; it also mandates the permitted use of the material and allocates patent and licence rights in new intellectual property that may be developed by the receiving party as a consequence of using the material. If proprietary information is also being transferred, as will often be the case, the agreement should separately clarify the permitted use of information.

Companies normally use boilerplate MTAs – although it is evident from US case law of recent years that boilerplate language might be exposed to interpretation in cases where the material is biological and the boilerplate clause entitles the provider to ownership of any derivatives developed by the receiver, in contrast to non-derivative materials. However, when materials are transferred from industry to academic research institutions, such agreements may have to be renegotiated because they can frequently contain language that conflicts with basic academic rights and impose restrictions on the academic freedom of the researchers. Among the conflicts that can arise between industry and academia are:

- issues relating to delay in (or ban on) publication to protect patentable inventions or confidential information;

- use of the results of investigation in further research and the proprietary rights in such derivative research;
- the interaction of researchers in future projects with other sponsors; and sometimes also
- conflicts with existing agreements.

(b) Commercial research and development agreements

Non-collaborative arrangements: In a funding transaction, the financier is often not interested in owning patent or licence rights but in having its investment returned with interest. Accordingly, the patentee (R&D company) would undertake to pay royalties or a share of the profits to the financier should the research and development prove successful. In addition, the option of pledging the patent or the income therefrom should be considered – for instance, a duly recorded pledge would be effective against other creditors of the patentee. Also, the patentee will have to obtain the financier's consent in case it should wish to license-out the patent.

A specific type of licence transaction may also be characterised as a nuanced non-collaborative arrangement. This is when the patentee-licensor has an interest in the licensee's success in bringing the product to the market, because the product (rather than royalty revenues) is important to its business. In this kind of transaction, the patentee actually outsources the development of the finished commercial product to a trusted third party that funds it and carries the entire risk of failure. Such a licence agreement may include, besides the licence terms, a framework for the future supply and purchase of the commercial product. It is important to draft in such an agreement, in addition to a standard due diligence clause and developmental milestones, a 'no-fault' deadline for commercialisation with optional termination in case of failure, in order to allow the patentee to seek an alternative in a timely manner.

Collaborative arrangements: Collaborative arrangements and joint ventures are synonymous, except that joint venture companies are often (but not necessarily) incorporated. An arrangement in which one party carries all R&D and the other party carries all funding and marketing is a collaborative arrangement. Even an arrangement in which one party carries only funding, in exchange for patent or licence rights from the fruits of the R&D, might be considered a collaborative arrangement. But a truly R&D collaborative arrangement is one where both parties contribute to the R&D itself.

As detailed above in the section dealing with joint ventures, the following principles are recommended to underlie R&D collaborative agreements:

- existing patent rights are preserved;
- newly developed technology is allocated to the inventor;
- jointly invented new technology is allocated to one party; and
- freedom to operate is secured by non-exclusive licences.

R&D collaborative arrangements have a better chance to prosper when the parties are not competitors and wish to co-develop complementary technologies.

However, in such cases close working relationships and even exchanges of personnel are not uncommon and existing confidentiality agreements of relevant employees of both parties may require updating (eg, to clarify that they are obligated to maintain both parties' trade secrets and know-how).

(c) Contract manufacturing agreements

Contract manufacturing agreements govern the outsourced manufacture of a finished product or its components. Inevitably, the outsourced contract manufacturer gains access to the commissioning company's trade secrets and know-how regarding the product. Therefore, it is recommended that a contract manufacturing agreement includes, in addition to non-disclosure provisions, a mechanism addressing how to safeguard the information provided for the contract manufacture, how to make sure it is not used post termination and how to monitor that no manufacturing takes place outside the contract. Specifying the employees of the contract manufacturer who will be authorised to gain access to proprietary information, the standard of information security and the specific steps that will be taken by the contract manufacturer to ensure deletion and retrieval of information upon termination should all also be considered.

Of course, sometimes the contract manufacturer will also be using its own proprietary information, and the contract manufacturing agreement will need to provide for adequate reciprocal obligations by both parties. Sometimes, too, the contract manufacturer (for example) will own patentable manufacturing processes which it would prefer to keep as a trade secret out of 'design around' concerns.

In addition to non-disclosure-related issues, difficulties may arise in cases where the patents covering the product to be contract-manufactured are jointly owned, or the party wishing to employ the contract manufacturer is not the patentee but a patent-licensee that is prohibited from sub-licensing. It has yet to be considered in Israel whether 'have made' rights are implicitly inherent in the bundle of patent rights. It is also important to bear in mind that where manufacture in Israel is found to infringe any patents, the party that commissioned the manufacture will be held liable (in concert with the contract manufacturer) even if it did not set foot in Israel.

(d) Product distribution agreements

Standard clauses in product distribution agreements include the distributor's acknowledgement of the supplier's sole ownership of all intellectual property rights, and non-disclosure clauses relating to the supplier's trade secrets and know-how. In addition, the distribution agreement will usually include an indemnity clause addressing scenarios in which the product is claimed to infringe intellectual property rights and the distributor is being 'dragged' to the proceedings. Of course, in cases where the distribution agreement has already been negotiated in the development phases of the pharmaceutical product (as part of what can be categorised as a joint venture/collaborative arrangement), the contractual terms might be nuanced in different ways.

A distributor, whether exclusive or not, will have no standing against third parties that are infringing the supplier's patents. Where a supplier-patentee wishes a distributor to have power to bring patent infringement proceedings against a local infringer without the patentee's involvement (for example, in order not to be exposed to costs in case of losing the trial), it must grant the distributor an exclusive patent licence and the licence must be recorded with the Patent Office.

In Israel, Ministry of Health regulations require that the marketing authorisation holder (MAH) of the pharmaceutical product be an Israeli resident or a corporation registered in Israel. Thus, sometimes the distribution agreement will provide that the local distributor is registered as the MAH with the Ministry of Health. This will not compromise the supplier's rights in or to the pharmaceutical product since the Ministry of Health's regulations expressly state that MAH status does not entail any such rights.

(e) *Co-promotion agreements*

In a co-promotion relationship a company uses another company's sales force, in addition to its own, to promote a product or range of products. Co-promotion agreements can be found in the pharmaceutical and biotech industries, in particular when the innovator does not have sufficient marketing capacity in a certain territory or in a particular therapeutic area. The stage of development at which co-promotion agreements are entered into can vary.

Co-promotion agreements must among other things regulate the precise co-promotion rights given to the respective parties and whether promotion rights are exclusive or, if not, how the promotional activity is to be divided between the parties. Responsibility for supply and publications regarding the product must be allocated between the parties. Co-promotion agreements are sometimes part of broader alliance agreements between the parties.

The termination of the co-promotion relationship can bring about complex issues regarding the maintenance of confidential information (scientific and commercial) and the transition of promotion to another entity. All possible termination scenarios must be carefully considered and adapted to the particular relationship in question.

(f) *Clinical trial agreements*

Israel is a thriving environment for clinical trials. Israel became a popular site for clinical trials after it received US Food and Drug Administration (FDA) recognition as an approved site for FDA clinical trials in 1997. Israel has a highly skilled and educated workforce and it is relatively easy to recruit patients for clinical trials on an expedited basis. This also led to Israel's popularity as a preferred site for clinical trials.

All clinical trials in Israel must be carried out in compliance with the principles of the Helsinki Declaration and additional regulations and guidelines published by Israel's Ministry of Health. In particular, *Guidelines for the Clinical Trials in Human Subjects*, published by the Ministry of Health, provide a comprehensive overview of the regulatory requirements. In addition to compliance with all the prevailing requirements, clinical trial agreements need to address issues relating to the design of the trial, including data collection, analysis and interpretation, access to data for research purposes, and publication of the results.

One of the potentially controversial issues needing to be addressed is whether and to what extent the confidentiality obligations in clinical trial agreements may relate to the publication of the results of the studies. Other than the possible publication of results and a duty to publicly disclose certain basic information with respect to certain clinical trials (through the US National Institutes of Health's site,[9] the drug owner will likely be able to preserve in the clinical trial agreement its core confidentiality interests in the information it discloses to the hospital. Generally, the parties will negotiate terms which will ensure that results of scientific interest arising from the clinical trial are appropriately published and disseminated, subject to maintaining the sponsor's interest in taking steps to protect its proprietary information.

Another key feature of clinical trial agreements is that all intellectual property rights or know-how arising from the clinical trials will vest in the sponsor.

3. Issues associated with patent transactions

3.1 Regulatory law and compliance

Pharmaceutical products and medical devices may theoretically be eligible in Israel for patent term extensions (PTEs) of up to five years if they obtained PTEs in the United States and/or the European Union (the period of regulatory approval in the Israeli Ministry of Health is not in practice relevant here). The Israeli PTE system is, however, notoriously complex in comparison with the US and EU regimes (primarily since the term of extension is determined based on multiple linkages to extensions granted in other countries). Thus, any assessment of potential PTE rights should be made on a case-by-case basis by a patent expert that specialises in this field of law. In addition, and in any event, it is important to bear in mind that any Israeli PTE rights will be dependent on the continued registration of the pharmaceutical product in the Israeli Ministry of Health's Drug Register.

Similarly to PTE rights, Israeli data exclusivity protection – which in fact amounts to only a limited marketing exclusivity – will only be available to substances that have no prior registration in the Drug Register. Because of a change in the data exclusivity legislation, the exclusivity period for drugs registered in Israel from August 7 2011 onwards is a year longer than for drugs registered before then (theoretically, up to 6½ years rather than 5½ years).

3.2 Commonly disputed matters

The competent courts to hear patent-related disputes are the District Courts. When only monetary compensation is sought, the Magistrate Courts will have jurisdiction up to a claim value of ILS2.5 million (approximately US$720,000), and the District Courts will have jurisdiction over higher-value claims.

Most disputes are normally related to the interpretation of the contractual relationship between the parties. In many cases, such disputes are triggered by licence agreements which were entered into at a preliminary stage of development between private individuals or small companies and which are either incomplete or

9 At www.clinicaltrials.gov.

do not address all possible commercial scenarios. Generally speaking, despite a more restrictive approach recently adopted by the courts, it appears that Israeli courts will tend to imply contractual terms into 'incomplete' agreements or poorly drafted agreements to the extent they deem such terms to be fair to the parties involved or reflect the parties' expectations upon entering the contractual relationship.

Post-patent-expiration royalties have also been litigated in Israel. In the *Blass* case,[10] the Supreme Court categorically held that post-expiration royalties are unlawful. However, de-scaled post-patent-expiration royalties are permitted in hybrid licensing agreements (covering both patents and trade secrets or know-how).

3.3 Competition rules

Essentially, Israeli competition laws generally follow the substantive framework of competition laws in Europe and in the United States. Patent and know-how licence agreements involving 'restraints' that are white-listed in the respective EU block exemptions are likely to be permitted in Israel; and, conversely, blacklisted restraints are likely to render the licence an anti-competitive practice.

Furthermore, vertical price fixing and no-challenge clauses will not be permitted. However, clauses permitting the licensor to terminate the agreement if the licensee challenges the validity of the licensed rights are likely to be regarded as legitimate. In addition, the Israeli Anti-Trust Authority has issued certain block exemptions to franchise agreements, R&D agreements and joint ventures, which will apply depending on several conditions (including whether the parties are competitors in the relevant product market, and the market share of the respective parties). Agreements that are not covered by these exemptions need to be cleared by the Restrictive Practices Tribunal.

Under the Israeli Patents Act, the Patent Registrar may grant a compulsory licence if the patentee abuses a monopoly position that he holds. Prior to TRIPS, compulsory licences were a powerful weapon in the hands of the local generic industry; however, the courts have banned the use of the compulsory licence provisions as a means of conferring business advantages to the local industry at the expense of foreign pharmaceutical companies. In addition, following TRIPS, the compulsory licence provisions were significantly revised and curtailed. Consequently, compulsory licences will only be available in the most extreme circumstances of abuse of patent rights and, in fact, since the legislative revisions no further petitions for compulsory licences have been filed.

3.4 Securitisation of patent rights

Like any valuable asset, intellectual property rights can be recognised as financial assets because they give rise to a stream of future revenues. They can therefore play a role in a company's funding and can be used as a cash-flow-generating asset by exploiting them as collateral for debt. One of the best-known securitisation deals is the David Bowie deal, in which rights for future royalties from Bowie records were securitised.

10 CA 427/86, *Blass v Kibbutz Dan* ILR 43 (3) 323.

Even so, notwithstanding the potential for raising funds, it is rather ambiguous whether patent securitisation is generally regarded as an effective alternative to traditional funding. Overall, patent securitisation is an advantageous cash-raising tool for companies with significant technology but limited access to capital markets. However, for such companies the regulatory obstacles and costs involved in securitisation may be prohibitive.

3.5 The effect of insolvency

When a licensee becomes insolvent, usually the licensor will be able to terminate the licence, subject to the duty to act in good faith. If the licensor becomes insolvent, usually the liquidator will not be permitted to terminate the licence, subject to two main exceptions:

- where the licence relationship requires the licensor to perform ongoing duties which are onerous given its current status; and
- where the liquidator is resolved to sell the licensor's intellectual property and the court is convinced that maintaining the licence would seriously jeopardise the sale.

Additionally, an Israeli court may under certain circumstances bar legal proceedings in Israel against a party that obtained a stay in the United States under Chapter 11 proceedings.

3.6 Taxation

A corporation resident in Israel is subject to Israeli corporate taxation. If the company's activities are managed and controlled from Israel or if it is organised under the laws of Israel, it will be considered as an Israeli resident. A company deemed to be resident in Israel is subject to taxes on global profits (including capital gains), but credit is granted for overseas taxation. A non-resident company will be subject to taxes in Israel only on profits accrued in Israel. Business losses can be offset against income in the same year or carried forward indefinitely. The basic company tax rate as of 2014 is 26.5%. Companies classified as approved industrial enterprises are entitled to reduced tax rates and, depending on eligibility, to grants from the Government Investment Centre. Royalties paid to non-Israeli residents are subject to 25% withholding tax, but the rate may be reduced if a tax treaty applies. Sales of patents will normally be subject to capital gains tax (sale by the inventor may be subject to income tax).

Conventions for the avoidance of double taxation either grant exclusive taxation rights to one or the other country; or provide that one country – generally the source country in which the income was derived – will have 'primary taxation rights', while the other country – the country of domicile of the person deriving the income – will have 'residual taxation rights' (ie, the right to tax the income alongside the duty to prevent double taxation by allowing credit on the tax paid in the country with the primary taxation rights). The Organisation for Economic Cooperation and Development model serves as the basis for Israel's tax conventions, which can be found at www.financeisrael.mof.gov.il/FinanceIsrael/Pages/en/EconomicData/

InternationalAgreements.aspx. Also, when income is earned in a country with which Israel does not have a double taxation avoidance agreement, there are provisions granting credit on foreign tax that was paid to another country.

3.7 Hot topics

Several contemporary issues define the landscape of the Israeli patent system; some have a direct impact on the Israeli patent transactions market, while others influence the local patent ecosystem in which the market operates. Among others, the impact of the policies of the chief scientist at the Ministry of Industry, Trade and Labour (OCS) on the terms of patent transactions involving Israeli start-ups that received OCS support is a hotly debated issue (eg, an Israeli start-up selling its intellectual property to a foreign company will pay the OCS between three and six times the value of support it received, depending on whether the start-up's activity will remain in Israel). In addition, in recent years Israeli regulators have taken a firm stance in enforcing the state's rights in service inventions developed by government employees, and several high-profile proceedings have been filed to reinstate the state's rights in important pharmaceutical inventions or obtain appropriate compensation for the misappropriation of the state's rights in service inventions.

In addition, the strong Israeli patent system relating to pharmaceuticals is continually undermined by the Israeli PTE mechanism, which is based on linkages to the shortest extensions granted in other countries and additional caps. Its latest development is Amendment 11, which reduces the number of reference countries (to include only the United States, Italy, the United Kingdom, Germany, Spain and France). While this in itself is at least in principle a positive development for innovative drug companies (fewer reference countries entails fewer potentially short reference extensions), Amendment 11 contains several provisions that more than offset the narrowing-down of reference countries (eg, a policy against deadline extensions during PTE prosecution and a new cap providing that patent revocation relating to the drug in reference countries may automatically lead to revocation of the Israeli PTE).

Italy

Luca Trevisan
Trevisan & Cuonzo

1. Patents, SPCs and know-how: foundations for transactions

1.1 Patents and supplementary protection certificates

(a) General

A patent is a set of exclusive rights granted by a state to an inventor for a limited period of time.

Italian patent law is contained in Articles 45–81*octies* and by Articles 117–146 of Legislative Decree 30/2005, also known as the Industrial Property Code, which has been effective since March 19 2005. The Italian Civil Code also contains general provisions on patents (Articles 2584–2591).

International treaties also apply. Specifically, Italy is a party to the Union Treaty of Paris of 1977, the Patent Cooperation Treaty of 1970 (including its associated regulations), the Strasbourg Agreement Concerning the International Patent Classification of 1971 and the Agreement on Trade-Related Aspects of Intellectual Property Rights (TRIPS) of 1994. Italy also signed the Patent Law Treaty of 2000, but it has not yet been implemented in Italy.

As to European legislation, Italy is bound by the 1963 Strasbourg Convention on the harmonisation of certain principles of patent law, by the European Patent Convention of 1973 (as revised in 2000) and by several EU regulations, such as Regulation 469/2009/CE on the creation of supplementary protection certificates (SPCs) for medicinal products. In particular, the European Patent Convention is a set of rules at a European level that aims to confer exclusive rights within all the 37 signing states and to provide a specific procedure for the achievement of a European patent through a single application to the European Patent Office (EPO). Any granted European patent does not, however, confer a unitary set of rights, but a bundle of rights equal to a national patent in each state.

As regards the Italian patentability requirements, Article 45.1 of Italy's Industrial Property Code, in accordance with the European Patent Convention, states that an invention requires novelty, inventive step and industrial application in order to be patentable. As to the first requirement, Article 46 of the Industrial Property Code stipulates that novelty exists when an invention is not comprised in prior art, or in general knowledge, made public before the filing date or the priority claimed by the application itself. Furthermore, under Article 48 of the same code, which recalls Article 56 of the European Patent Convention, an invention is considered to have an inventive step if it is not obvious from the prior art to an expert in the field – that is, an individual with ordinary skills in the field of which the invention is part. The third requirement, industrial application, is set forth in Article 49 of the Industrial Property Code, which recalls Article 57 of the European Patent Convention: an invention is considered to be industrially applicable if its subject matter can be manufactured or used in any kind of industry, including agriculture. Finally, Article 50 of the Industrial Property Code, reproducing Article 53a of the European Patent Convention, states that what is against morality and public order cannot constitute the subject matter of a patent.

The list of what can be the subject matter of a patent is not mandatory but can embrace new types of inventions except for those expressly indicated by law. Article 45 and following of the Industrial Property Code list what cannot be considered as an invention: discoveries; scientific theories and mathematical methods; business methods; software; and information presentations. Furthermore, regardless of their qualification, it is not possible to patent surgical, therapeutic and diagnostic methods applied to humans or animals; plants varieties; animal breeds; essentially biological procedures and biotechnological inventions (which, however, benefit from patent protection under a specific section of the Industrial Property Code that implemented in Italy the EU Biotechnology Directive (EC/98/44).

Pursuant to Article 78 of the European Patent Convention and Article 51 of the Industrial Property Code, a patent, which can only concern one invention at a time, consists of the following elements:

- a summary, which aims at briefly defining the subject matter and the aims of the invention, in order to allow third parties, including the examiners, to immediately understand the relevance of the document;
- a specification (see Article 51 of the Industrial Property Code), which must be sufficiently clear and complete as to allow an expert in the field to implement the invention without further documents (except for the quotations made by the patent itself and common knowledge) – and for a description to be considered sufficient, case law has established that it must include a specific explanation of the technical problem solved by the invention;
- drawings, subject to specific formalities, which can be used in order to allow a complete and detailed description of the invention; and
- the claims (see Article 52 of the Industrial Property Code and Article 69 of the European Patent Convention) which define the scope of the patent; claims can be independent (if autonomous from one another and able to be the subject matter of a patent on their own) or dependent (if justified by those which are independent).

(b) *Supplementary protection certificates*

Specific attention must be given to SPCs. It is known that inventions in the pharmaceuticals field require specific administrative authorisations in order to be commercialised. However, the procedure to obtain such authorisations can be very long, thereby substantially reducing the time in which the owner of the patent is able to benefit from the exclusive rights. Being aware of those circumstances, the

Italian legislature introduced, by way of Law 349/1991, a specific set of rules for SPCs, allowing recovery of the time 'lost' while waiting for the authorisations to be issued. Indeed, the SPC allows the extension of the patent for the period between the filing date and the date when the first marketing authorisation was granted by the competent administrative authorities. In any event, though, an SPC cannot extend the patent for more than five years.

Those that can benefit from an SPC are the owners of patents whose subject matter is:

- an ingredient or the composition of ingredients of a drug;
- the manufacturing procedure for an ingredient or the composition of ingredients of a drug; or
- the use of an ingredient or the composition of ingredients in a drug.

In addition, an SPC can be granted for a product if:

- the product is protected by a basic patent in force;
- a commercialisation authorisation was granted for the product;
- the product has not yet been the object of an SPC; or
- the commercialisation authorisation on which ground the SPC is granted is the first commercialisation authorisation for the product as a drug.

(c) *Patent applications and the rights granted*

In order to obtain a patent, an application must be filed with the Italian Patent and Trademark Office (IPTO), which makes a merely formal examination of the application and of the presence of the requirements. In the event of a positive outcome, the patent is granted and its effects accrue from the publication of the application (ie, 18 months after the filing date). Before the grant, the applicant can withdraw his application, submit observations to an objection of an examiner and amend the application – however, its content cannot be extended beyond the scope of the original application.

The rights conferred by the patent are subject to the principle of territoriality, expressed by Article 66 of the Industrial Property Code, which states that patent rights for industrial inventions grant to their owner the exclusive power to implement and benefit from the invention within the territory of the state.

The rights granted by the patent are cancelled in any of the following cases:

- expiration of the rights after 20 years from the filing date (or longer when an SPC has been granted);
- renouncement notified by the owner to the ITPO, with no retroactive effects; and
- forfeiture of the rights (after failure to pay the relevant fees, failure to implement the invention or insufficient implementation thereof).

(d) *Infringement*

As far as infringement is concerned, the statutory provision in this regard can be found in Article 66 of the Industrial Property Code, which sets out the primary rights conferred to a patentee. In particular, a patent confers on its owner the exclusive right to make, use, put on the market, sell or import the patented products, to apply the patented process and to use, put on the market, sell or import the product directly obtained. Therefore, with very few exceptions (such as, for example, the so called 'Bolar' clause – ie, the exception concerning the activities necessary to obtain a marketing authorisation for a drug), every behaviour in violation of such rights amounts to direct infringement, even if such use is only partial.

In addition, in Italy indirect or contributory infringement is not expressly regulated by statutory provisions. However, according to the relevant case law, contributory infringement applies when the following requirements are met:

- the activity of the contributor is essential for the direct infringement, and
- there is no lawful use of the activity of the contributor other than for the carrying out of the direct infringement; or
- when there are other uses, the contributor is aware that his activity will be used in an act of direct infringement.

In addition, patent infringement may qualify as an act of unfair competition under Article 2598 of the Italian Civil Code. Therefore, in patent litigation it is normally possible to claim both infringement and unfair competition.

1.2 Who owns the patent?

According to Articles 63 and 118(1) of the Industrial Property Code, the right to patent is conferred on the inventor and his successors in right, and they have the right to file a patent application. Article 63 of the same code expressly establishes that the inventor's right to patent can be assigned to third parties, while the moral right to be named as the inventor is not assignable.

The filing of the patent application by a non-entitled person results in a ground for invalidity of the patent (Article 76(1)(d) of the Industrial Property Code). However, an entitled person may avail himself of the remedies foreseen by Article 118 of the Industrial Property Code, which allows the filing of court proceedings aimed at acknowledgement of the entitlement.

The patenting system contemplates cases in which inventors are *ab origine* not entitled to the filing of the patent application, namely when the inventions were made in the framework of employment agreements and certain conditions are met (Articles 64(1) of the Industrial Property Code). In these cases, the right to file a patent is on the employer, without the need of any assignment.

In certain cases, however, employees maintain a right to the filing of a patent application, but with either the right of the employer to obtain part of the profits from the exploitation of the invention (in the case of universities or public administrations; Article 65 of the Industrial Property Code) or the right of the employer to obtain an assignment of the patent in exchange for a fee (otherwise; Article 64(3) of the code).

Should an invention be obtained within the framework of a research agreement with a university or research institution, financed by a private entity, the rights on the invention can be in the name of the financing entity (Article 65.5 of the Industrial Property Code). Similar rules apply when the invention is obtained by private research institutions, or any other economic undertakings, whenever the invention is a direct result of the research requested within the framework of a contractual agreement. However, special provisions apply for subcontracting agreements, forbidding that such agreements may foresee that any intellectual property (IP) rights are automatically assigned, without compensation, to the principal by the contractor, whenever the latter, in the execution of the agreement, obtains an invention or any other IP protectable result that was not expressly requested (and paid for) by the principal.

As regards which law must be applied in order to determine which person is entitled to the patent, Article 60 of the European Patent Convention provides that "if the inventor is an employee, the right to the European patent shall be determined in accordance with the law of the State in which the employee is mainly employed". This provision is in line with the general principles of Italian patent law, according to which when an Italian patent is filed by a foreign entity, and although the patent is of course governed by Italian law, the issue of ownership – and in particular whether the rights on an invention made by an employee or a researcher reside with the employer – is governed by the law of the employment contract.

Lastly, Article 67(3) of the Industrial Property Code provides that an implied licence is considered to have been granted for any patent protecting new processes whenever the patent owner supplies to a third party – without any different agreement in this respect – the means and tools specifically aimed at implementing the invention.

1.3 Inventor compensation

Should the employment agreement contemplate inventive activities as a possible task of an employee, and in the employee's salary a specific amount is foreseen as compensation for such an activity, no further compensation is due by the employer to the employee for any invention made. In contrast, should the invention being made in the execution of his duty be assigned by the employer, without being a specific task contemplated by the employment agreement and specifically paid for, a fair compensation – to be determined based on equity (so-called *equo premio*) – is due to the employee.[1] The economic value[2] of the invention (the higher the value, the higher the compensation), as well as the tasks the employee was assigned (the lower and more distant from the field of the invention, the higher the compensation), as well as the employee's salary (the lower the salary, the higher the compensation) and the support to the invention obtained by the employee from the employer (the

---

1 Both the above-mentioned solutions require the employee to carry out researches aiming at the development of the invention. However, the distinctive element between the former and the latter solution is the presence of an explicit contractual provision of specific compensation for the inventive activity. Lacking the said provision, the employee is entitled to a fair compensation (the *equo premio*). The burden to prove the existence of specific compensation promised for the inventive activity is on the employer, as established by case law, with a view to giving the employee as much protection as possible (see, eg, in this respect Supreme Court case 6367 of March 21 2011).

2 According to the Supreme Court (case 7161/1998, *Nuovo Pignone v Zanobini*), 'economic value' has to be assessed by means of a forecast of the profits expected to be obtained by the employer by exploitation of the invention, having reference to its field of activity, and without making reference to the actual profits obtained.

higher the support, the lower the compensation) are taken into account in the determination of the compensatory amount (similarly to the so-called 'German formula').[3]

1.4 Know-how and trade secrets

The main difference between protecting information and know-how by keeping it in confidence or by patenting it is that the first alternative is only possible whenever information can be actually kept confidential. Typically, information which will be incorporated in products or processes to be made accessible to third parties without restrictions on possible reverse-engineering cannot be kept secret or confidential. A further difference is that no formalities are required for protecting confidential information, and the burden of proof as to its extent lies in full with the owner.

Furthermore, patents are enforceable against any third parties using the invention without licence, regardless of their good faith or the fact that the third party has reached the same solution independently. In contrast, know-how is not enforceable against any third party which obtained the same result by independent research, or who received the information in good faith without having cooperated with the discloser.

Lastly, patent protection requires the fulfilment of formalities (eg, annual fees for maintaining the registration), and has an expiry date (normally 20 years from the filing), whereas confidential information remains protected while not disclosed or obtained through a different source.

Confidential information and know-how are regulated by Articles 98–99 of the Industrial Property Code, according to which information is protectable whenever it is:

- confidential – that is, the information is not generally known nor easily accessible to skilled persons in the field, as a whole or in specific combination;
- economically valuable by being secret; and
- kept protected by the owner in a way reasonably sufficient for maintaining confidentiality.

1.5 Employees, consultants and protecting know-how

Employees are expressly bound by a duty of confidentiality in respect of any information obtained in the fulfilment of their employment duties, without the need of any specific provision in their employment agreements. This obligation is in light of Article 2105 of the Civil Code, which sets forth, among other duties for the employee, that of not disclosing or using information that refers to the undertaking and its production methods.

---

3 The so-called 'German formula' is the following: $I = V \times P$, where I is the fair compensation determined based on equity (the *equo premio*); V is the value of the invention itself; and P is a percentage, calculated taking into account the employee's contribution to the initiative that led to the invention, the enterprise's contribution in the solution to the problem, and the position and tasks assigned to the employee within the enterprise's organisation. The given formula has been applied by Italian courts in many instances – as, for example, by the Supreme Court (decision 7161/1998, referred to above, and decision 2646/1990) and by lower courts, for example in the *Technicolor* case (Court of Rome, October 23 2007),where the Court of Rome applied the German formula in a case between the Technicolor company and one of the inventors (Mr GM) of an improvement in the devices used to develop movie films.

A general principle of implied confidentiality is considered by the settled case law to also bind independent consultants or any business partner of an economic undertaking in respect of any information gathered during the relationship.

However, it is advisable – before disclosing any confidential information to external consultants, business partners or (whenever information is of particular importance) employees – to sign a confidentiality agreement or an engagement of confidentiality, in which:

- the information is in some way defined or identified, in order to avoid issues on what the obligation referred to as to the confidentiality of such information;
- the duty of confidentiality is expressly set out; and
- a penalty is provided for any breach.

Italian Courts have dealt several times with the concept of confidentiality in respect of information gathered by an employee during their working relationship. It is worth recalling in this respect a notable decision by the Court of Florence on November 26 2008, where company A claimed a breach of confidentiality by one of its former employees, who had allegedly transferred to company B information, data and contacts gathered at his workplace at company A. The court ruled that part of the information and data, being economically sensitive information that was subject to specific security measures and was not generally available to all employees, was to be considered confidential and thus the employee had breached his duty of confidentiality. On the other hand, as to other information allegedly stolen, the court ruled that it was not confidential, being a mere list of contacts and addresses which could have been retrieved from other (public) sources.

2. Context and features of patent transactions

2.1 The principal dealings in a patent

(a) Patent assignments

The assignment of a patent results in the transfer of ownership of the same. There are several reasons for this kind of operation, which may vary depending on the assignor.

One reason is simply to monetise the patent whenever the patentee is not capable of, and/or willing to, exploit it directly or through a licence (for example, because the inventor is an individual, or because the invention falls outside the scope of business of the company, or simply because the undertaking does not intend to exploit the invention). A second reason is that the patent may be transferred within the framework of a joint venture agreement, a merger or acquisition, or a spin-off operation. In this case it is not rare that the assignment refers only to a part of the ownership, so as to constitute a joint ownership of the patent. A third case is an assignment as warranty or within the framework of other financial deals (such as a venture capital investment).

The formalities to be followed for assigning a patent are to:

- draft a written deed of assignment;
- register the said deed with the tax office; and
- register the said deed at the IPTO.

Generally, the assignment is a relatively simple deed, which indicates:

- the details of the patent and better specifies the subject matter thereof;
- the price (or at least the consideration) to be paid for the assignment;
- the effective date of the assignment; and
- the duties for the assignor to cooperate with the assignee for registration at the IPTO.

Further clauses – particularly in cases of joint ventures or venture capital transactions – could refer to further obligations for the assignor not to further develop new inventions in the same field (which could diminish the value of the invention assigned) or a pre-emption right for the assignee in respect to further inventions/patents in the same field.

The usual anatomy of an assignment agreement is:

- identification of the parties;
- recitals;
- definitions;
- assignment;
- compensation (price);
- obligations of the parties;
- term and termination;
- conflict resolution;
- applicable law; and
- miscellaneous.

Whenever the assignment refers to inventions not yet patented or not fully patented (ie, for which corresponding patents can still be filed claiming priority), it should define which party has the right to file patents claiming the same priority.

The main differences when drafting an assignment for an assignor or for an assignee are related to warranties and indemnities. In fact, although the assignee will try to obtain a warranty that the patent is not only fully owned by the assignor, but also that it is valid, the assignor will try to effuse such warranty, being only ready to warrant that, to the best of his knowledge, the invention has been solely obtained by the assignor, who has full ownership of the patent, that any fair compensation due has been paid when the invention has been made by an employee, that the formalities for filing the patent application have been correctly carried out, and that no prior art or other grounds for nullity (eg, pre-disclosure) exist. The difference between the two cases is that, should a warranty on validity be granted, in a case of nullity the assignee could claim the payment of an indemnification or even full reimbursement of the price paid for the assignment. When a contract does not contain any warranty in respect of validity, under Italian law no indemnification is due (see Article 77.1(b) of the Industrial Property Code, according to which the contract is still valid but the assignee may only request that, under certain circumstances, a part of the amount paid is reimbursed).

(b) *Patent licences*

Licences are usually granted with the aim by the patentee of monetising the invention, similarly to assignments.

According to Italian law, two kinds of licence exist: voluntary licences (ie, licences granted by the patent owner) and compulsory licences (ie, licences provided by the law as a consequence of some particular set of circumstances, eg, that the patent has not been exploited, the scope of protection or the duration has been modified or the exploitation of the invention is dependent upon the use of a third-party patent (which could also lead to a cross-licence)). A further specific compulsory licence is set forth in Article 81*octies* of the Industrial Property Code for biotechnological patents for the avoidance of overlaps in protection of plant varieties.

In particular, under Article 71 of the Industrial Property Code the proprietor of a later patent who cannot properly exploit his patent without using the information from a previous patent can be granted a compulsory licence on the basis of dependency. A compulsory licence on the basis of dependency can, however, only be granted provided that the more recent invention involves a key technical advance of economic significance in relation to the invention claimed in the older patent. The receiver of a compulsory licence on the basis of dependency is obliged to grant a cross-licence to the patentee of the older invention.

Voluntary licences are usually granted after a negotiation and are usually based on one-to-one agreements, but they can also be granted as a public offer.

A licence can be exclusive (if the licensee is the only party authorised to exploit the patent) or non-exclusive (if more than one licensee exists or the patentee has retained the right to exploit the patent in addition to the licensee).

Exclusive licences tend to be very similar to assignments, since all the rights of using the invention are transferred to the licensee and the licensor is prevented from using it (in this respect differing from a sole licence, in which the licensor is prevented from granting further licences, but not from using the licensed invention), the main difference being that the consideration is paid as an ongoing royalty. Exclusive licences must indicate whether this also implies a right to sub-license, and under what conditions.

Non-exclusive licences can be tailored to the specific needs of the parties, and the agreements can contain indications as to the specific field of activity or territory in which the invention can be exploited, as well as the number of uses made, and further obligations as to the quality of the resulting products. Such licences can also be combined with know-how and trademark transfer agreements.

Exclusive licensees are considered by settled case law as legitimised to enforce the licensed patent, even in court. However, claims or counterclaims relating to nullity need to involve the actual patent owner. Non-exclusive licensees are, to the contrary, not legitimised to instigate infringement actions by themselves (ie, not having the patent owner as joint plaintiff), whilst they are indeed legitimised to instigate claims for damages in the same infringement proceedings with the patent owner.

As to formalities, there is no obligation that a patent licence contract has to be entered in the Patent Register – although under Article 138 of the Industrial Property Code only by means of such an entry can the licence becomes opposable to third parties. For instance, the licence can be opposable to a later assignee of the patent only if the patent has been entered in the Patent Register.

The usual anatomy of a licence agreement is:

- identification of the parties;
- recitals;
- definitions;
- licence grant;
- compensation (price);
- obligations of the parties;
- term and termination;
- conflict resolution;
- applicable law; and
- miscellaneous.

As to warranties and indemnities, they are regulated similarly to assignments. We refer to the previous section 2.1(a) in this respect.

As stated above, royalties are the main characteristics of a licence agreement. Royalties may be construed only as a periodic payment or, instead, as a combination of a periodic continuing payment and an initial lump-sum or upfront licence fee. Sometimes royalties can be substituted by the payment of a single overall amount (creating a fully paid-up licence).

Royalties can be paid monthly or annually, and their amounts may be fixed (a specific amount for each unit produced or for each use of the invention) or variable (proportional to the price or to the turnover – in this case there should be an indication of whether costs, taxes and other expenses are included). The amount of the royalty to be paid can also be construed as decreasing over time, or proportionally to quantities.

Although it is a common principle, it is advisable for the avoidance of doubt to express that a lack of payment of a royalty makes further use of the invention an infringement; it doesn't just create a debt. This is of importance in order to have access to the effective measures of judicial protection offered by Italian law for protecting IP rights.

2.2 Patent rights in corporate transactions

(a) Mergers and acquisitions

In a share sale-and-purchase agreement, only the ownership of a company is transferred – it remains the same legal entity. Therefore, ownership and existence of all the assets, including intangible ones, is not affected by the agreement; their existence and ownership only have relevance for establishing the price of the transaction. Indeed, these kinds of agreements are preceded by a due diligence exercise, and the price is subject to adjustments in light of possible differences between the due diligence results and the closing date situation, as well as of possible new circumstances which can affect the value of the company.

In the business sale and purchase, only a defined set of assets are transferred, among which are likely to be patents. Patents therefore have to be expressly mentioned and referred to in the agreement. Furthermore, the relevant formalities for the transfer of ownership of the patents must be completed, as with assignments.

As to patents and know-how, due diligence should verify that all formalities and deadlines have been met (particularly for still-pending patent applications). As to patents already granted, it is important to verify that all formalities to maintain ownership have been carried out (eg, in case of European patent, that the relevant national translations have been filed in a timely manner – in Italy, within three months from publication) and that annuities have been paid regularly. It is important to verify whether nullity actions have been brought against the patents, and whether other proceedings involving the patents exist. Moreover, verification needs to occur that all fair compensations due to employees who are indicated as inventors have been regularly paid. Where a patent has been obtained by the owner through a prior assignment or purchase, it must be verified that the first assignor had full ownership of the patent. The existence of any licence should be verified.

As to know-how alone, the due diligence exercise should in addition verify its confidentiality and establish whether the measures and policies adopted by the owner are sufficient in order to maintain the secrecy of the information involved. In particular, there should be verification that adequate technical protection measures and confidentiality agreements/engagements with employees and/or external consultants, suppliers and sub-contractors have been in existence.

(b) Financing

As intangible assets, patents are rarely inserted in the balance sheet of a company, in particular because of the difficulty in assessing their value, which needs to be determined in a non-disputable way (taking into consideration the liability for directors when the value reported in the balance sheet is not consistent with the real value of the assets, as well as the need for an adjustment of the balance sheet where the value of the patent changes). Thus, IP rights are indicated in balance sheets only with a figurative value, and in any case their value is not detailed patent by patent (this is in any case not required by law).

In Italy there is a public stock exchange (the Borsa valori), which is divided in different sections in which the requirements differ on size and financial capacity of the companies listed therein.

Where a flotation of shares in a public exchange is planned, and because the value of the shares in the flotation does not reflect the balance sheet value alone but, rather, the real value of the business itself, the value assigned to any patents can be taken into consideration, particularly when launching an initial public offering. When preparing a prospectus for flotation or an IPO, it is therefore possible also to place a value on IP rights and particularly on patents. However, since in the preparation of a prospectus a specific liability exists in relation to false declarations, a due diligence exercise has to be carefully conducted, and the value of the IP rights should be confirmed by external auditors and certified by an independent expert. During all of these activities, the company has a duty to disclose all relevant information.

(c) *Joint ventures*

A joint venture as such does not set up a new legal entity. Therefore, the creation of a joint venture does not imply an automatic assignment of IP rights by one participant to the others. However, since joint ventures correspond to the pooling of means and resources, including technical knowledge and intangible assets, this goal can also be reached by contributing patents. Such a contribution can be put in place by assigning the ownership of the patents (done by mean of an assignment agreement, which does not differ from an 'ordinary' assignment as illustrated above), normally in order to create joint ownership among all participants to the joint venture. However, the same result can be obtained more easily by granting a licence to the other participants to allow them to exploit the associated invention. In this case, ownership formally remains with the original participant, acting as a licensor, who is legitimised with regard to the enforcement of the patent. The participant(s) acting as a licensee can join the proceedings instigated by the patent owner for damages claims.

In a joint venture agreement, the ownership of possible new rights deriving from the common activity performed has to be carefully regulated – in particular, whenever new inventions are made possible by the exploitation of a patent owned by one of the participants.

(d) *Divestment*

There are many reason for divesting a patent – for example, because the patents to be divested are no longer of interest due to a change in production or economic strategy, or because the operation to which the patent refers is no longer considered to be a core business for the company.

Divestments are normally obtained through assignments, but they can also be obtained by exclusive licences (see section 2.1(b) above in this respect). When divestment of a patent is obtained within the framework of an assignment of part of a business, know-how and employee confidentiality related to the patent must also be regulated in the agreement regulating the assignment. However, in contrast, whenever patents are transferred independently from the business, which remain in the ownership of the assignor, know-how and employee confidentiality must be regulated in order to ensure and warrant that the assignee is not damaged and has full protection on the one side, as well as full access to the relevant information on the other.

When patents are divested separately from the business as a whole, it is also important to regulate the further use of the invention by the assignor, in case it is needed (eg, for spare parts to be provided to existing customers, or whenever a secondary production process remains active). A possible solution in this respect is to grant a non-exclusive licence to the assignor.

As to warranties and indemnities related to divestments, they are very similar to those of assignments/licences, as described in section 2.1 above.

(e) Consortium agreements

Patents are of relevance whenever a consortium of parties is created to act together for conducting research on products.

In some cases the research project might already exist, and a consortium is needed in order to develop the invention and increase its value; this is typically what happens with venture capital projects. The consortium can act to provide joint funding, development, marketing or technical capacities. The respective shares of resulting profit or increases in inherent value, as well as the possible derivation of new IP rights, must be carefully regulated.

Alternatively, patents can be the result of the consortium itself, which is set up with the aim of creating new inventions. The share in the ownership of the deriving assets and/or the respective rights of use thereof must again be carefully regulated.

One particular issue to be managed is when members of the consortium do not have the same core activity (ie, some are financial entities and some are industrial entities). In this case, some participants may be interested in directly exploiting the IP rights, whilst others may be interested in assigning or licensing them. Since joint ownership is regulated by Italian law only partially, it is advisable to stipulate precisely which kind of use is reserved to the co-owner who intends to exploit the patent, and under what conditions a licence can be granted to third parties.

2.3 Particular business transactions

(a) Material transfer agreements

Relevant provisions for material transfer arrangements (MTAs) under Italian law are in Article 81*sexies* and 81*septies* of the Industrial Property Code.

In MTAs, it is of particular importance to regulate for what purpose the material is transferred, in order to be able to precisely define whether the material is transferred for allowing further reproductions thereof. This regulation is achieved by carefully drafting the relevant agreement, which should include any appropriate detail for determining whether the organic material is strictly transferred and/or which specific further use is allowed.

In particular, it is advisable, if the patent concerns a biotechnological method or a new use of it, whenever this method is strictly embodied in the material, to draft the agreement such as to avoid the assignee claiming that the transfer of the material implies the transfer of the tools needed for implementing the method, thereby creating an implied licence in favour of the assignee in accordance with Article 67 of the Industrial Property Code.

(b) Commercial research and development agreements

Non-collaborative arrangements: When entering into commercial research and development agreements – and in particular non-collaborative ones – the main issue to be precisely defined is ownership of the resulting inventions. Thus, the agreement should specify whether the rights deriving from the resulting invention will be shared among all parties involved and, if so, the respective shares (usually corresponding to the respective shares of the initial investments needed to make the invention). A further option could be to provide for the resulting invention to be assigned to a sole owner (which may be the entity granting most of the investments or research efforts, or the entity willing to actually implement the resulting invention on the market), under payment to other parties of a fair consideration.

Collaborative arrangements: In commercial collaborative R&D agreements, the issue of which uses of the inventions will be allowed to each participant to the agreement must be taken into account, in addition to the issues already dealt with in relation to joint ventures and consortiums. Furthermore, the relevant European antitrust provisions (where applicable), set forth by EC Regulation 1217/2010, should also be considered to see whether they apply in the given circumstances.

(c) ***Contract manufacturing agreements***

When setting a contract manufacturing agreement, it is of importance to regulate the confidentiality of the information disclosed by the principal, as well as the rights of the parties if a new invention results from the execution of the agreement. If an invention or other protectable result is obtained by the subcontractor, a consideration should be in place as recompense for the assignment to the principal of the full ownership.

(d) ***Product distribution agreements***

Product distribution agreements involving patented products should be carefully assessed from a competition law perspective, particularly when they involve both distribution and production agreements. Provisions included in the agreement shall be assessed in respect of their compliance with EU Regulation 330/2010 on the application of Article 101(3) of the Treaty on the Functioning of the European Union to categories of vertical agreements and concerted practices.

In particular, the regulation prohibits, as restrictive of competition, any direct or indirect non-compete obligation, such as undertakings not to manufacture, purchase or resell goods or services which compete with the contract goods or services, or any direct or indirect obligations on the buyer to purchase from the supplier or from another undertaking designated by the supplier which apply indefinitely or for more than five years. Additionally, the regulation prohibits the inclusion in product distribution agreements of any direct or indirect obligation which prevents the buyer, after termination of the agreement, from manufacturing purchasing, selling or reselling goods or services, as well as any direct or indirect obligation which prevents the members of a selective distribution system from selling the brands of particular competing suppliers.

(e) ***Co-promotion agreements***

Under Italian law, the main issue related to co-promotion agreements is again compliance with antitrust regulations, since they could affect competition.

Guidelines on the applicability of Article 101 of the Treaty on the Functioning of the European Union to horizontal co-operation agreements (2011/C 11/01) have to be considered in this assessment.

(f) ***Clinical trial agreements***

By entering into a clinical trial agreement, on one side patent-related issues and on the other side know-how protection must be considered.

As to patent-related issues, they can refer to the party requesting that clinical trials are conducted (in this respect, since clinical trials have the scope of assessing, among others, the effectiveness and safety of a new drug, they could be of importance for developing new inventions, particularly as to improved formulations). It is important clearly to express a confidentiality agreement, and to regulate the rights of the party requesting the trials to be provided with all possible information deriving from those trials.

As to patent issues, they could refer to third-parties rights, and in this respect the 'Bolar' exemption applies, which in Italy is implemented by Article 68.1(b) of the Industrial Property Code. It is advisable that the purpose of a clinical trial is clearly defined, in order to confirm that the research conducted falls within the scope of Article 68.1(b).

3. Issues associated with patent transactions

3.1 Regulatory law and compliance

Specific provisions are set forth by Italian and EU law in respect of drugs regulation. Before the launch of a newly patented drug onto the market, a marketing authorisation is needed. In Europe, this can be obtained, depending on the marketing authorisation required and on the drug at issue, through a centralised procedure, a decentralised procedure, a mutual recognition procedure or a national procedure. The relevant EU provisions are contained in Directive 2001/83/EC on human medicinal products, as amended.

As to the national procedure in Italy, marketing authorisations are regulated by Legislative Decree 219/2006.

In order for a marketing authorisation to be granted for a new drug, applicants must prove the quality, safety and efficacy of their medicinal product, and do so by submitting the results of pharmaceutical tests as well as of clinical trials conducted on the medicinal product to be placed on the market. This very costly evidence is protected as confidential information (in Italy, through a special provision set forth in Article 98.2 of the Industrial Property Code) and by the granting of data protection and market exclusivity for a period of 10 years after the first marketing authorisation of the product – although this regime has only applied for medicinal products authorised after October 31 2005.

In this context, patent protection, data protection and market exclusivity act jointly in the protection of the product and its owner. But since new drugs cannot be marketed until a marketing authorisation has been granted – which can take a lengthy period of time, thus resulting in a reduction of the period of protection for the pharmaceutical product – a supplementary protection certificate (SPC) may be provided under Council Regulation 469/2009, allowing the patent owner to recover the time lost in the regulatory process up to a maximum of five years. A further SPC extension of six months is granted in a case where a marketing authorisation for paediatric use is obtained for the drug.

3.2 Commonly disputed matters

In Italy, there are 20 courts having jurisdiction on patent matters. However, they are not necessarily the same as those having jurisdiction for contract law matters, which number more than 100. In this respect, while a court having jurisdiction on patent matters can also conduct a hearing on contract law, the opposite is not true. Therefore, in drafting an agreement regarding patent rights (assignment, licence, joint ventures, etc), it is important to specifically regulate where any claim should be brought to court, by selecting a court having jurisdiction in both matters.

3.3 Competition rules

Under certain circumstances the exploitation or the enforcement of patents may be contrary to competition law. Articles 101 and 102 of the Treaty on the Functioning of the European Union (the Treaty) are relevant in this respect. As regards Italy, Articles 2 and 3 of Law 287/1990 contain the same principles established by Articles 101 and 102 of the Treaty, but Articles 2 and 3 only apply for cases that have a merely national relevance, as the articles do not have the effect of restricting competition between EU member states but apply only within the Italian market.

Article 101(1) of the Treaty and Article 2 of Law 287/1990 prohibit all agreements between undertakings, decisions by associations of undertakings and concerted practices which have as their object or effect the prevention, restriction or distortion of competition within the common market or national market, respectively. A breach of Article 101(1) or Article 2 automatically causes the nullity of the agreement concerned. Agreements can nonetheless be exempted from these restrictions if they contribute to improving the production or distribution of goods or to promoting technical or economic progress.

According to Article 102 of the Treaty and Article 3 of Law 287/1990, any abuse by one or more undertakings of a dominant position within the common or national market, respectively, is prohibited in so far as it may affect trade between member states or within the Italian national market.

3.4 Securitisation of patent rights

According to Article 140 of the Industrial Property Code, patents can be pledged in respect of money credits. A pledge on a patent is established by a deed and is effective only when registered with the IPTO. It is argued that the entry in the Patent Register represents an absolute requirement of validity and not just a requirement imposed for the pledge to be effective against third parties.

3.5 The effect of insolvency

Insolvency procedures are aimed at paying creditors proportionally to their credit amounts by distributing the revenues of the divestments of the undertaking. IP rights can also be divested, and in a case of insolvency a bankruptcy receiver will act as described in section 2.2(d) above. However, it is much more likely that divestments will only be obtained by assignment rather than the granting of a licence, unless a licence agreement was already in existence, and considering the remaining duration the bankruptcy receiver may decide to exploit the agreement until its expiry, rather than terminate it for divesting the IP right.

In order to ensure protection against the consequences of the insolvency of the other party, the agreement should clearly regulate the relevant circumstances, for example by providing the other party with the right to terminate the contract.

However, if a declaration of bankruptcy is issued by a competent court, Italian bankruptcy law provides that the performance of all contracts to which the company was party is suspended until a decision is made by the bankruptcy's official receiver to succeed in the contract. If there is a lack of a timely decision, the other party may request the court to assign a deadline to the official receiver to declare its intention to succeed in the contract, or otherwise to consider it terminated.

3.6 Taxation

Patent transactions are subject to the payment of a 3% registration fee on the declared value of the transaction, unless VAT applies.

Companies and economic undertakings carrying out R&D activities have available to them several kinds of tax incentive offered by both the national government and by local authorities, with a view to promoting innovation and scientific research. Those incentives are usually established on a year-by-year or multi-year basis, and renewed at the end of each period by the competent authority.

3.7 Hot topics

One hot topic in Italy (and in other countries) is that of patent pools and FRAND (fair, reasonable and non-discriminatory) licences. In particular, the issue concerns whether patent pool members can obtain preliminary injunction relief and under what conditions.

As to hot topics related to the life sciences, recent decisions have discussed the issues of preliminary reliefs and actions for obtaining declaratory judgements in respect of pharmaceutical patents.

Additionally, some recent cases have concerned the issue of whether a request for a marketing authorisation for the introduction onto the market of a generic drug may be considered an activity infringing the patent rights of the originator. This discussion has led to amendments to the relevant law provisions, which regrettably still do not clarify the matter.

Furthermore, SPCs have been extensively discussed in the light of recent decisions of the Court of Justice of the European Union that have been rendered in pharmaceutical cases in respect of the validity of limitations to the basic patents.

The author would like to express his gratitude to Lorenzo Battarino and Elena Tellini for their assistance in the preparation of this chapter.

Japan

Frederick Ch'en
Eiichiro Kubota
Hogan Lovells

The Japanese pharmaceuticals market is the second-largest in the world and, according to a report published by EIU Japan in August 2013, represents about 12% of the global market (the United States is the world's largest market, which has about 37% of the global market). Even so, many consider Japan a challenging market to enter.

In this chapter, we describe some of the features of patent transactions in the life sciences sector in Japan and hopefully thereby provide practical guidance to those who may be considering such activity. For non-Japanese companies in particular, this may be an opportune time (2014) to do so because the Japanese government is currently contemplating various changes in the life sciences sector, including the pricing of products, healthcare spending, and generic drugs. Indeed, it seems that companies are actively assessing their product pipelines and portfolios, and may soon enter into various transactions to anticipate and capitalise on such changes. So, despite the potential risks and challenges, now could be the moment to heed the words of a famous Japanese proverb, "If you do not enter the tiger's cave, you will not catch its cub"; put another way, nothing ventured, nothing gained.

1. Patent, patent-term extensions and know-how: foundations for transactions

1.1 Patents and patent-term extensions

(a) Invention

The term 'invention' is defined in Japan's Patent Act (Law 121 of 1959) to mean the highly advanced creation of technical ideas utilising the laws of nature (Article 2(1); references to Articles are to articles of the Patent Act unless otherwise stated). There are three categories: 'invention of a product', 'invention of a manufacturing product' and 'invention of a process' (Article 2(3)).

Under the current practice of the Japan Patent Office (JPO), a gene can fall within the definition of an invention (although it is a product of nature) if it is specified by its amino acid sequence and function.

(b) Conditions for patentability

The main requirements for patentability are as follows (the full list of conditions is set out in Article 49):

- The invention for which a patent is sought falls within the statutory definition of 'invention' (Article 2.1).
- The invention for which a patent is sought is industrially applicable, novel and inventive (Article 29).
- The invention for which a patent is sought is clearly defined (Article 36(6)(ii)).
- The detailed explanation of the invention (its specification) is clear and sufficient so as to enable a person ordinarily skilled in the field of art to which the invention belongs to work the invention (Article 36(4)(i)).

In connection with the novelty requirement (Article 29(1)), as in many other jurisdictions, an inventor is prevented from obtaining a patent for the following types of invention:

- inventions that were publicly known in Japan or a foreign country before the filing of the patent application;
- inventions that were publicly worked in Japan or a foreign country before the filing of the patent application; and
- inventions that were described in a distributed publication, or inventions that were made publicly available through an electric telecommunication line in Japan or a foreign country, before the filing of the patent application.

In connection with the inventiveness requirement (Article 29(2)), a patent will not be granted for an invention if, at the time of filing the patent application, a person skilled in the art would have been able easily to make the invention based on a prior art or publicly known invention.

Under the current practice of the JPO, it is not possible to obtain a patent for a method of surgery, therapy or the diagnosis of human diseases because such a process lacks industrial application. As in many other jurisdictions, this is on policy grounds to ensure that medical practitioners' freedom to treat patients is not constrained by patent rights. However, medicines (pharmaceutical products) and medical devices themselves are not included within this exception to patentability. Thus, in principle, it is possible to obtain a patent in respect of an invention of a medicine and a medical device, and the invention of a process to manufacture them. Furthermore, an invention of process to operate a medical device, such as a method of controlling a medical device, is patentable. And an invention of a process to obtain materials from the human body is industrially applicable and patentable, unless the process includes a method of surgery, therapy or the diagnosis of human diseases.

(c) The effect of a patent right

The owner of a patent (the patentee) has the exclusive right to work the patented invention as a business. This, however, does not apply to the extent that the patentee has granted an exclusive licence of the patent right to a third party (the exclusive licensee) entitling it exclusively to work the patented invention (Article 68).

The 'working' of an invention is defined by statute to mean the following acts (Article 2(3)):

- in the case of an invention of a product: the act of making, using, assigning, leasing, exporting or importing, or offering for assignment (including displaying for the purpose of assignment) the product;

- in the case of an invention of a process: the act of using the process; and
- in the case of an invention of a process for making a product, in addition to the act of using the process, the acts of using, assigning, leasing, exporting or importing, or offering for assignment the product produced by the process.

A patent right becomes effective on registration and lasts for a period of 20 years from the filing date of the patent application (Articles 66.1 and 67.1).

The term of a patent right may be extended if there is a period of time during which it was not possible to work the patented invention. The extension may be for up to five years (Article 67.2), on application by the patentee only (Article 67-3.1(4)). Multiple extensions are available provided the relevant criteria are met for each extension.

The inability to work the invention must be because of the need to obtain approval (or other 'disposition') that is governed by legislation intended to ensure that the working of patented inventions is safe, and which is provided for in a Cabinet Order as being such that, in the light of the purpose of the relevant disposition, a considerable period of time is needed for the proper execution of the disposition (Article 67(2)). The application for an extension needs to be filed within three months after the patentee obtains the disposition prescribed by a Cabinet Order. However, the application may not be filed after the patent right has expired (Article 67-2.3). Examples of such a disposition include the registration of agricultural chemicals based on the Agricultural Chemicals Regulation Act and the approval of drugs based on the Pharmaceutical Affairs Act; both are designated as dispositions by Cabinet Order (Article 3 of the Order for Enforcement of the Patent Act).

Note that if the term of a patent right is extended, the effect of such a patent right does not extend to acts other than the working of the patented invention in respect of the product that was subject to the disposition (Article 68-2).

Under the Pharmaceutical Affairs Act, marketing authorisation must be obtained for each product, specified by dosage, administration, efficacy, active component, etc. Conventionally, once a certain pharmaceutical product (the 'first medical product') has received marketing authorisation, the term of any patents covering other pharmaceutical products with the same active ingredient and efficacy as the first medical product (known as 'second medical products') cannot be extended. This is true even if the second medical products need a separate marketing authorisation from that for the first medical product.

However, the Supreme Court has recently noted that if a particular first medical product does not fall within the technical scope of the patent covering a second medical product, it is necessary to obtain a separate marketing authorisation for that second medical product in order to work the patent covering that second medical product. As a result, the Supreme Court held that the term of patents covering second medical products may also be extended separately from any patents covering first medical products.[1]

1 Supreme Court decision Heisei 21 (*Gyo-Hi*) 326 of April 28 2011.

Where a trial decision to the effect that a patent is to be invalidated has become final and binding, the patent right shall be deemed never to have existed (Article 125).

(d) *Application procedure*

Detailed requirements for a patent application are set out in Article 36 of the Patent Act. Once a period of one year and six months from the date of the filing of a patent application has elapsed, the patent application is laid open for publication (Article 64). A patent application will be examined if requested within three years from the filing date (Article 48-3).

A patent application is examined by an examiner, the qualifications of whom are prescribed by Cabinet Order (Article 47). If an examiner intends to refuse an application, the examiner must provide reasons and give the applicant an opportunity to submit a written statement of counterarguments within an adequate time limit (Article 50). If an examiner has no reason to refuse a patent application, the examiner will render a decision that a patent is to be granted (Article 51).

1.2 Who owns the patent?

Only an inventor, or a person who succeeds to the right to obtain a patent, can apply for a patent. If the patent applicant is not the inventor or does not otherwise have the right to obtain a patent, the patent application will be refused (Article 49(vii)). If such patent is wrongly granted, in such circumstances the patent will be invalidated (Article 123.1(vi)).

If the right to obtain a patent is jointly owned, a patent application must be made by all joint owners. No joint owner may assign his respective share without the consent of all the other joint owners.

1.3 Inventor compensation

Japan's regime for inventor compensation is designed to encourage employee-inventors to invent and thereby create an innovative economy. An 'employee invention' is defined by statute to mean an invention which, by the nature of the invention, falls within the scope of the employer's business and was achieved by an act or acts that were part of the present or past duties of the employee performed for the employer (Article 35.1).

Unless an employee's invention falls within the statutory definition of an employee invention, any prior agreement purporting to assign to the employer the right to obtain a patent, or the patent right itself, is null and void (Article 35.2). Thus, in such circumstances, an employer needs to obtain the employee's consent to be assigned the right to obtain a patent, or the patent right itself, after the invention has been made.

For an employee invention, an employer may enter into a prior agreement with an employee providing that the right to obtain a patent, or the patent right itself, shall vest in the employer (this is usually set out in the company's employee invention regulations). However, in such circumstances the employee shall have the right to receive reasonable remuneration (Article 35.3), and the amount is governed by the company's employee invention regulations.

What is 'reasonable' is hard to assess. Rules for determining the amount of remuneration are not considered unreasonable in circumstances where the employer and employees have negotiated the criteria by which the amount is determined, the criteria have been disclosed, and the employees' opinions of the criteria have been received (Article 35.4). If there are no provisions for determining the amount of remuneration, or if the amount to be paid in accordance with the relevant provisions is considered unreasonable, the amount is calculated by taking into account the employer's profit from the invention, the employee's contribution, the employer's treatment of the employee, and any other circumstances relating to the invention (Article 35.5).

In practice, determining 'reasonable' remuneration is difficult, as illustrated by the 2004 case of *Nakamura v Nichia Corp*,[2] in which the Tokyo District Court awarded an employee-inventor more than USD180 million before the case settled (the final amount was USD8 million). Note that Japan is currently contemplating a revision of its inventor compensation regime, especially in the light of the collective transaction and litigation costs to Japanese companies; this area should be monitored carefully in the coming few years.

1.4 Know-how and trade secrets

(a) *Protection by the Patent Act*

If know-how meets the conditions for patentability (Article 29), it can be registered and protected as a patent right. The advantage is that a patentee will have the exclusive right to work the patented know-how. However, the disadvantage is that the know-how is disclosed, and protection is limited to a period of 20 years from the filing date of the patent application.

(b) *Protection by the Unfair Competition Prevention Act*

Even if know-how does not meet the conditions of patentability or if the creator of the know-how does not wish to patent it, depending on the nature of the know-how it may be protected as a trade secret under the Unfair Competition Prevention Act.

'Trade secret' is defined by statute to mean technical or business information useful for business activities, such as manufacturing or marketing methods, that is kept secret and that is not publicly known (Article 2.6 of the Unfair Competition Prevention Act). The act of acquiring a trade secret by theft, fraud, duress or other wrongful means, or the act of using or disclosing a trade secret through an act of wrongful acquisition etc, constitutes unfair competition under Article 2.1(4)–(9) of the same act.

A person whose business interests have been infringed, or are likely to be infringed, due to unfair competition has a right to demand an injunction (Article 3 of the Unfair Competition Prevention Act), seek damages (Article 4 of that act) and seek necessary measures to restore business reputation (Article 14 of that act).

Protection in this way is advantageous because the confidentiality of know-how is protected as long as it is kept secret, and non-patentable know-how can be

2 Heisei 13 (*Wa*) 17772 of January 30 2004.

protected. However, if other companies develop the know-how independently, the original creator of the know-how does not have an exclusive right.

(c) ***Protection by a confidentiality agreement***

If know-how falls within the definition of confidential information that is protected by a confidentiality agreement, the know-how is protected in accordance with the terms of that agreement. Confidentiality agreements may be between companies and between an employer and employee. Typically, such agreements set out the scope of confidential information, the obligation of confidentiality, exemptions, the duration of confidentiality and the remedies for a breach of the contractual obligations.

2. Context and features of patent transactions

2.1 The principal dealings in a patent

A key question before assigning a patent right is whether a licence of the patent right (rather than an assignment) may in fact be preferable in terms of achieving the patentee's commercial objectives. The main advantage for the patent owner of an assignment is that it can receive a relatively high amount in compensation for the assignment while also being released from having to pay patent maintenance fees. Moreover, from the assignee's perspective, it can mean much greater flexibility in exploiting the patent right. However, one of the advantages for the patentee of licensing instead of assignment is the possibility of control over the licensee's use of the patented technology through the provisions of the licence. In addition, the royalties payable under the licence can be calculated and tied to the use of the patented technology; this might be fairer to both parties than the price that is negotiated and agreed for an outright assignment, which must be determined at the time of the assignment when the full commercial potential of the patent is not known and can only be estimated. Nevertheless, there is a risk for the patentee that licensing revenues will be lost if the validity of the patent is challenged and the patent right is invalidated.

For these reasons, it is important to consider carefully which type of transaction is most appropriate in the context of the patentee's broader commercial objectives and strategy. Both assignment and licensing are described further next.

(a) ***Patent assignments***

A patent right is a property right, and the owner of a patent right may assign all or a part of the right. Generally, a company sells and assigns its patent right to another in order to turn it into cash, and may do so for a number of reasons. For instance, the company may be closing down a particular part of its business such that the patent right is no longer necessary or useful for the remaining business. Alternatively, a company may wish to sell part of its business as an asset sale that includes patent rights, or to entrust part of its business in some countries to a third party. In the latter case, there may be some circumstances where a patent owner may consider assigning a share of the patent right to the third party that runs the business in the relevant countries; this results in a jointly owned patent right.

In Japan, the assignment of a patent right must be registered at the JPO to take effect (Article 98 of the Patent Act). If a patent right is jointly owned, a joint owner must obtain the consent of all the other joint owners to assign its share of the patent to a third party (Article 73).

There are a number of key considerations when granting an assignment:

- *Grant-back clause:* In principle, an assignor (ie, a former patent owner) is prohibited from using the patented invention after it has assigned its patent right. If the assignor intends to use the patented invention after the assignment, the assignor should ensure that the assignment agreement grants a licence of the patent right back to the assignor via a 'grant-back' clause.
- *Existing licensees:* Following the implementation of revisions to the Patent Act in 2012, a non-exclusive licensee that exists at the time the patent right is assigned, but which is not registered at the JPO, is now protected from any enforcement action by the assignee (ie, the new patent owner). Therefore, a potential assignee should conduct careful due diligence and investigate the existence of any non-exclusive licences of the relevant patent: a search of the patent registry will not reveal non-exclusive licensees of the patent right.

In Japan, it is often difficult for a potential assignee to obtain accurate information about existing licences of the patent right because many licence agreements, or their associated confidentiality agreements, prohibit the licensor (ie, the potential assignor) from disclosing the existence of the licence to third parties, including to a potential assignee. Even if there is no such confidentiality agreement, potential assignors may be reluctant to disclose information about existing licences. Nevertheless, it is of course important for a potential assignee to minimise the risk that it fails to achieve the commercial purpose of acquiring the patent right. For example, if the reason for acquiring the patent right is exclusively to use the patented technology, the potential assignee should seek a warranty from the assignor that there are no existing licensees of the relevant patent. If the purpose is to enforce the patent right against competing companies, the potential assignee should consider obtaining a warranty that those companies have not been granted a licence. Further, if there is an existing licensee of the relevant patent, it is advisable for the potential assignee to clarify to whom the royalty should be paid after the patent has been assigned (ie, to the assignor or the assignee) and to notify the existing licensees accordingly.

Warranty of patent validity: In some cases, an assignment agreement includes a provision that represents and warrants the validity of the relevant patent, or imposes on the assignor an obligation to indemnify the assignee for loss caused in the event that the patent is invalidated. From the assignor's perspective, such a provision should be avoided because the assignor would otherwise be exposed to the risk that it breaches the contract, or that it will be obliged to indemnify the assignee, for a quite long period – that is, at least until the expiration of the patent term. Even if a potential assignee is concerned about patent validity and demands a warranty that the patent is valid, assignors very often try to address such concerns by reflecting them in the price for the assignment, rather than providing such a warranty.

(b) Patent licences

In Japan, as in other parts of the world, the conventional licensing model in the pharmaceutical industry is one where a major pharmaceutical company grants an exclusive licence of rights in respect of a chemical compound to a relatively small company. However, an increasing number of major pharmaceutical companies have recently started to license-in from venture businesses the rights to use chemical compounds that were developed by those venture businesses, and then to sell drugs containing such chemical compounds under the company's own brand. This model produces benefits for both sides: while the major company can save development costs, the venture business can exploit and commercialise its technology and overcome the disadvantage of a low profile.

In addition, a growing number of pharmaceutical companies are seeking the route of co-development with a third party, rather than granting an exclusive licence, in an attempt to reduce the costs of development. However, where multiple companies share a licence as a result of co-development, there are often a variety of associated problems. In particular, typical issues that are frequently encountered and the subject of heavy negotiation are the extent and permissible scope of the licence, each company's percentage costs and percentage profit, and the necessary procedures that need to be followed if the co-development ends in failure.

If companies are about to license their products or processes to other businesses, either for marketing purposes or manufacturing, certain requirements must be taken into account as follows:

- A business that involves the marketing of drugs, quasi-drugs, cosmetics or medical devices in Japan must obtain a marketing business licence from the relevant prefectural governor. The licensing requirements for a business involved in the marketing of a drug include the appointment of a general marketing compliance officer, who is a pharmacist, and compliance with good quality practice (GQP) for quality control and good vigilance practice (GVP) for post-marketing safety surveillance.
- A business that involves the manufacture of drugs, quasi-drugs, cosmetics or medical devices in Japan must obtain a manufacturing business licence in accordance with the manufacturing category specified by Japan's Ministry of Health, Labour and Welfare (MHLW). If a foreign company grants a patent licence to a Japanese company that starts a business involving the manufacture or marketing of a drug covered by a licensed patent, the Japanese company (ie, the patent licensee) must also obtain the relevant manufacturing or marketing business licence.
- A business that wishes to manufacture drugs, quasi-drugs, cosmetics or medical devices that are exported to Japan from overseas must receive accreditation from the minister of the MHLW. Thus, if such an overseas manufacturer grants a patent licence to a Japanese company that starts a business involving the marketing of drugs that are manufactured by the overseas manufacturer, as well as the Japanese company (ie, the patent licensee) obtaining a marketing business licence the overseas manufacturer (ie, the patent licensor) must receive the aforementioned accreditation.

Drug 'marketing authorisation' means governmental permission for a drug with appropriate quality, efficacy and safety, or a drug that is manufactured by a method in compliance with standards for manufacturing control and quality control, and which is generally distributed and used clinically under an appropriate quality and safety management system. The minister of the MHLW or prefectural governor (in practice, the Pharmaceuticals and Medical Devices Agency (PMDA)) reviews the name, ingredients, composition, dosage, administration, indications, adverse drug reactions etc of the proposed drug product in an application submitted by a person with a marketing business licence. Thus, when granting a patent licence to a Japanese company that also includes the right to distribute drug products covered by the licensed patent, the Japanese company (ie, the patent licensee) must separately obtain an authorisation for marketing such drugs.

It is advisable for a potential licensor to take the following key preparatory steps when considering whether to enter into a patent licence agreement:

- Search for suitable potential licensees.
- Prepare a business strategy for the subject technology or products in the territory to be licensed.
- Estimate licence royalty rates and amounts.
- Consider the applicable regulations in the territory to be licensed (eg, antitrust laws, tax regulations, intellectual property laws).
- Prepare the information to be disclosed to potential licensees.
- Organise a negotiation team.
- Draft an appropriate terms sheet.
- Consider the location, venue and schedule for licensing negotiations.
- Determine the criteria for walking away from any potential licensee or licensing deal.

Similarly, there are a number of preparatory steps a potential licensee may wish to consider before entering into a patent licence agreement:

- Consider the strength of the patent for the subject technology, the remaining term of the patent and the possibility of infringing a third party's patent by using the patented invention.
- Conduct market research on the subject technology, review the competitive power of the product and make a suitable sales plan.
- Consider the reasonableness and feasibility of royalty obligations.
- Review the financial state of the potential licensor.
- Consider the applicable regulations in the territory to be licensed (eg, antitrust laws, tax regulations, intellectual property laws).
- Organise a negotiation team.
- Review the terms sheet.
- Determine the criteria for walking away from the licensor or any licensing deal.

There are some key aspects that should be considered when negotiating a typical patent licence. Set out next are points to bear in mind from the perspective of Japanese law and practice.

Parties: A common issue is how to deal with the parties' subsidiaries and group companies. Although subsidiaries and affiliates are often subsumed into the definition of the contracting parties, it is not always completely clear, for instance, that the licensee's subsidiaries and affiliates would necessarily receive the contractual benefits and be subject to the contractual burdens of the licence. A better way is to provide in the licence the right for the licensee to sub-license a subsidiary as appropriate and on suitable terms, and to have a separate sub-licence agreement governing that relationship.

Types of licence: Article 77 of Japan's Patent Act defines the term 'exclusive licence'; but this definition is not exactly the same as the concept of an exclusive licence as used in the United States, United Kingdom and elsewhere, so we refer to it in this chapter as a 'Japanese exclusive licence'. The key characteristics are threefold:

- A Japanese exclusive licence is only effective if it is registered at the JPO.
- Once a Japanese exclusive licence is granted and effective, even a patent owner is prohibited from working the patented invention.
- The licensee of a Japanese exclusive licence has the right to demand an injunction against a patent infringer.

Aside from the Japanese exclusive licence, there is a type of licence called an 'exclusive normal licence' or 'contractual exclusive licence'. Here, exclusivity is expressly conferred on a 'normal licence' in a licence agreement. A contractual exclusive licence does not need to be registered to become effective, and whether a patent owner can or cannot work the patented invention depends on the terms of the licence agreement. A licensee of a contractual exclusive licence is not allowed to demand an injunction but only to claim damages against a patent infringer (in this regard, this type of exclusive licence is different from an exclusive licence as used in the United States or the United Kingdom).

Given these different types of licence, it is advisable for a patentee to ensure that any patent licence it grants expressly states whether or not the patentee also has the right to work the licensed patent.

Clear definitions: From the perspective of a licensee, it is often advisable to include in the definition of 'licensed patents' any derived or additional patents (including patent applications) along with foreign counterpart patents and their derived or additional patents. It is also not uncommon for a licensee to request the inclusion of patents and patent applications owned by the licensor's subsidiaries.

The definition of 'licensed products' will be the subject of negotiation. It may be defined as the products covered by the licensed patents or may directly describe and specify the product itself. The term needs to be clearly defined because the amount of royalty payable will be directly linked to the scope of the licensed products.

Royalties: The term 'royalty' is used to describe any consideration paid for the rights granted to a licensee. Exclusive licences typically require an upfront, non-refundable licence fee that may cover the costs of obtaining the patent, as well as running royalties and minimum annual royalty payments (see also below). Non-exclusive licences usually have more flexible terms. According to a research study published by the Japanese Patent Office in 2009, the average royalty percentage payable for a licence of a patent related to pharmaceutical products and biotechnology is about 6% (Japanese domestic transactions only).

As in many other jurisdictions, in Japan there are a wide range of royalty payment structures, including initial payments, milestone payments, upfront payments, minimum annual royalties, earned royalties (running royalties) and reach-through royalties. Initial payments are generally coupled with periodic royalty payments. From the perspective of a licensor, the licence agreement should stipulate that any initial payment is non-refundable even if the development is cancelled for any reason.

In order to secure the payment of a royalty, a licensor may impose on its licensee an obligation to work the invention of the licensed patent, breach of which gives the licensor the right to terminate the licence.

A minimum annual royalty provision ensures that the licensor will receive a royalty payment no lower than a specified amount. This sort of provision is especially prevalent in exclusive licence agreements and protects a licensor's interest where the licensee is not diligent in developing or marketing a product based on the licensed technology. This sort of provision may stipulate that if the royalty does not reach a predetermined minimum amount after a certain period:

- either the licensor may terminate the licence agreement or convert the exclusive licence agreement to a non-exclusive one; or
- the licensee must pay the difference between the actual royalties paid and the minimum annual royalty amount.

For licences in the Japanese pharmaceuticals industry, upfront payments, milestone payments and/or running royalties are typically used. Milestone payments normally become payable as a pharmaceutical product progresses through various stages of development. This is advantageous for the licensee because there is usually a lower upfront payment in such a case, which saves the licensee initial costs. There are advantages for the licensor, too, as it can spread the risks associated with drug development and obtain a certain amount of consideration even if the development ends in failure.

In milestone payments, a licensor may also require certain 'success milestones' coupled with obligations on the licensee to develop the technology in a certain manner within a certain time – for example, by performing clinical trials by a certain date, or getting the product on the market by a certain later date. However, success milestones are particularly difficult to negotiate for technology in its very early stages of development – where parties are conscious of many unpredictable technical hurdles in the product's development – making it hard and often unreasonable to demand specific dates for success.

Best efforts: 'Best efforts' clauses are common in exclusive licences and preferred by many licensors, especially in the pharmaceuticals industry. These clauses oblige the licensee to "use its best efforts" to develop the invention to commercial utility, or to use "all reasonable efforts" to maximise the benefits to be realised by the licensor. In other words, the licensee must perform its obligations with the effort that companies in a similar situation would exert in respect of their own research programme. As in many other countries, best efforts clauses are prone to litigation and the Japanese courts tend to enforce them by giving them their literal meaning. Therefore, a better way to avoid a dispute in Japan is to set measurable goals with the licensor. Absent any express provisions, it is unclear whether the licensee has an implied obligation to the licensor.

Due diligence: 'Due diligence' clauses impose on the licensee an obligation to proceed diligently with the development, manufacture and sale or lease of the licensed product, and to develop markets diligently to meet certain specified goals. A licensor will usually try to set specific timelines for achieving certain milestones, including a deadline for drug development and clauses dealing with delays in performance. As delay is common in drug development, the inclusion of provisions governing the notice requirements for expected delays, and on dealing with such delays, is recommended.

Infringement: The right to bring an action against any infringers of licensed patent rights is often heavily negotiated. The parties may add a provision regarding whether the licensor, or the licensee, or both in cooperation, are allowed to control the defence of infringement actions brought against them by third parties, and control proceedings against third party infringers. This provision may include the division of costs and any recovered damages for the infringement.

2.2 Patent rights in corporate transactions

(a) Types of M&A transactions under Japanese law

In Japan, a merger and acquisition (M&A) transaction usually takes the form of:

- an absorption merger by a company of another company, under which all rights and obligations of the merged company are transferred to the surviving company by operation of law;
- a business or asset transfer, under which a company sells the whole or part of its business to another company;
- a share sale, under which the shares of the target company are sold by the seller to the purchaser; or
- a company split, under which a portion of a company is split off to become a new company by operation of law, where only a portion of the company is to be sold to the purchaser.

It is possible to have a combination of these structures – for instance, a company split linked with an absorption merger.

(b) How patent ownership is treated under each type of M&A transaction

Sole ownership of a patent is treated somewhat differently depending on what type of M&A transaction is occurring:

- In an absorption merger, the patents that are owned by the merged company will be transferred to the surviving company by operation of law. It is unnecessary for the parties to enter into a sale-and-purchase agreement for each patent. The transfer takes effect without registration of the transfer at the JPO, but the surviving company must notify the JPO of the transfer without delay.
- In the case of a business transfer, the patents owned by the seller may be assigned based on the agreement between the seller and the purchaser. In order for the assignment to take effect, the assignment for each patent must be registered at the JPO. In other words, until registration is completed, patent ownership remains with the seller.
- For a share sale, the target (acquired) company continues to be the owner of any patents it owned prior to the transaction.
- In the case of a company split, patents owned by the original company will be transferred to the new company (or, in the case of a linked company split and absorption merger, to the absorbing company) in accordance with the company split plan (or, in case of a linked company split and absorption merger, the company split agreement between the original company and the absorbing company) by operation of law. The transfer takes effect without registration of the transfer at the JPO, but the new company (or the absorbing company) must notify the JPO of the transfer without delay.

If a patent is jointly owned by more than one company, and one of the joint-owner companies is involved in an M&A transaction, how that patent is treated will depend on the type of transaction:

- In the case of a merger, a share sale or a company split, the jointly owned patent is treated in the same way as if it were fully owned (see above). In other words, a company may deal freely with its partial ownership share of the jointly owned patent.
- In the case of a business transfer, a company may not assign its partial ownership share of a jointly owned patent to another company unless all of the other joint owners consent.

When a licensor is involved in an M&A transaction, the following applies:

- In the case of a merger, the status of licensor is transferred from the merged company to the surviving company (together with ownership of the patents) by operation of law. The licensee's consent is not necessary.
- With a business transfer, the status of licensor may be assigned from the seller to the purchaser by agreement between the parties, and this requires the licensee's consent unless the licence agreement provides otherwise. The seller might, however, try to sell only the legal title to the patent (ie, patent ownership) to the purchaser. If such a sale completes, the patent licensee would be able to assert a licence defence against the purchaser in respect of any claim of patent infringement. (Note that Japanese case law is unclear as to whether the licensee must pay royalties to the seller or to the purchaser in such a scenario.)

- For a share sale, the target (acquired) company continues to be the licensor. The licensee's consent is not necessary.
- In the case of a company split, the status of licensor is transferred from the original company to the new company (or the absorbing company). The licensee's consent is not necessary.

When it is the licensee that is involved in an M&A transaction:

- In the case of a merger, the status of licensee is transferred from the merged company to the surviving company by operation of law. The licensor's consent is not necessary.
- For a business transfer, the status of licensee may be assigned from the seller to the purchaser by agreement between the parties. This does not require the licensor's consent, as long as the status of licensee is assigned together with the transfer of the business that uses the patented invention (Article 94(1) of the Patent Act).
- For a share sale, the target (acquired) company remains the licensee. The licensor's consent is not necessary.
- With a company split, the status of licensee is transferred from the original company to the new company (or the absorbing company). The licensor's consent is not necessary.

As is clear from the foregoing, the licensor's consent is generally unnecessary. However, the licence agreement might still contain a non-transfer clause or a change-of-control clause. Such a clause might provide that any purported transfer or assignment of the status of licensee in violation of such a clause is void. However, in the case of a merger or company split, as the transfer of the status of licensee occurs by operation of law, it is unlikely that the transfer becomes void merely because there is such a clause in the licence agreement.

In contrast, if a clause provides that the transfer or assignment of the status of licensee in violation of such a clause entitles the licensor to terminate the licence, it is likely that such a termination clause is valid and enforceable.

(c) Compensation to employee-inventors under different M&A transactions

As mentioned in paragraph 1.3 above, if an employee creates an invention in the course of his employment, and if the right to obtain a patent for the invention is assigned to the employer (the employment rules of a Japanese company usually provide so), the employee is entitled to seek reasonable compensation from the employer (Article 35 of the Patent Act). If the patent is valuable, the amount of compensation could become very high.

In the case of an M&A transaction, a company's obligation to pay reasonable compensation to the employee-inventor alters as follows:

- In the case of a merger, the obligation of the merged company is assumed by the surviving company by operation of law. The employee's consent is not necessary.
- In a business transfer, the obligation of the seller may be assumed by the

purchaser by agreement between the parties. However, this requires the employee's consent.

- In a share sale, the target (acquired) company continues to have the obligation. The employee's consent is not necessary.
- For a company split, the obligation of the original company may be assumed by the new company (or the absorbing company) in accordance with the company split plan (or the agreement with the absorbing company) by operation of law.

(d) Matters on which to focus in M&A due diligence

When a company is willing to acquire another company (through one of the M&A transactions mentioned in section 2.2(a) above) and is conducting due diligence on the target company, there are several key patent-related matters on which to focus, as set out next.

Ownership of patents and patent applications: If the target company provides a list of patents (identified by a patent number) that it allegedly owns, it is possible to verify the ownership by obtaining a copy of the JPO register entries for those patents. Note that under the present system in Japan the JPO database does not provide a way to search for patents that are currently owned by a particular entity. (It is possible to search for patents owned by a particular entity at the time of issuance by searching the patent publication, but the list of such patents would not cover patents that the entity obtained after issuance.) Therefore, the target company should be asked to provide a list of patents.

It is important to ask the target company to provide a list of patent applications, and to check the status of each application. If the application is published, it is possible to obtain the file wrapper of the patent and to check the current status. In the case of an unpublished application, it is necessary to ask the target company to confirm the status of the application. Moreover, it is advisable to check whether there is an invention for which a patent application has not yet been filed, or whether there is research and development activity that is likely to result in such an invention.

Whether a patent is owned solely or jointly with other companies can usually be verified by obtaining a copy of the JPO register entry. As mentioned previously, a company may not assign its share (partial ownership) of a jointly owned patent to another company unless all other joint owners consent. Therefore, it is important to check whether the target company fully or partially owns a patent.

It is also important to check whether there are any agreements (such as joint research and development (R&D) agreements) that grant special rights or impose special restrictions on the joint owners (as mentioned in section 2.3(a) below).

Security interest: An owner of a patent is able to grant a security interest (pledge) over a patent for the benefit of a creditor (see also section 3.4 below). Whether there is a security interest over a particular patent can usually be verified by obtaining a copy of the JPO register entry.

Existing licensees when the target company is the patent owner: As described in section 2.1(b) above, there are a number of different patent licences in Japan, which can broadly be divided into the Japanese exclusive licence and the ordinary or 'normal' licence, on which exclusivity can be conferred to create an exclusive normal licence.

Whether there is a Japanese exclusive licence in respect of a particular patent can usually be verified by obtaining a copy of the appropriate JPO register entry, as a Japanese exclusive licence becomes effective only upon registration at the JPO. In contrast, a normal licence is not verifiable merely by obtaining a copy of the JPO register entry because such licences are not recorded. Further, if a company purchases a patent from the target company (in the case of a business transfer), the normal licensee would be able to assert a licence defence against the purchaser in respect of any claim of patent infringement.

Therefore, it is normally necessary to ask the target company to provide information as to whether there is a licensee of the patent. However, it should be noted that in the case of a comprehensive licence (eg, a broad, general licence of a company's patents or intellectual property), the licence agreement may not necessarily specify which patents are covered by the licence.

Compensation to employee-inventors: It is possible to find out from the patent publication the name of the inventors. However, it is often quite difficult to predict in advance the amount of reasonable compensation that may be payable to them. If the target company holds a patent from which the target company gains a significant amount of profit, an expert's opinion might become necessary to assess the potential risk in respect of the amount of reasonable compensation that may be payable to employee-inventors.

Scope and validity of patent: The value of a patent typically depends on the scope of the patent, how much time is left before the patent term expires, whether annuities are duly paid, and also whether the patent could be invalidated for some reason (such as lack of inventive step). It should be noted that it is not rare for a Japanese patent to be invalidated after grant: according to the JPO, in 2012 about 30% (75 cases out of 254) of first-instance invalidation trial decisions held that the patent-in-suit was invalid. If a certain patent is an important asset of the target company in the M&A transaction, an expert opinion might become necessary to assess the scope and validity of the patent.

Target company a licensee: If the target company has been granted a patent licence from another company, it is important to check the scope, terms and conditions of that licence. In particular, it is important to check whether the licence agreement:

- contains a non-assignment or change-of-control clause that could be triggered by the contemplated M&A transaction (where, if so, it may be necessary to secure a waiver or consent from the licensor);
- is a comprehensive cross-licence and, if so, whether the company that is acquiring the target company would be obliged to license its patents under the terms of the cross-licence;

- is a sub-licence agreement and, if so, whether there is a valid (head) licence agreement between the patent owner and the sub-licensor (although such a licence agreement might not be accessible during the due diligence process, as the target company is not a party to any head licence agreement); and
- is a Japanese exclusive licence and, if so, whether the licence is duly registered at the JPO (which may be verified by obtaining a copy of the JPO register entry) since it only becomes effective on registration at the JPO.

Other agreements: Other agreements may affect the ownership of relevant patent rights and should be reviewed and considered. For example, the contract for a joint R&D programme often has a clause governing how patents that arise from the joint research and development should be treated (such as licensed or jointly owned).

In the case of joint ownership, the agreement might further contain a clause that restricts a party's use of the invention, or a clause that allows a party freely to grant licences to third parties. It is important to be vigilant over clauses such as these that alter the parties' usual statutory rights and obligations.

Third-party involvement: In some cases – for example, if there is a competitor of the target company, or if the target company has received a warning letter from another company – it may be important to check the ownership details of third-party patents, and further assess the risk of being sued for patent infringement by third parties. If there is such a risk, it would also be important to consider possible countermeasures, such as whether the target company owns a sufficient patent portfolio to enable it to counterclaim or bring its own action against the competitor.

Representations and warranties: As it is often difficult to verify the patents owned by the target company in the due diligence process, a purchaser often seeks a representation and warranty to the effect that the target company owns the patents stated in a certain list and that that list is an exhaustive list of the patents owned by the target company. The purchaser may also seek a warranty regarding the status of the patents and patent applications.

It is also common for a purchaser to seek representations and warranties in respect of the licences held by the target company. For example, if a licence is a sub-licence, the purchaser may request a warranty that there is a valid (head) licence agreement between the patent owner and the sub-licensor (especially if no access to the head licence agreement is possible during the due diligence process).

2.3 Particular business transactions

(a) Material transfer agreements

A material transfer agreement is an agreement under which the first party (usually a research institution) provides certain material to the second party (also a research institution) to allow research using the material by the second party.

Similar to clinical trial or research agreements, how a material transfer agreement treats patent ownership and licences of patents arising from the provided material varies greatly between different research institutions and their standard forms.

(b) Research and development agreements

Research and development agreements often contain provisions that deal with how intellectual property rights arising from joint R&D should be treated among the parties. Often, they specify that the IP rights arising from the joint R&D shall be jointly owned by the parties in a certain percentage. An R&D agreement is one of the common instances where joint ownership of a patent arises.

Under the Patent Act, the rights and obligations of joint patent owners are as follows:

- A joint owner is not able to assign its share (partial ownership) of the patent to a third party unless all the other joint owners consent to the assignment. This also applies to the creation of a security interest in such a share.
- A joint owner is able to work the patented invention by itself and without the consent of the other joint owners.
- A joint owner is not able to grant licences to third parties to use the patented invention unless all the other joint owners consent to the licence.

However, an R&D agreement may alter these provisions. For example, the R&D agreement might provide that a certain joint owner may not use the patented invention (even by itself), or might provide that a certain joint owner may grant licences to third parties without the other joint owners' separate consent. In the latter case, the R&D agreement might further provide for the sharing of royalties among the joint owners.

(c) Contract manufacturing agreements

A contract manufacturing agreement is an agreement under which a first party (a manufacturer of a product) outsources the manufacturing work to a second party. In the case of pharmaceutical products, the marketing authorisation holder must manufacture the pharmaceuticals by itself, import the pharmaceuticals from outside Japan or outsource the manufacturing work of the pharmaceutical products to another company pursuant to the contract manufacturing agreement.

The first party usually holds the patent covering the product to be manufactured, and licenses the patent to the extent necessary for the second party to manufacture the product. In the meantime, the second party may also hold a patent covering certain manufacturing methods/processes. Further, in the course of preparing to manufacture the product, a new invention might be created that could result in a new patent. How these patents are handled can vary greatly between different contract manufacturing agreements, but one typical way of handling them is that the pre-existing patents (owned by either the first or the second party) remain with the current owner but any newly created patents are assigned to the first party (the outsourcing party).

(d) *Product distribution agreements*

A product distribution agreement is an agreement under which a first party (often a manufacturer of products) appoints a second party as the distributor of the products in a certain area. The distributor purchases the products from the first party and sells them in the area, or acts as an intermediary in relation to the sale and purchase of the products between the seller (the first party) and the purchaser.

A product distribution agreement usually does not deal with patent ownership, since it is often taken for granted that patents covering the products (if any) would remain with the first party. However, under the doctrine of patent exhaustion, once the product is put on the Japanese market the first party may not enforce its patent rights against resale of the product. This applies to resale not just by the distributor, but also by any downstream wholesalers and retailers.

(e) *Co-promotion agreements*

A co-promotion agreement is an agreement under which two companies promote products sold by one of the companies under a single brand name. In the case of pharmaceutical products, a co-promotion agreement is often used to take advantage of the sales channels of one party where the other party does not have a strong sales channel in Japan.

A co-promotion agreement usually does not deal with patent ownership. However, as the products promoted under a co-promotion agreement are typically branded with a brand that is owned by one of the parties, co-promotion agreements may often include the grant of trademark licences.

(f) *Clinical trial or research agreements*

A clinical trial or research agreement is an agreement under which one party (usually a pharmaceutical company or a medical device manufacturer) requests another entity (usually a hospital) to conduct clinical trials or clinical research. Such an agreement often contains provisions regarding patent ownership and the licensing of patents arising from the clinical trial or research.

The general rule under the Patent Act is that the person who creates an invention owns the right to obtain a patent for that invention. Consequently, the right to obtain a patent is likely to belong to the hospital (whose employees actually conduct the research). However, a clinical trial or research agreement might vary this general rule.

Usually, each hospital has its own standard form of clinical trial or research agreement (which is often found on the hospital's website). However, how patents are treated varies greatly depending on which hospital's standard form is used. The standard forms of some hospitals provide that any patents arising from the clinical trial or research belong to the hospital, while the standard forms of other hospitals vest the patents in the requesting company. There are also standard forms that stipulate that the parties will separately discuss and determine the issue of patent ownership. The standard form issued by the Japanese Association of Pharmaceutical Medicine provides that patent ownership is determined in accordance with the rules of the relevant research institution, but also that the requesting company will have a royalty-free licence to use the invention.

3. Issues associated with patent transactions

3.1 Regulatory law and compliance

(a) *Extension of a patent term under the Patent Act*

The term of a patent in Japan is 20 years from the date of its application. However, as mentioned above in section 1.1(c), the term of a patent can be extended for a period not exceeding five years, to reflect the period of time during which the patentee was unable to use the patented invention because of the need to confirm the safety of the patented invention under applicable law (Article 67(2) of the Patent Act). As in many other jurisdictions, in Japan a new medicine must be examined and not be marketed until its safety is confirmed (under the Pharmaceutical Affairs Act – see section 1.1(c)). Thus, this provision on extending the patent term essentially applies to all inventions relating to medicines.

(b) *Re-examination period under the Pharmaceutical Affairs Act*

Other than by the Patent Act, a new medicine is in effect protected by the so-called 're-examination' system required under Article 14(4) of the Pharmaceutical Affairs Act. Under this system, for generally eight years from the time a new medicine is approved by the relevant authority, the originator (brand-name) pharmaceutical company must have its new medicine re-examined in order to re-confirm its effect, efficacy and safety based on data collected from medical institutions that have used that new medicine.

During this typical eight-year period from first regulatory approval until the end of re-examination, if another pharmaceutical company (a generic pharmaceutical company) wishes to apply for approval in respect of its own generic version of the original medicine – ie, for which the brand-name pharmaceutical has already obtained marketing approval and which is currently the subject of re-examination – the generic pharmaceutical company must prepare documents such as clinical test data in the same way as the originator pharmaceutical company has done. Of course, this requires a long time and an enormous expense. In other words, during the re-examination period, other generic pharmaceutical companies are effectively prevented from applying for approvals for their generic medicines and marketing them at low prices, and the originator (brand-name) pharmaceutical company has the exclusive right in respect of the branded medicine and its equivalents.

This protection under the re-examination system is independent of patent protection and is similar in effect to the regulatory data protection and marketing exclusivity systems found in the United States and in Europe (where the rationale is more to compensate originator companies for the time and expense it takes to develop a new drug, rather than focusing on assessing the safety of the new drug). What is important is that, even though the term of a patent in respect of a new branded medicine has expired, if the new medicine is still within the regulatory re-examination period, a generic pharmaceutical company is prevented from marketing generic versions of it.

3.2 Commonly disputed matters

Matters in patent transactions in Japan that are commonly negotiated, or that can become the subject of dispute, are probably not greatly different from those negotiated or disputed in most other countries. They include:

- the structure of the transaction (see section 2.2), including tax and transfer pricing considerations (see section 3.7);
- the ownership of patents and patent applications, including joint ownership issues (see section 2.2);
- the scope of relevant patents (see section 2.2);
- representations, warranties and indemnities (see sections 2.1 and 2.2);
- the risk of infringement (see section 2.2);
- the issue of inventor compensation (see sections 1.3 and 2.2); and
- in the case of licences: exclusivity, the definition of 'licensed patents' and 'licensed products', and royalties (see section 2.1(b)).

3.3 Competition rules

(a) Introduction

Article 21 of Japan's Anti-monopoly Act provides that "The provisions of this Act shall not apply to such acts recognisable as the exercise of rights under the Copyright Act, the Patent Act, the Utility Model Act, the Design Act, or the Trademark Act." Therefore, in principle, the Anti-monopoly Act is not applicable to acts that are considered to be an exercise of patent and other intellectual property rights and, as a result, such acts do not violate the Anti-monopoly Act.

However, the Anti-monopoly Act is applicable to any act that may appear to be an exercise of a relevant right but that is not "recognisable as the exercise of the rights". In other words, even if an act is an exercise of a relevant right, if such an act is found to deviate from, or run counter to, the intent and objectives of the Japanese intellectual property system, in the light of the intent and manner of the act and its degree of impact on competition the Anti-monopoly Act is applicable to that act.

When an owner of a technology grants other entrepreneurs a licence to use that technology subject to certain restrictions, such a licence is normally seen as an exercise of relevant intellectual property rights. However, there are some cases that might not be considered as "recognisable as the exercise of the rights" and thus potentially fall within the Anti-monopoly Act. Japan's Guidelines for the Use of Intellectual Property under the Anti-monopoly Act issued by the Fair Trade Commission (the Guidelines) indicate the following examples of such cases.

(b) Examples given in the guidelines

There are a number of examples quoted in the Guidelines that may not be considered as being "recognisable as the exercise of rights" and thereby likely to constitute unfair trade practice. Several of these are described further next, where relevant to the life sciences sector.

Restrictions on raw materials and components: A licensor may impose limits on licensees as to the quality or suppliers of raw materials, components and other items needed to supply the product using the licensed technology (including services and other technologies; collectively 'raw materials and components'). Such limits could be considered necessary to ensure the function and effect of the technology, to maintain safety and to prevent the disclosure of confidential information, and hence are recognised as reasonable to some extent.

However, because the supply of products that incorporate the licensed technology is deemed part of the licensees' business activities, restrictions on raw materials and components have the effect of constraining the means of competition used by licensees – or, in other words, the freedom to choose the quality of raw materials and components and the suppliers of them. The restrictions have an additional effect of depriving entrepreneurs who supply alternative raw materials and components of trading opportunities. Such restrictions therefore constitute an unfair trade practice if they exceed the extent necessary in the light of these considerations and may tend to impede fair competition.

Restrictions on sale and resale prices: If a licensor restricts licensees' sale or resale prices of products incorporating the licensed technology, such a restriction limits the most fundamental means of competition for licensees and for the distributors purchasing such products from them. This necessarily reduces competition and such a restriction is in principle recognised as constituting an unfair trade practice in Japan.

Restrictions on the manufacture and sale of competing products or on transactions with competitors: If a licensor restricts licensees in relation to the manufacture or sale of any product that competes with the licensor's products, or in relation to the acquisition of a licence for a competing technology from a competitor of the licensor, such conduct has the effect of impeding licensees from effectively using technology, and obstructing technology transactions, with the effect of depriving competitors of trading opportunities. Such a restriction therefore constitutes an unfair trade practice if it has the tendency to impede fair competition.

Nevertheless, it is thought that such restrictions are likely to be recognised as not tending to impede fair competition under certain conditions; likewise restrictions that remain effective for a short period after the termination of the licence agreement. The conditions are that:

- the restrictions are only to an extent necessary to maintain confidentiality;
- the licensed technology is concerned with know-how; and
- there exists no other means of preventing disclosure or unauthorised use of the technology.

Calculation of royalties using criteria unconnected to the use of technology: When a licensor imposes royalty payment obligations that are calculated using criteria unrelated to the use of the licensed technology (for example, by imposing an obligation to pay royalties according to the quantity of products manufactured or sold without the licensed technology), licensees may be hindered from using competing products or technology. This conduct therefore constitutes an unfair trade practice if it tends to impede fair competition.

In principle, however, such an approach will not constitute an unfair trade practice if the licensed technology is used in part of the manufacturing process, or is associated with any component, and it is a reasonable way to calculate royalties. For instance, calculating royalties using the manufactured or sales quantity, or the value of the final product using licensed technology or components, or the quantity of raw materials and components used, is recognised as reasonable for the convenience of computation.

Restrictions after the extinction of rights: When a licensor restricts a licensee from using a technology, or imposes an obligation to pay royalties even after the rights to the technology have become extinct, such a restriction generally hinders the free use of technology. The restriction will constitute an unfair trade practice if it tends to impede fair competition.

The Fair Trade Commission of Japan has rendered an advisory decision stating that where Company A granted know-how relating to an epoxide-based plasticising agent to Taiwanese Company B pursuant to an international agreement, prohibiting Company B's import of epoxide-based plasticising agent after expiry of the international agreement violated the Anti-monopoly Act.

Restrictions on research and development activities: Restrictions by the licensor relating to licensees' R&D activities, such as preventing licensees from independently or jointly with any third party conducting R&D activities concerning the licensed technology or any technology that competes with it, generally affects R&D competition and ultimately reduces future competition in the technology or product market. Such restrictions are recognised as having the tendency to impede fair competition and are in principle recognised as an unfair trade practice.

However, when the licensed technology is protected and controlled as know-how, restricting licensees from jointly performing research and development activities with any third party to the extent necessary to prevent disclosure of the know-how (or its use for unauthorised purposes) is generally not recognised as having the tendency to impede fair competition and does not constitute an unfair trade practice.

In the past, the Fair Trade Commission held that when a pharmaceutical manufacturer A grants Company B the right to distribute a patented product, prohibiting Company B from conducting research and development in respect of products that compete with the patented product during the term of agreement and five years after its termination had a tendency to violate the Anti-monopoly Act.

Obligations to assign or to grant exclusive licences in respect of improved technology: If a licensor obliges a licensee to hand over to the licensor, or any designated third party, the rights to improved technology that is developed by the licensee, or to grant the licensor an exclusive licence in respect of such improved technology, this conduct enhances the licensor's position in the technology or product market. It also discourages the licensee from working on research and development by obstructing it from using and exploiting the improved technology. Normally it is not thought that there is any justifiable reason for imposing such restrictions and, in principle, it constitutes an unfair trade practice to impose any such obligation.

An obligation that forces a licensee to co-own with the licensor the rights in respect of the improved technology they invent is a restriction of the licensee's freedom to use or dispose of the results of its own improvements or applied research. However, the degree to which the obligation discourages a licensee from undertaking R&D activities is less than the restrictions stated above (ie, the obligation to assign the rights to improvements to the licensor). Even so, it may constitute an unfair trade practice if it has the tendency to impede fair competition.

If a licensee's improved technology cannot be used without the licensed technology, an obligation imposed on the licensee to assign the rights in respect of the improved technology in exchange for fair consideration could be recognised as essential to promoting technology transactions. Moreover, such an arrangement is not recognised as being detrimental to a licensee's motivation to conduct research and development, and is generally found not to have a tendency to impede fair competition.

Obligations to grant non-exclusive licences in respect of improved technology: When a licensor obliges a licensee to grant it a non-exclusive licence in respect of improved technology developed by the licensee, such an obligation may not constitute an unfair trade practice in principle as long as the licensee may still freely use its own improved technology. The obligation has little impact on a licensee's business activities and is not recognised as being likely to discourage a licensee from undertaking research and development.

However, if this obligation accompanies a restriction on which parties can be licensed to use the improved technology (for instance, by imposing an obligation to not grant a licence to any competitor of the licensor or to any other licensee), such an obligation may reduce a licensee's motivation to undertake research and development and possibly enhances the position enjoyed by the licensor in the technology or product market. Such an obligation therefore constitutes an unfair trade practice if it has the tendency to impede fair competition.

Obligations to report obtained knowledge and experience: Obliging a licensee to notify the licensor of the knowledge or experience it obtains in the process of using the licensed technology will enhance the licensor's incentive to offer further licences and will not reduce the licensee's motivation to undertake research and development. Such an obligation does not therefore constitute an unfair trade practice in principle.

There has been some discussion that the Anti-monopoly Act may apply to acts that inhibit upstream technology that has low substitutability, such as research tool technology. However, there are no actual cases in which the Anti-monopoly Act has been held to apply.

3.4 Securitisation of patent rights

(a) *Legal framework: a pledge and a* jouto-tampo

With recent innovations of technology by many venture companies, the importance of taking security over patent rights has been increasing. In particular, many venture companies do not have sufficient assets other than their intellectual property rights, and need to grant security in order to obtain finance for their activities.

The Patent Act provides that an instrument of a pledge may be used to obtain a security interest in a patent right, an exclusive licence and a non-exclusive licence (Article 95). In the case of a pledge over a patent right or an exclusive licence, the pledge will not become effective unless it is registered at the JPO (Article 98(1)(iii)). In contrast, there is no need to register a pledge over a non-exclusive licence right.

When exercising a pledge, a pledgee (eg, a bank) must first file a petition at a district court to obtain an attachment order. The pledge can be exercised not only against the pledged patent right itself, but also against any consideration that arises from the patent right, such as the licence fees arising from a licence of a patent that is subject to the pledge.

As another instrument for obtaining a security interest in a patent right, a *jouto-tampo* (security by way of assignment) is also available where the patent right of a patentee (not necessarily a debtor) is assigned to a creditor subject to an assignment back to the original patentee if the debtor pays all of its debts to the creditor. If the debts are not paid in full, the creditor will have the right to hold its rights in respect of the patent. A *jouto-tampo* is thus a kind of patent assignment; but it is of no effect without being registered at the JPO (Article 98(1)(i)). In practice, a *jouto-tampo* is more often used than a pledge because the fees for creating a pledge are high.

(b) *Practice on having a security interest in patent rights in Japan*

Unlike other property that is tangible, such as real estate, it is not easy to assess the value of intellectual property rights, including patent rights. Therefore, commercial banks in Japan do not commonly provide finance in return for security interests over patent rights.

Nonetheless, the Development Bank of Japan, a government-controlled bank, has a good track record of providing finance in return for taking security over intellectual property rights: JPY21 billion (approximately USD200 million) was financed in 310 cases in total from 1995 until March 2008.

3.5 The effect of insolvency

The effect of insolvency is important, especially in the case of the insolvency of a licensor of a patent licence agreement.

Under Japan's Bankruptcy Act a trustee in bankruptcy who has the authority to control and dispose of the remaining assets of a bankrupt entity can terminate a bilateral agreement between the bankrupt entity and its counterparty, including a licence agreement (Article 53 of the Bankruptcy Act). However, if the counterparty has registered its right, or meets any other requirement for duly asserting such right against any third party based on the licence agreement, the trustee in bankruptcy cannot terminate the bilateral agreement (Article 56 of the Bankruptcy Act). This also applies to other kinds of insolvency, namely civil rehabilitations and corporate reorganisations under the Japanese Civil Rehabilitation Act and the Corporate Reorganisation Act, respectively.

Before the revision of the Patent Act in 2011 (effective from April 2012), a licensee in a non-exclusive patent licence agreement could not assert its right to use the patented invention against third parties if the non-exclusive licence was not registered at the JPO. In practice, there are many licence agreements that are not registered. The two main reasons for this are that parties to licence agreements do not wish to have information relating to the licence agreements (eg, details of the licensed patents and the names of the parties) disclosed to the public by registering them; moreover, that registration is costly, especially in case of licences of hundreds or thousands of patents. However, this has meant that if a licensor were to become insolvent, a licensee of a non-exclusive licence that had not been registered at the JPO would face a real risk that the licence agreement could be terminated by the licensor (or its trustee in insolvency).

The amended Patent Act eliminated this disadvantage for licensees: now, a licensee of a non-exclusive patent licence agreement can assert its right to use the patented invention against third parties without registration (Article 99 of the Patent Act). Consequently, the previous risk of the licence terminating on insolvency is no longer an issue.

The foregoing also applies to a non-exclusive licence of an invention in respect of which a patent has not yet been granted (Article 34-5 of the Patent Act).

It should be noted that, as mentioned in section 2.1(b) above, the Japanese exclusive licence still needs to be registered for a licensee to assert its right against third parties (Article 98(1)(i) of the Patent Act) and to avoid being terminated by a licensor or its trustee in insolvency when the licensor becomes insolvent.

3.6 Taxation

As a patent right is one kind of intangible asset, taxes are imposed on transactions relating to patents. For example, in Japanese domestic transactions, consumption tax is imposed on the consideration for a patent transaction such as a patent assignment or licence. In addition, a person who prepares a written agreement of assignment of a patent right in Japan needs to pay documentary stamp tax, the amount of which is based broadly on the amount described in the agreement. In the case of a patent licence agreement, documentary stamp tax is not imposed. In contrast, income tax and normal tax conventions are applicable as for cross-border transactions.

In terms of cross-border licence agreements, which are increasingly common, the withholding tax rate imposed on the consideration is 20%; however, this rate is reduced to 10% or 15% depending on the tax conventions applicable between Japan and the relevant counterpart country. Furthermore, a taxpayer is exempt from withholding tax under certain tax conventions such as that between Japan and the United States, between Japan and the United Kingdom, and between Japan and France. In order to benefit from reduced withholding tax rates, an application regarding the appropriate tax convention must be submitted to the relevant tax authority.

Transfer price taxation is important in cross-border patent transactions. If a company located in Japan concludes a deal with a company outside Japan at a price (the so-called 'transfer price') different from the 'normal' (ie, arm's length) price applied to the same type of transaction, the company in Japan effectively transfers the profits that the company should have gained to the other company outside Japan. Then, the company in Japan can reduce its corporation tax payable in Japan. This kind of transaction can be seen especially between affiliates within group companies, which transfer profits to other countries that impose a lower corporation tax. For example, when manufacturing products outside Japan, a company in Japan may grant a patent licence to its subsidiary outside Japan with a royalty rate that is lower than the normal rate in Japan; this would thereby reduce the corporation tax payable by the company in Japan.

In order to prevent such a loophole, there is transfer price taxation. In summary, by virtue of transfer price taxation, a cross-border transaction will be deemed to be made at a normal (arm's length) price. With the expansion and globalisation of companies, transfer price taxation issues are expected to arise more often than has occurred so far, including in cases of patent transactions. However, as there is no clear or suitable valuation method for patent rights, and patent transactions differ greatly from each other in comparison with other kinds of transactions, it is difficult to determine what price should be the normal (arm's length) price to be used for a relevant patent transaction.

3.7 Hot topic

An originator (brand-name) pharmaceutical company that develops a new patented medicine will usually lose its exclusive right in respect of it after the expiration of the patent term. At that stage, generic pharmaceutical companies will typically start to sell generic versions of the patented branded medicine. In order to prevent sales from plummeting on the expiry of the patent, originator (brand-name) pharmaceutical companies are increasingly granting licences to other pharmaceutical companies to manufacture and sell medicines equivalent to the patented medicine before the expiry of the patent. These medicines are so-called 'authorised generics'.

The originator (brand-name) pharmaceutical company can benefit from licence fees from the licensed pharmaceutical company selling authorised generics. The licensed pharmaceutical company also benefits as it can market generic medicines earlier than any other generic pharmaceutical companies (namely, before the expiry of the patent term). Additionally, authorised generics are usually the same as the brand medicine, not only in terms of their active ingredients but also in their inactive ingredients, such as additives. Thus, authorised generics may appear more credible and perhaps reliable from the perspective of doctors and consumers compared with other generic medicines (which are required to contain only the identical active ingredient as the branded product).

Japan's first authorised generic medicine was launched in June 2013, following an agreement between Sanofi and Nichi-iko Pharmaceutical Co Ltd. According to press releases, Nichi-iko has been licensed by Sanofi to work relevant patents and is permitted to sell exclusively in Japan an authorised generic version of Sanofi's Allegra (fexofenadine hydrochloride tablets) under the brand name Sanik.

We would like to express our thanks for the cooperation of our colleagues, Mr Yusuke Inui, Ms Kiyoko Nakaoka, Mr Masahito Imai, Mr Hirotaka Noguchi and Ms Iku Kumagai, in the production of this chapter.

Netherlands

Carreen Shannon
Deterink NV

1. Patents, SPCs and know-how: foundations for transactions

1.1 Patents and supplementary protection certificates

A patent provides the patent owner with the right to exclude others from using the claimed invention (a novel and inventive technical solution to a problem) for commercial purposes. It should be noted that a patent does not grant the owner the right to use the claimed invention. Earlier patents may after all limit the application of the invention that is claimed in a later patent. Pursuant to Article 53 of the Dutch Patent Act, and subject to certain exceptions, a patent relating to a product more specifically confers on its owner the exclusive right to make, use, put on the market or resell, hire out or deliver the patented product, or otherwise deal in it in or for his business, or to offer, import or stock it for any of those purposes. In the case of a process patent, the patent owner has the exclusive right to use, put on the market, or resell, hire out or deliver the product obtained directly as a result of the use of the patented process, or otherwise deal in it in or for his business, or to offer, import or stock it for any of those purposes.

The geographical scope of a patent is limited, in that its owner is only granted exclusive rights for those countries or regions for which the patent is granted and in force. Dutch patents can therefore only be invoked against infringing activities taking place in the Netherlands.

A Dutch patent and the Dutch part of a European patent provide protection for a maximum period of 20 years, provided that an annual maintenance fee is paid to the Netherlands Patent Office. If those fees are not paid, the patent expires as of right. In the case of medicinal and crop protection products, the maximum period of protection may, however, be extended by a maximum of five years by means of a supplementary protection certificate (SPC). Payment of an annual fee is also required to maintain an SPC. The duration of a supplementary protection certificate may in turn be extended with a fixed six months period in the case of orphan medicinal products.

The first step in obtaining a Dutch patent is the filing of a patent application with the Netherlands Patent Office. Such an application must be followed by a request for a novelty search no later than 13 months after filing the application. After obtaining the search results, including a written opinion as to the novelty, the inventive step and industrial applicability of the claimed invention, the applicant may amend the patent application. Such amendment is, however, not necessary for the grant of a Dutch patent, even if the search results (clearly) show that the patented invention is not new and/or inventive. A Dutch patent is granted automatically without any examination of (eg) novelty or inventive step. From that perspective, the validity of a Dutch patent is rather uncertain and should always be analysed before, for example, entering into a licence or purchase agreement in respect of a Dutch patent.

In the Dutch Patent Act a distinction is made between direct and indirect (or contributory) infringement of a patent. Anyone who, without the patentee's consent, performs the acts listed in Article 53 of the Dutch Patent Act (as given above in the first paragraph of this section) is a direct infringer. A patent is indirectly infringed if someone, without the patentee's consent, supplies or offers to supply to non-authorised persons in the Netherlands essential means for putting the invention into effect in the Netherlands, knowing that, or it being obvious that, those means are suitable and intended to put the invention into effect.[1]

Pursuant to Article 64 of the Patent Act, the ownership of a patent, a patent application and the right to obtain a patent can be assigned (ie, sold) to a third party or otherwise transferred to a third party (eg, resulting from a (de)merger or inheritance). The proprietor of a patent or patent application can also grant one or more licences in respect of its patent (application).[2] Patents and patent applications can furthermore be the subject of security rights, more specifically a pledge.[3]

1.2 Who owns the patent?

Primarily it is the inventor who is entitled to apply for a patent and who can claim ownership of the patent (application). However, Article 12 of the Patent Act provides for the following exceptions to that principal rule:

- If the invention is made by an employee and the nature of his employment is to make inventions of the same kind as the invention at issue, the employer is entitled to apply for a patent and claim ownership of the patent (application).[4] The employer has the responsibility to prove that those requirements are met.
- If the invention is made by a trainee, the person for whom the trainee performs the services is entitled to apply for a patent and claim ownership of the patent (application), unless the invention is not connected to the subject of the trainee's services (which is for the trainee to prove).[5]
- If the invention is made by a person carrying out research in the service of a university, college or research establishment, the university, college or research establishment is entitled to the patent.[6]

The above-mentioned statutory exceptions may, however, be departed from by

1 Article 73(1) of the Patent Act.
2 Article 56 of the Patent Act. Compulsory licences are also possible under certain circumstances.
3 Article 67 of the Patent Act.
4 Article 12(1) of the Patent Act.
5 Article 12(2) of the Patent Act.
6 Article 12(3) of the Patent Act.

written agreement.[7] For example, in employment agreements a waiver clause is often included, stipulating that each invention that is made by an employee passes to the employer.

Finally, it should be noted that freelance consultants and contractors do not fall within any of the above exceptions, meaning that, if not agreed upon otherwise in writing, they will – in accordance with the primary rule – be entitled to patent protection for any inventions they make, even if such inventions are made in the course of their consulting or contractor services for a client.

Where an invention has been made by two or more persons working together by agreement, they shall have a joint right to apply for a patent and have joint ownership of the patent (application).[8] The mutual relationship of such joint owners is primarily governed by the terms of the agreement between them.[9] Where there is no agreement or where the agreement does not provide otherwise, each of the joint owners shall have the right to use and apply the invention (ie, perform the acts listed in Article 53 of the Patent Act) and to act against third-party infringers; but a licence may only be granted with the common consent of the joint owners.[10] Unless an agreement between them provides otherwise, each joint owner can assign its share in the joint ownership to a third party. Assignment of the patent (application) in its entirety requires the consent of all joint owners. Furthermore, the joint owners are severally liable for payment of the annual maintenance fees.[11]

Also relevant in respect of the ownership of a patent (application) is that Article 8 of the Patent Act stipulates as a principal rule that the applicant is deemed to be the person entitled to the patent. An exception to this rule is provided for in Article 11 of the Patent Act, pursuant to which the applicant is not entitled to a patent in so far as the contents of his application have been obtained unlawfully from subject matter made or applied by another person or from descriptions, drawings or models belonging to another person without that person's consent. The latter person remains entitled to file a patent application himself to the extent that the subject matter thus unlawfully obtained is patentable (the application filed by the unlawful applicant not being taken into consideration as prior art) or he may claim (partial) entitlement to the patent (application) that was unlawfully obtained.

1.3 Inventor compensation

According to Article 12(4) of the Patent Act, the employee-inventor who cannot be deemed to have been compensated – by the salary or any extra remuneration he receives – for not having been granted a patent pursuant to Article 12(1) of the Patent Act (ie, his employer is granted a patent on the basis of that statutory provision) is entitled to an equitable remuneration from his employer "related to the pecuniary importance of the invention and the circumstances under which it was made". The same goes for the trainee-inventor and research-inventor who are not granted a

7 Article 12(5) of the Patent Act.
8 Article 13 of the Patent Act.
9 Article 66(1) of the Patent Act.
10 Article 66(2) of the Patent Act.
11 Article 66(3) of the Patent Act.

patent pursuant to Articles 12(2) and 12(3) of the Patent Act. The inventor's right to claim such remuneration lapses three years after the date of the grant of the patent. The aforementioned rules constitute mandatory law, and thus cannot be contracted out of.

As the right to compensation is rarely litigated in the Netherlands, there is only limited guidance in case law on how the principle of compensation is to be applied. According to case law of the Dutch Supreme Court,[12] as a general principle – particularly in respect of large, sophisticated companies – the salary agreed in the inventor's employment contract is deemed to include compensation for patents not granted to the inventor. This is particularly the case if the inventor was specifically hired to carry out research and development.

Generally, the Dutch courts consider the following factors when determining whether to award additional compensation and the amount thereof:

- the employee's position and function within the employer's organisation;
- the employee's remuneration and other terms of employment;
- the nature and the financial significance of the invention;
- the extent of the employee's contribution to the invention;
- the role played by other employees in making the invention;
- the extent to which the employer has provided the necessary facilities and research opportunities; and
- the extent to which the employer has assisted with the exploitation of the invention.

The lack of a specific salary component representing compensation for the employer's ownership of the patent is neither necessary nor sufficient for the additional compensation to be awarded. Furthermore, the Supreme Court[13] has emphasised that in determining the amount of compensation, the relationship between the employee and employer should not be equated with a relationship between a licensor and a licensee.

In summary, in current Dutch legal practice any requests for compensation are granted only in very exceptional circumstances and, if granted, the awarded compensation is likely to be relatively low.

1.4 Know-how and trade secrets

Not only patents but also know-how (also called confidential information or trade secrets) relate to (technical) knowledge upon which a proprietor can base certain technical achievements and competitors cannot. Both know-how and patents can be the subject of a licence agreement. An important difference is that know-how is only valuable as long as it remains secret, whereas the patent system requires publication and thus disclosure of the patented knowledge. Also, from a legal point of view, know-how and patents are totally different. Dutch law does not provide for exclusive

---

12 Supreme Court, May 27 1994, *Hupkens v Van Ginneken*; and Supreme Court, March 1 2002, *TNO v Ter Meulen*.

13 Supreme Court, May 27 1994, *Hupkens v Van Ginneken*.

and absolute rights of know-how protection as opposed to patents and other intellectual property rights. Know-how mainly provides its proprietor with a 'head start', due to the fact that, and for as long as, competitors do not possess that knowledge. Furthermore, the duration of protection of know-how is not limited to a specific period set by law or to a specific territory, as opposed to the exclusive rights of a patentee.

In general, protection of know-how under Dutch civil law is mainly awarded under:

- the general principles of unfair competition derived from the general tort clause embodied in Article 162 of Book 6 of the Civil Code; and
- contract law, by entering into non-disclosure agreements and other contractual obligations of confidentiality.

Under Dutch unfair competition law, benefiting from another's know-how is generally allowed. The simple use of know-how by third persons does not necessarily fall under the law of tort, but is dependent on the way in which the secret information was obtained. A person who uses improper methods in discovering another's know-how – eg, by means of industrial espionage, theft or the bribery of employees – may be held liable. The use of information disclosed by employees in violation of their obligation of secrecy, as well as the abuse of confidential information acquired during a pre-contractual stage in negotiations, may also be held unlawful. The aggrieved party may recover damages for loss suffered as a result of the unlawful acquisition or use of its know-how. The aggrieved party may furthermore obtain a (temporary) injunction to prohibit the disclosure and/or the (continued) use of unauthorised obtained information.

In contractual relationships the content of the (non-disclosure) agreement primarily determines the scope of the confidentiality obligations and the sanctions that apply when such obligations are violated. For example, the agreement may provide for contractual penalties if the secrecy obligation is violated. If no contractual sanctions are included in the agreement, the aggrieved party may – based on Article 6:74 of the Civil Code – be able to claim damages which it has suffered as a result of the breach of contract. The aggrieved party may furthermore (partly) terminate the agreement or demand specific performance of the secrecy obligation included in the agreement.

Although Dutch civil law does not encompass statutory provisions specifically relating to the protection of know-how (other than a specific provision relating to employees – discussed under section 1.5 below), this situation is likely to change once the proposed Trade Secrets Directive is adopted and implemented by the Dutch legislature. The Dutch Penal Code does, however, already include several specific know-how-related provisions:

- Pursuant to Article 272 of the Dutch Penal Code, the deliberate betrayal of a trade secret which one knows or reasonably should have known had to be kept secret because of an employee's (former) office or profession or any statutory provision, is a criminal offence, sanctioned by imprisonment or a criminal fine.

- Article 273(1) of the Dutch Penal Code makes it a criminal offence for a person to deliberately disclose information, in respect of which he has sworn secrecy, relating to an organisation in which he is or was employed.

Furthermore, a person who incites an employee of a competitor to disclose a trade secret of that competitor is also guilty of a criminal offence pursuant to Article 47 of the Dutch Penal Code.

1.5 Employees, consultants and protecting know-how

There are generally three ways of acting against an employee or consultant who discloses the know-how of his employer: under the law of tort, under contract law or under criminal law. As regards employees, employment law can also form a basis for acting against disclosure of the know-how of employers.

In general, under Dutch law the disclosure of know-how by an employee or consultant (whether during or after employment) is considered an act of wrongdoing towards the employer. Besides such an unlawful act giving rise to a claim for damages under the law of tort, it is also considered to form a valid ground for immediate termination of employment or consultancy agreement (as appropriate). As regards employees, Article 7:678(2)(i) of the Civil Code even explicitly stipulates that disclosure of information which the employee is supposed to keep secret is considered a valid ground for dismissal.

In most employment and consultancy agreements, a clause will be found stipulating that the employee/consultant is supposed to keep secret any and all company confidential information (know-how), both during his employment and after termination of his employment. The employee/consultancy agreement will often also include an obligation to return to the employer all (digital) documents containing the know-how of the employer. If such contractual obligations are indeed entered into, the disclosure of the employer's know-how by the employee will amount to a breach of contract. The agreement may itself provide for certain sanctions if the confidentiality obligations are breached – for example, contractual penalties and/or immediate termination of the employment agreement. If no contractual sanctions are included in the agreement, the employer may (based on Article 6:74 of the Civil Code) be able to claim for the losses it has suffered as a result of the breach of contract.

Article 273 of the Dutch Penal Code provides that where secrecy of information has been imposed upon an employee or consultant, the disclosure of such secret information during or even after termination of employment is considered a criminal act. Such criminal offence is sanctioned by imprisonment or a criminal fine.

2. Context and features of patent transactions

2.1 The principal dealings in a patent

(a) Patent assignments

A patentee will generally decide to assign his patent or patent application if he is no longer interested in exploiting the patented invention himself or by means of licences, or if he is no longer able or willing to pay the costs of prosecuting a patent application and maintaining a patent. Instead of simply dropping the patent (application) by not complying with prosecution or maintenance formalities and thereby saving on future costs, the patentee could in fact even gain financially if he assigns the patent (application) to an interested party who is willing to pay for the assignment.

For the assignment of a patent (application), a written document (whether in the form of a notarial deed or a private agreement) signed by both parties is required. That document must explicitly state that the assignor assigns the patent (application) to the assignee and that the assignee accepts such assignment.[14]

The patents (applications) which are transferred should be specifically identified in the deed of assignment. Generally, the deed of assignment explicitly provides that all rights and claims relating to the assigned patent (application) are transferred to the assignee, including the right to sue for and to recover and retain damages and profits for past infringements of the patent by a third party.

Furthermore, the deed of assignment should include a provision on whether the assignment takes place for free or against payment (consideration is not required under Dutch law for a contract to be enforceable), in the latter case specifying the amount to be paid and the payment conditions. In the case of intercompany assignments, an arm's length transfer price must be agreed upon for tax reasons, unless the assignee and assignor form a fiscal unity, in which case the assignment can take place without any compensation.

Deeds of assignment typically include warranties providing that the assignor has full ownership of the patent (application) and is entitled to assign the patent (application), and that the assigned patent (application) is not the subject of a right of pledge or seizure. The assignee may also request the assignor to warrant that no licences have been granted in respect of the assigned patent (application), or if a licence has been granted, to warrant that the licensee cannot or does not object to the assignment of the patent (application). In some cases the assignor may even be willing to warrant that the use of the patented invention will not amount to an infringement of intellectual property rights of any third party. Any of these warranties may be combined with an indemnification and hold-harmless of the assignee in the event of third-party claims relating to the subject matter of such warranties.

For the assignment to have effect with regard to third parties (but not for the assignment to be valid and binding between the assignor and the assignee), the assignment should be registered in the Patent Register controlled by the Netherlands Patent Office.[15] For the purpose of registering the assignment, an additional summary version of the deed of assignment is generally prepared and signed by the parties. As

14 Article 65(1) of the Patent Act.

15 Article 65(3) of the Patent Act. In the event that a certain third party is otherwise notified of the patent assignment, it is argued that such a party cannot claim that the assignment does not have effect towards him if the assignment is not registered in the Patent Register. Nevertheless, registration in the Patent Register is advisable to avoid any discussion on this point.

the registered document will be made available by the Patent Office to the public upon request, registration of the entire deed of assignment, including possibly confidential information (eg, the transfer price), is not advisable.

(b) Patent licences

Article 65 of the Patent Act relates to licences granted in respect of a granted patent. Although the Patent Act does not mention the possibility of patent application licences, it is generally accepted in Dutch legal practice that applications can also be licensed.

Generally, a patentee will decide[16] to grant one or more patent licences simply because of the financial gain resulting from such licences. The patentee may not be able or willing to manufacture, market or apply the patented invention in certain territories or for certain applications for which a patent is granted, but a licensee may be willing to pay good money to be able to do so.

A patent licence is created by agreement, whether in writing or orally, or by accepted testamentary disposition.[17] Clearly, a written licence agreement is preferable from an evidentiary point of view, but a written document is not required for a licence to be valid.

Assuming that the licence is granted by a written agreement, the agreement should first of all identify the patents (applications) which are licensed. Secondly, the agreement should specify the scope of the licence rights: is the licensee entitled to manufacture, sell and/or apply etc and are such acts limited to a specific product or process? If the licence agreement does not include any explicit limitations in that respect, the licensee will be deemed to be entitled to perform all acts referred to in Article 53 of the Patent Act – ie, all acts prohibited to any person other than the patentee. Similarly, the licence agreement should specify whether there are any limitations as to geographical territory, market sector or technological field of use for which the licence is granted.

The licence agreement should also stipulate whether an exclusive or a non-exclusive licence is granted, and in the event that an exclusive licence is granted, whether or not the patentee himself is still allowed to apply the patented invention. The agreement should further provide whether the licensee is allowed to grant sub-licences to affiliated companies and/or other third parties, and if so whether prior (written) consent by the licensor is required.

Often, the licence agreement also includes provisions relating to any improvements made to the licensed product or process and, more specifically, stipulating which of the parties is the owner of those improvements and entitled to file a patent application for such improvements. Generally, this will depend on which of the parties made the improvements. Mostly, the party who is appointed as the (IP rights) owner of improvements in turn agrees to grant the other party a non-exclusive royalty-free licence to apply such improvements.

---

16 Under certain circumstances a patentee may be forced by a court or the Netherlands Government to grant a licence to a third party. However, such compulsory licences fall outside the scope of this chapter.

17 Article 56(2) of the Patent Act.

The licence agreement should also provide which party is in charge of or should take the lead in prosecuting, maintaining, defending and enforcing the licensed patent (application) and which party will bear the costs relating to each of those actions. Related thereto is a right of first refusal for the licensee in the event that the licensor no longer wishes to maintain or wishes to sell one or more of the licensed patents (applications).

Generally, the term of the licence agreement will be specified; if the term is not specified, the licence is deemed to have been granted for the entire duration of the licensed patent.[18] The grounds for termination of the licence agreement will generally not be different from the termination grounds of other types of agreement (eg, termination due to the other party breaching its contractual obligations or being declared bankrupt or insolvent).

Licence agreements typically include a warranty providing that the licensor is entitled to license the patent (application) to the licensee, and that the licensed patent (application) is not the subject of a right of pledge or seizure. A licensee will generally not be willing to warrant that the manufacturing, use or sale of the licensed product does not constitute an infringement of any third-party intellectual property rights.

The licence agreement should stipulate whether the licensee is entitled to assign his licence rights to a third party (without the consent of the licensor) and whether the licensor is entitled to assign the licensed patent (application) to a third party (without the consent of the licensee).

Another important provision in licence agreements relates to the remuneration that is to be paid for the licence. Remuneration can be agreed upon in various forms (or a combination thereof), the following forms being more or less common:

- *A lump sum or upfront payment:* a specific sum to be paid once or periodically.
- *Milestone payments:* a specific sum to be paid upon reaching certain milestones. In the life sciences practice, such milestone payments are very common. Milestones may, for example, consist of the realisation of a proof of concept or prototype, the entering into or completion of various phases of clinical trials, the successful completion of a funding round, obtaining EMEA/FDA approval, or the launch of the product.
- *Royalties:* payment of a specified percentage of the (net) sales or profit realised by the licensee in respect of the licensed product. In the event that the licensee is entitled to grant sub-licences, the licensor will commonly also receive a percentage of the royalties and/or other remuneration which the licensee receives from its sub-licensees. A royalties clause should always be combined with a clause pursuant to which the licensor will be periodically provided with a royalty report (including sales/profits realised) and will be entitled to (have an accountant) verify the royalty reports by means of an inspection of the licensee's administration.
- *Shares:* the licensee obtains a specified percentage of the shares of the licensee's company, upfront and/or upon reaching certain milestones, the total percentage of shares of the licensor mostly being capped.

18 Article 56(2) of the Patent Act.

- *Reimbursement of past patent costs:* especially in the case of an exclusive licence, the parties may agree that the licensee will reimburse the licensor for all patent prosecution and maintenance costs borne by the licensor before the licence agreement was entered into.

In the event that parties agree upon milestone payments and/or royalty payments, it would be beneficial for the licensor to include an 'anti-shelving' clause in the licence agreement, dealing with the possibility that the licensee may not put sufficient effort into, or may simply not be able to realise, the agreed-upon milestones or to realise sufficiently high sales figures. Such an anti-shelving clause could take the form of a periodically guaranteed minimum amount of royalties, or a ground for termination of the licence agreement by the licensor in the event that certain milestones are not met by a certain date or if specified minimum sales figures are not realised.

For the licence to have effect in relation to third parties (but not for the licence to be valid and binding between the licensor and licensee), the licence should be registered in the Patent Register of the Netherlands Patent Office.[19] For the purpose of registering the licence, an additional summary version of the licence agreement is generally prepared and signed by the parties. As the registered document will be made available by the Patent Office to the public upon request, registration of the entire licence agreement, including possibly confidential information (eg, the transfer price), is not advisable.

2.2 Patent rights in corporate transactions

(a) Mergers and acquisitions

In the case of a share sale-and-purchase transaction, only the shares of the company at issue are sold and assigned to the purchaser. With the transfer of the shares and thus control over the company, the new shareholder will in principle also gain control over the assets – including patent rights – of the company, without the seller and purchaser having to enter into any specific transfer agreements in respect of such assets. This is different with a business sale-and-purchase transaction, in which case specifically appointed assets – eg, patent rights – of a company are assigned to the purchaser and in respect of which the seller and purchaser must enter into one or more deeds of assignment (see section 2.1(a) above).

Regardless of whether the deal is structured as an asset transaction, a stock transaction or a merger, such transactions will generally be preceded by a due diligence exercise. By means of the documents disclosed in the (virtual) data room and any public registers (such as various national patent registers), the party conducting the due diligence will, in respect of the patent rights that will be transferred as part of the deal, want to know and verify:

- which patents (applications) are part of the deal and what their status is (eg,

19 Article 56(2) of the Patent Act. In the event that a certain third party is otherwise notified of the patent licence, it is argued that such party cannot claim that the licence does not have effect towards him if the licence is not registered in the Patent Register. Nevertheless, registration in the Patent Register is advisable to avoid any discussion on this point.

(co-)ownership, patent application or granted patent, application still pending, patent still in force, remaining patent duration, validity objections of patent office);
- which licence agreements have been granted in respect of the patents (applications) that are part of the deal and the terms and conditions of such licence agreements;
- whether the patents (applications) that are part of the deal are the subject of a right of pledge, seizure or any other demand by third parties;
- whether employees have any claim for ownership or compensation in respect of the patents (applications) that are part of the deal; and
- whether any proceedings are pending or whether there is a threat of proceedings relating to the validity or infringement of the patents (applications) that are part of the deal.

The warranties in the share or asset purchase agreement will generally include warranties relating to:
- the full and unencumbered ownership of the company's patent rights;
- the company having those third-party patent licences that are necessary for continuing the business
- the sale of the share or business having no effect on the continuance of those third-party patent licences;
- no consultants, contractors or employees of the company having any right in respect of the company's patent rights; and
- the non-existence of (pending) litigation relating to the company's patent rights.

(b) Financing

The main Dutch public exchange is Euronext Amsterdam, which forms part of the pan-European exchange of NYSE Euronext. Apart from Euronext, there are some other foreign exchanges (eg, Nymex and Comex) that have permission to organise regulated markets in the Netherlands. Another well-known market is Alternext, a multilateral trading facility for small and mid-sized companies, which is also run by NSYE Euronext. The text below focuses on companies whose shares are traded on Euronext Amsterdam.

A company that wishes to float its shares (or other securities) on a public exchange located in the Netherlands will, leaving aside more complex matters such as dual listings, passporting etc, need to draw up a prospectus and include in that prospectus the information prescribed by the Prospectus Regulation (EC Regulation 809/2004), have its prospectus approved by the Dutch Financial Markets Supervision Authority (AFM in Dutch) and make available the prospectus to the public, in order to be allowed to float its shares.

Apart from the fact that the (value of) the patents will be reflected in the financial statements which are generally included in the prospectus by reference, complying with the Prospectus Regulation implies that, when patents or licences are material to an issuer's business or profitability, a summary regarding the extent to which the issuer is dependent on patents, licences or new manufacturing processes is to be included in the paragraph providing the business overview.

Furthermore, in a separate paragraph regarding research and development, patents and licences, the company will have to describe the issuer's research and development (R&D) policies for each financial year for the period covered by the historical financial information taken as a basis for the prospectus, as well as the amount spent on issuer-sponsored R&D activities.

If any material patent- or licence-related litigation is pending or threatening, this should be disclosed in the financial paragraph, and more specifically in the subparagraph regarding pending and threatening legal and arbitration proceedings.

Important licence agreements may also need to be disclosed in the paragraph on material contracts. These contracts may encompass conflicts of interest (eg, directors of the issuing company being also inventors, (indirectly) holding patents or receiving licence fees). Such conflicts will also have to be disclosed in the prospectus.

Finally, in a case of dependability on patents or certain licences, this dependability or the change of termination of licences or opposition to patents or discussions on licences may be qualified as risks that are to be disclosed in the paragraph regarding risk factors.

Accordingly, a legal due diligence exercise or audit prior to a listing will concentrate on such issues as:

- (co-)ownership of the patents;
- whether the patents are already granted or still in the application phase;
- remaining terms for the patents;
- validity objections or pending oppositions against patents;
- pending or threatened litigation; and
- other risks related to the validity or infringement of patents, any right of pledge or seizure in respect of the patents.

Where licences are concerned, issues such as term, possible termination, litigation and other risk factors are to be considered.

A prospectus is valid for 12 months after the AFM has approved it, provided it is supplemented with relevant information (also to be approved by the AFM and presented to the public) when a significant new factor, a material mistake or an inaccuracy relating to the information included in the prospectus arises that is likely to affect the assessment of the securities or that is noted between the time when the prospectus is approved and the final closing of the offer to the public or, as the case may be, the time when trading on a regulated market begins.

Previously, there was an ongoing requirement for issuers to publish annually a document which, briefly summarised, had to refer to documents and information that the issuers had published anywhere in the world on the basis of security laws. This provision was regarded as having become obsolete due to the incorporation of publication obligations under the EU Transparency Directive and has been revoked.

(c) Joint ventures

The goal of the types of joint venture that are to be considered in this chapter is to further develop and/or commercialise intellectual assets belonging to two or more independent organisations, through a separately incorporated legal entity or in a project collaboration. More specifically, the parties to the joint venture share risks and contribute with their intellectual property towards R&D, production and commercialisation.

Because intellectual property plays a central role in conducting such joint venture, it is of great importance that the joint venture agreement clearly defines:

- which party will initially contribute existing (so-called 'background') IP, which party is the owner of such background IP and the terms and conditions of the contribution of such background IP into the joint venture;
- which party is the owner of IP created during the joint venture, but not for the purposes of the joint venture (so-called 'sideground' IP);
- which party or parties will be the owner of the new IP created by the joint venture (so-called 'foreground' IP), which party or parties will have the right to file patent applications in respect of such foreground IP and which parties will have to bear the costs of such patent applications;
- each party's rights to use the background, sideground and/or foreground IP of the other party (ie, access rights) pending and after termination of the joint venture, and the terms and conditions of such access rights;
- the rights for a party who joins the venture at a later date or who exits the venture before termination of the joint venture, in respect of the background, sideground and/or foreground IP; and
- the confidentiality obligations in respect of background, sideground and foreground know-how pending and after termination of the joint venture, including the rules applicable to any publications by one or more of the joint venture parties in respect of such know-how (also with a view of preventing prior public disclosure from standing in the way of applying for a patent).

(d) *Divestment*

In the case of a divestment, certain assets – regularly all assets relating to a certain business activity of a company – are sold and transferred to a third party. The considerations stated in section 2.2(a) above in respect of mergers and acquisitions are equally applicable to a divestment in which patent rights are involved.

(e) *Consortium agreements*

A consortium agreement is entered into in respect of collaboration between two or more parties in a certain project, without the incorporation of a separate joint venture entity. The considerations stated above in respect of joint ventures are equally applicable to consortium agreements.

2.3 Particular business transactions

(a) *Material transfer agreements*

Material transfer agreements (MTAs) are used in connection with the transfer of tangible research materials between two organisations, for use by the recipient for its own research purposes. The MTA defines the rights of the provider and the recipient with respect to the materials and any derivatives of the transferred materials. MTAs are most frequently used for the transfer of biological materials (eg, reagents, cell lines, plasmids and vectors), but other materials (eg, chemical compounds) can be the subject of an MTA.

Issues which should be dealt with in an MTA are:

- for which specific purpose the materials can be used and a prohibition on any other use;
- whether the recipient may transfer the materials to third parties;
- ownership of the materials and any patent rights relating to the materials;
- ownership of modifications or derivatives made from the materials or (other) inventions arising from the use of the materials by the recipient, and the right to apply for a patent in respect of such modifications, derivatives and (other) inventions, and any licence rights granted to the other party in respect of such inventions; and
- confidentiality obligations and terms and conditions relating to the publication of research results.

(b) Commercial research and development agreements

The considerations stated in section 2.2(c) above in respect of joint ventures are equally applicable to collaborative or joint R&D agreements.

In the situation where a company (the client) contracts with another company (the contractor) to conduct R&D work (non-collaborative R&D), the agreement will generally state that all background information provided to the contractor by the client, all R&D results and any IP rights (eg, patent rights) pertaining thereto will be exclusively owned by the client. Such an agreement should also provide for strict confidentiality obligations on the part of the contractor.

(c) Contract manufacturing agreements

In the situation where a company (the client) contracts with another company (the contractor) to manufacture a patented product, the agreement will stipulate that all background information provided to the contractor by the client, all manufactured products and any IP rights (eg, patent rights) pertaining thereto will be exclusively owned by the client. Such an agreement should also provide for strict confidentiality obligations on the part of the contractor.

(d) Product distribution and co-promotion agreements

A product distribution agreement is entered into in respect of the distribution of products which – for the purpose of this chapter – fall within the scope of a patent. A co-promotion agreement manages the relationship where one party engages the services of another party to help sell a product that has already been developed. What is stated in section 2.3(c) above in respect of contract manufacturing agreements equally applies to product distribution and co-promotion agreements.

(e) ***Clinical trial agreements***

A clinical trial agreement manages the relationship between a party which has developed a pharmaceutical product that needs to be evaluated by means of clinical trials (generally named the sponsor) and the institution which has facilities and personnel required to support the performance of the clinical trial by the principal investigator. What is stated in section 2.3(b) above in respect of non-collaborative R&D agreements is equally applicable to clinical trial agreements.

3. Issues associated with patent transactions

3.1 Regulatory law and compliance

The Dutch Medicinal Products Act implements EU Directive 2001/83/EC on the Community code relating to medicinal products for human use. Under the Medicinal Products Act a marketing authorisation must be obtained from the competent authority before medicinal products can be marketed.

Holders of a marketing authorisation for innovative medicinal products obtain at least eight years of data exclusivity after the date on which the marketing authorisation is granted. After those eight years, generic manufacturers may submit a marketing authorisation application for a generic product and refer to the pre-clinical information of the innovative product.

Article 43(2) of the Medicinal Products Act provides that, apart from the eight years of data exclusivity, the holders of a marketing authorisation for innovative medicinal products can obtain 10 years of market exclusivity (the '8+2' rule). This means that the generic product cannot be launched on the market before the expiry of a period of 10 years after the date on which the marketing authorisation for the innovative medicine was granted.

This period of 10 years can be extended by one year (the '8+2+1' rule) if, during the first eight years, the marketing authorisation holder obtains a marketing authorisation for one or more new therapeutic indications that bring a significant clinical benefit in comparison with existing indications.

Clearly, if patent or SPC protection is still available after the data and market exclusivity period has expired, the patentee can object to the manufacture and sale of the medicinal product notwithstanding the expiry of the exclusivity period. Pursuant to the Bolar-type exemption implemented in Article 53(4) of the Patent Act, the performance of studies and tests that are necessary to determine that a medicinal product is a generic product of an innovative product cannot, however, be objected to by the holder of a patent or SPC.

3.2 Commonly disputed matters

Generally, all elements of patent transactions can give – and have in the past given – rise to disputes, which if not settled amicably have been brought before the Dutch courts – disputes as to whether or not contractual obligations have been complied with, which patents and know-how are covered by an agreement, the amount of royalties to be paid by a licensee, whether an agreement is validly terminated, whether a party can claim sole or joint ownership of IP rights, etc. Such disputes often arise from the fact that the agreement does not contain sufficiently clear provisions on the disputed subject matter.

In the Netherlands, the District Court and Court of Appeal in The Hague have exclusive jurisdiction in all matters relating to patent infringement and patent validity.[20] The courts in The Hague do not have exclusive jurisdiction to hear matters arising from patent-related contracts. Whether a Dutch court and, if so, which Dutch court has jurisdiction to hear matters relating to patent-related contracts, is generally agreed upon by the contractual parties in the agreement itself. If this is not the case, court jurisdiction will be determined in accordance with the applicable national law (Dutch Code of Civil Procedure) or EU regulations and international treaties.

3.3 Competition rules

The Dutch competition rules are embodied in the Dutch Competition Act of 1997, as subsequently amended, and are enforced by the Netherlands Authority for Consumers and Markets and the Dutch courts.

The Dutch competition rules largely correspond with the EU competition rules. The three pillars of both EU and Dutch competition law are the prohibition of cartels,[21] prohibition of the abuse of a dominant position[22] and merger control.[23] Pursuant to Articles 12 and 14 of the Competition Act, EU/EC block exemption regulations – including the Technology Transfer Block Exemption Regulation and R&D Block Exemption Regulation – are equally applicable to cartel matters assessed under Dutch competition law.

3.4 Securitisation of patent rights

Patents and patent applications can be the subject of security rights, more specifically a right of pledge (*pandrecht*).[24] The proprietor of a right of pledge (the pledgee) does not obtain ownership of the encumbered asset but is granted a title to recover his financial debt-claim from the sale proceeds of the encumbered asset prior to all other creditors of the person to whom the encumbered asset belongs (the pledgor). The pledgee is entitled to sell the encumbered asset only if the pledgor does not pay off its debts to the pledgee in a timely manner.

A right of pledge is vested by means of a written document. For the right of pledge to have effect in relation to third parties, it should be registered in the Patent Register of the Netherlands Patent Office.[25]

---

20 Article 80 of the Patent Act.

21 Articles 6–16 of the Dutch Competition Act of 1997, as amended; Article 101 of the Treaty of the Functioning of the European Union.

22 Articles 24–25 of the Competition Act; Article 102 of the Treaty of the Functioning of the European Union.

23 Articles 26–49 of the Competition Act; the EC Merger Regulation (Regulation 139/2004 of January 20 2004).

24 Article 67 of the Patent Act.

25 Article 67(1) of the Patent Act. In the event that a certain third party is otherwise notified of the patent assignment, it is argued that such party cannot claim that the assignment does not have effect towards it if the assignment is not registered in the Patent Register. Nevertheless, registration in the Patent Register is advisable to avoid any discussion on this point.

3.5 The effect of insolvency

As mentioned in section 2.1 above, for an assignment and licence of patent rights to be effective towards third parties it is in principle necessary to have the assignment or licence registered in the Patent Register of the Netherlands Patent Office. The trustee or receiver in a company which is declared bankrupt is considered to be such a party.

However, whether or not a patent licence is registered in the Patent Register, it is assumed in Dutch legal practice that the trustee/receiver in the bankruptcy of the licensor is entitled to disregard the licence agreement, leaving the licensee empty-handed. This is the consequence of a decision of the Dutch Supreme Court in a case regarding a tenancy agreement,[26] which decision is considered to apply equally to IP licence agreements – at least for as long as the Supreme Court (or a Court of Appeal) does not explicitly decide otherwise. This risk clearly results in requests of licensees to be provided with additional security against the licensor going bankrupt. Such security could be provided by qualifying the licence as a right of usufruct (which right a trustee/receiver cannot disregard), by (conditionally) assigning (co-)ownership of the licensed patent to the licensee, or by granting the licensee a (conditional) right of pledge in respect of the licensed patent.

3.6 Taxation[27]

If a Dutch company (eg, a private limited liability company (BV) or public limited company (NV)) has the legal and economic ownership of patents, any income derived from these patents is included in the taxable base and is subject to Dutch corporate income tax. The Dutch corporate income tax rates are currently (2014) 20% on profits up to €200,000 and 25% on the excess.

To stimulate R&D activity in the Netherlands, a special elective regime known as the 'innovation box' applies to income from self-developed qualifying intangible assets. Under this regime, 80% of the qualifying income is exempt from tax, resulting in an effective corporate income tax rate of 5%. Intangible assets generally qualify for the regime if they are patented and/or derived from R&D that benefits from an employment-cost-focused R&D payroll tax subsidy. The innovation box may extend to, for example, software, production methods, product development or improvement, and certain types of business secrets. It does not apply to marketing intangibles such as trademarks and logos. Subject to certain conditions, intangible assets that are partially derived from outsourced R&D or developed in cooperation with one or more (foreign) group members by means of a cost contribution agreement may also qualify. An acquired intangible asset that was patented before acquisition by the transferor is generally not eligible for innovation box benefits; however, the income attributable to its further development does qualify for the innovation box, provided the R&D payroll tax subsidy is granted for the further development and these development activities essentially result in a new qualifying intangible asset (not just maintenance of the old asset).

---

26 The *Nebula* case, Supreme Court, November 3 2006.

27 Contributors of this section 3.6 are Ronald Honings and Michael Dekkers, both tax advisors at and part of the Innovation Box and Intellectual Property Group of KPMG Meijburg & Co.

The innovation box regime covers all types of income attributable to qualifying assets, including capital gains. In the case of patents, not only licensing income is eligible for the innovation box, but also income derived from the sale of products or services based upon the innovations. In calculating the amount covered by the regime, generally accepted transfer pricing rules apply.

3.7 Hot topics

Currently there are no hot topics relating to patent transactions in life sciences under Dutch law.

South Korea

Mikyung Choe
H Joon Chung
Yoon Young Huh
Ji Eun Kim
Jongguk Pak
Kim & Chang

By gross domestic product, South Korea is the fourth-largest economy in Asia and the 12th-largest in the world.[1] The country's pharmaceutical market grew by 5.2% in 2010, to USD17 billion.[2] South Korea has a tremendous growth potential in the life sciences industry with its strong government support, highly educated population, growing infrastructure and burgeoning innovation from the academic and private sectors. In line with its development in the biotech and pharmaceutical sectors, South Korea is also strengthening its intellectual property rights, indicating yet another correlation between economic improvement and growth in intellectual property.

1. Patents, SPCs and know-how: foundations for transactions

1.1 Patents and supplementary protection certificates

Of the many forms of intellectual property (IP) protection available in South Korea, the most significant in the life sciences sector is patent protection under the Patent Act (since 1961). Other IP-related laws include the Trademark Act (since 1949), the Copyright Act (since 1957) and the Unfair Competition Prevention and Trade Secret Protection Act (since 1961).

(a) Patentability

To be patentable, an invention must be industrially applicable,[3] novel[4] and inventive.[5]

South Korea is a 'first-to-file' or 'absolute novelty' country in which the novelty of an invention is destroyed once it is made available to the public (anywhere in the world) prior to the filing of an application.[6] However, a safe harbour provision offers a reprieve from such disclosures provided that the disclosure was made by the inventor or against the will of the inventor, and that the patent application is filed within 12 months of the disclosure.[7]

---

1 CIA, *The World Factbook*, available online at www.cia.gov/library/publications/the-world-factbook/geos/ks.html (data retrieved April 30 2014).
2 See Ribbink Kim, "The Changing Face of South Korean Pharma", *PharmaVoice*, September 2013, p50.
3 Articles 29(1) and 32 of the Patent Act. A patent application claiming unpatentable subject matter is typically rejected for lack of industrial applicability. In particular, medical activities applied to a human being, including surgical/therapeutic/diagnostic methods, are considered lacking industrial applicability.
4 Articles 29(1), 29(3) and 30(1) of the Patent Act.
5 Article 29(2) of the Patent Act.
6 Article 36 of the Patent Act.
7 Article 30(1) of the Patent Act.

An invention directed at a behavioural or research product that is prohibited under South Korea's Bioethics and Biosafety Act (eg, artificially fertilised embryos for purposes other than pregnancy, or the use of unused and expired embryos for purposes other than research relating to disease treatment) is unpatentable. Furthermore, methods of treating or preventing diseases in humans are considered therapeutic in nature and thus industrially inapplicable.

(b) Patent applications

An applicant for a patent must submit the following documents with the Korean Intellectual Property Office (KIPO):

- an application stating the name and address of the inventor and the applicant, the title of the invention, and priority data (if the right of priority is claimed);
- a specification setting forth the title of the invention, a brief description of the drawing(s) related to the invention, a detailed description of the invention, and claim(s);
- drawing(s);
- an abstract; and
- a power of attorney.[8]

Special requirements are imposed on life sciences inventions:

- *Sequence listing:* All patent applications seeking protection of one or more sequences are required to submit sequence listings or listings as an electronic file meeting World Intellectual Property Organization (WIPO) Standard ST25.
- *Deposit of micro-organisms:* A patent application seeking to claim an invention involving a micro-organism must deposit the micro-organism, at either a local KIPO-designated depository or a recognised international depository. However, if the production or isolation of the micro-organism would not involve undue experimentation, a deposit may not be necessary.

For an international patent application filed under the Patent Cooperation Treaty (PCT) designating South Korea to enter into the national phase, a South Korean national phase application should be submitted to KIPO within 31 months from the priority date.[9]

A South Korean patent application will not be substantively examined until a formal request for substantive examination is made. A request for examination must be made within five years from the filing date of the application.[10]

Except for patent applications under a secrecy order, all pending applications will be automatically published in the official gazette of KIPO, called the *Patent Laid-Open*

8 Articles 29(3) and 42 of the Patent Act.

9 See WIPO, *Time Limits for Entering National/Regional Phase under PCT Chapters I and II (in Months from Priority Date)* available online at www.wipo.int/pct/en/texts/time_limits.html (data retrieved May 1 2014).

10 Article 59(2) of the Patent Act.

11 Articles 29(3) and 42 of the Patent Act.

Gazette, after 18 months from the South Korean filing date (or from the priority date, if claimed).[11]

(c) ***Scope of patent rights***

Once the patent right is granted, the patentee has the right to exclude others from practising the patented invention in South Korea.

However, there are some exemptions to this rule within the country, including the so-called 'Bolar' exemption by which an action by the accused infringer may be exempted if the action is reasonably related to developing and submitting information pursuant to a law regulating the manufacture or sale of drugs.[12] In particular, certain experiments to be conducted on a patented pharmaceutical during the lifetime of the patent are allowed so as to enable generic manufacturers to demonstrate bioequivalence prior to the expiry of a patent.

(d) ***Patent duration and patent term extension***

The patent term lasts 20 years from the effective filing date of the South Korean application or the effective filing date of the PCT application.[13]

Patent term extensions are available for medicinal and agrochemical inventions for time lost to the regulatory approval process, up to five years.[14] The extendable time period is the sum of the number of days spent on conducting clinical trials (if any) in South Korea and the number of days spent by the Ministry of Food and Drug Safety (MFDS) reviewing a drug approval application, but with any delay attributable to the applicant being deducted.

Further, for a patent application filed on or after March 15 2012 the applicant may request a patent term adjustment for unreasonable delay by KIPO, but only delays within the period beginning the day after four years from the filing date of the patent application or three years from the date of examination request, whichever is later, and ending on the patent grant date when the patent was registered. Again, any period of delay attributable to the applicant shall be deducted.[15]

(e) ***Infringement***[16]

To constitute direct infringement, at least one of the infringing activities – such as manufacturing, using, or importing a patented product – should take place in South Korea before the patent expiration. If a patent is related to the invention of an article, an act of producing, transferring, leasing, importing or offering for sale or lease, as a

---

12 See eg, *Syngenta v LG* (Seoul High Court decision 2001 Ka Hap 1074 of June 15 2001), where manufacturing and using a patented agrochemical for conducting a field test in South Korea for obtaining agrochemical manufacturing approval to sell the agrochemical after the patent expiration were judged not to be infringing activities; and *Sanofi-Aventis v Ahn-Kook* (Patent Court case 2008 Heo 5168 of December 30 2008), where activities such as applying for a generic drug approval, manufacturing test samples and conducting bioequivalence tests, including storage of the surplus test samples, were exempted.

13 Article 87(1) of the Patent Act.

14 Article 7 of the Enforcement Decree to the Patent Act, and the relevant sections of the Pharmaceutical Affairs Act.

15 Article 92-2(1) of the Patent Act.

16 See also Article 126 of the Patent Act.

business, articles 'used exclusively' for the production of the patented article is considered as an infringement.[17]

Regarding indirect infringement, the South Korean Supreme Court clarified the meaning of 'exclusive use' by declaring that theoretical, experimental or temporarily feasible uses do not qualify as alternative use.[18]

A patent infringement may be found under the doctrine of equivalents in South Korea, but this is seldom successful.

(f) *Transfers of patents*

A patent right is well understood to be alienable. If a South Korean patent is jointly owned, no owner may transfer or license his interest in the patent without the consent of other joint owners – although each owner may freely practise the patent. Furthermore, no transfer of patent rights becomes effective against third parties unless the event (eg, a merger, acquisition or grant of an exclusive licence) is recorded in KIPO's patent register.[19]

1.2 Who owns the patent?

(a) *Initial ownership and assignment of ownership*

In South Korea, the right to a patent belongs to the inventor or his successor in title.[20] The right to a patent may be transferred by assignment in whole or in part (see also sections 1.2(b) and 1.3 below).

An inventor has the right to be identified in a patent application. Conversely, the applicant for a patent application need not be the inventor, as the right to a patent may be transferred – for example, by contract or inheritance.

(b) *In-service invention*

An 'in-service invention' (otherwise known as a 'work-for-hire invention') is defined as an invention resulting from present or past duties of an employee when the invention is within the scope of the employer's business.[21] Unless otherwise stipulated, the ownership of an in-service invention belongs to the inventor/employee. A small or medium-sized enterprise (SME) is automatically granted a royalty-free and non-exclusive licence (a 'shop right') to use the invention.[22] For a non-SME, the employer must have a contract or employment policy executed or adopted after consultation with the employee under which the

---

17 Article 127 of the Patent Act.

18 See Case 2007 Hu 3356 of September 10 2009. The controversy in this case was over-interpreting the requirement stated in the indirect infringement provision that "an article is exclusively used for the production of the patented article". Only a few cases exist relating to indirect infringement.

19 Article 87 of the Patent Act.

20 Article 33(1) of the Patent Act.

21 See the Invention Promotion Act.

22 See the Framework Act on Small and Medium Enterprises. See also Small and Medium Business Administration, *Criteria of Korean SMEs*, available online at www.smba.go.kr/eng/smes/scope.do?mc=usr0001146 (data retrieved May 1 2014). According to that website, a company engaging in the "human health" sector qualifies as an SME if (i) the company employs less than 300 workers and (ii) its sales are less than KRW20 billion (approximately USD19 million).

employee transfers the patent ownership of an in-service invention to the employer in advance.

An employee has a duty to report the details of any inventive work to his employer; and when two or more employee-inventors contribute to the invention, all of the employee-inventors must report together. In response, the employer must notify, within four months from receipt of such a report, the employee(s) in writing of its decision whether to acquire the in-service invention. If the employer fails to notify its employee(s) of its intention to acquire the invention, the employee(s) will retain ownership of the invention.

1.3 Inventor compensation

In exchange for an in-service invention acquired by the employer, the employee is entitled to "reasonable compensation".[23] The compensation must be determined in consideration of the profits that the employer gained or expects to gain from the in-service invention, the relative degree of contribution between the employer and the employee in creating the invention, and the relative degree of contribution among inventors if there are multiple inventors.

Even if a patent application on an in-service invention is not filed by the employer (because, for instance, it involves trade secrets), the employee is still entitled to reasonable compensation.

1.4 Know-how and trade secrets

Confidential information is protected under the Unfair Competition Prevention and Trade Secret Protection Act (hereinafter the Unfair Competition Act) and the Act on the Prevention of Divulgence and Protection of Industrial Technology. Trade secrets under the former statute and 'industrial technology' under the latter are also protected against misappropriation.

Under the Unfair Competition Act, 'trade secret' is defined as technical or business information unknown to the public that possesses independent economic value. To substantiate the value of a trade secret, the owner must demonstrate that substantial efforts were made to maintain its secrecy.

Trade secret misappropriation is a growing problem in South Korea. In 2012, there were more than 100 instances of trade secret misappropriation in the country. Companies conducting business and research and development (R&D) in South Korea should adopt clear guidelines to educate employees on handling confidential information and to comply with the law and its associated regulations.

1.5 Employees, consultants and protecting know-how

The Unfair Competition Act provides the following remedies in cases of trade secret infringement:

- injunctive relief;
- compensation for loss; and
- restoration of goodwill or business reputation of the trade secret holder.

23 See Article 15 of the Invention Promotion Act.

The acquisition, usage or disclosure of trade secrets for the purpose of making illegal profit or causing damage to the trade secret holder is subject to criminal sanctions, as are acts of attempt, preparation or conspiracy to illegally acquire usage or disclosure of a trade secret. The maximum penalty for trade secret infringement is imprisonment for up to 10 years or a fine of up to KRW100 million (about USD100,000).

2. Context and features of patent transactions

2.1 The principal dealings in a patent

(a) *Patent assignments*

In South Korea, no transfer of a patent right becomes effective against third parties unless it is recorded within the patent register at KIPO. Documents required for the registration process are:

- a deed of assignment executed by the assignor;
- a notarised corporate nationality certificate executed by the assignor; and
- power of attorney granted by the assignor and assignee.

For the purpose of the registration of an assignment at KIPO, an assignment form indicating the exact names and addresses of the assignor and assignee and the subject patent being assigned is needed.

(b) *Patent licences*

There are two ways in which licences can be granted: as exclusive licences and as non-exclusive licences, as described further below. When patent rights are licensed under a licence agreement in either category, the patent laws require licence registration with KIPO for the licensee to obtain the full breadth of benefits.

Exclusive licences[24]: An exclusive licence is tantamount to ownership in South Korea once the licence is registered with KIPO. Therefore, recording a grant of rights with KIPO as soon as possible is advisable – especially, for multinational companies operating in South Korea – to enforce the patent rights under the licence.

Typically the patent is filed by a parent company, which would grant its South Korean branch or subsidiary an exclusive licence so that the South Korean branch or subsidiary may exercise the licensed patent – for example, by marketing the relevant product. Once the licensing arrangement is registered at KIPO, the branch or subsidiary is permitted, by statute, to bring a suit against any alleged infringer.

If registration does not occur, the effectiveness of the exclusive licence will only be recognised between the licensor and licensee and not in respect of any third party. If the licensor violates the agreement and grants a licence to a third party, the unregistered exclusive licensee cannot assert infringement claims against the third party – its only recourse is with the licensor, for example by seeking damages.

24 See Articles 100 and 101 of the Patent Act.

Non-exclusive licences[25]: To be enforceable against a third party, a non-exclusive licence is recommended to be registered with KIPO as well. For example, in the event that a licensor transfers ownership of a patent, if the non-exclusive licence is registered, the licensee may assert its non-exclusive right to use the licensed property against the new owner.

There are no statutes or regulations setting a maximum royalty rate or payment term. The amount and payment must comply with South Korean law, including fair-trade (antitrust) regulations. These regulations impose some restrictions on the enforceability of patent licences, as follows:

- The term of a licence extending beyond the expiration dates of the licensed patents can be found unenforceable.
- Restrictions on resale price, restrictions on transaction of competing products, and the tying of non-patented products can be found unfair and unenforceable.
- A patentee is generally not allowed to prevent a licensee from contesting the validity of one or more licensed patents.

2.2 Patent rights in corporate transactions

(a) Mergers and acquisitions

IP due diligence is essentially an audit to assess not only the quantity and the quality of intellectual property assets owned by, or licensed to, a company, business or individual, but also the liabilities from operating the business or commercially exploiting intellectual property assets. A due diligence exercise should also include an assessment of how intellectual property is captured and protected by the relevant company or business. Typically, IP due diligence is carried out by a prospective purchaser in relation to the IP assets of the target company or business.

Determining the structure of a business sale as a business/asset sale or a stock sale is complicated and depends on the various needs of the parties. An asset sale is the purchase of individual assets and liabilities, whereas a stock sale is the purchase of the owners' shares a corporation. If the target business is a sole proprietorship, a partnership or a limited liability company, the transaction cannot be structured as a stock sale because none of these entity structures would issue stocks. Instead, owners of these entity types can sell their partnership or membership interests as opposed to the entity selling its stocks. If the business is incorporated, the buyer and seller must decide whether to structure the deal as an asset sale or a stock sale. Each is described below.

Asset sales: In an asset sale, the seller retains possession of the legal entity and the buyer purchases individual assets of the company, such as equipment, fixtures, leaseholds, licences, trade secrets, trade names and inventory.

Asset sales may represent problems for buyers. Certain assets, including intellectual property, contracts and leases, are more difficult to transfer due to issues of assignability, legal ownership and third-party consent.

25 See Article 102 of the Patent Act.

Profits realised by the target company from the transfer of its business or assets are subject to ordinary corporate income tax, which is currently (2014) 22% including a resident surtax.[26] The tax base would be the actual sale price less the net book value of the assets sold by the target company.[27] If the target company has a net operating loss, such net operating loss can be utilised to offset any profit in calculating the taxable income for the year in which the asset or business transfer takes place. The purchaser would be liable for acquisition tax and/or registration tax depending on the properties transferred (primarily real property).

For an asset sale, value added tax (VAT) will be levied at the rate of 10% of the purchase price allocated to those assets subject to VAT.[28] Assets that are subject to VAT include inventory, machinery and equipment, buildings and goodwill; land or accounts receivable, however, are not subject to VAT. If the target company is engaged in a VAT-payable business, the target company would generally be able to credit such VAT paid against its VAT payable (or obtain a refund later if there is a shortfall) when it files its VAT return.

Stock sales: A buyer may purchase the selling shareholders' stock directly in stock sales to obtain ownership in the seller's legal entity. Unlike an asset sale, stock sales do not require numerous separate conveyances of each individual asset because the title of each asset lies within the corporation. If the target business has significant intellectual property assets that are difficult to assign, a stock sale may be the better option because the business, not the owner, retains ownership.

Sellers often favour stock sales because all the proceeds are taxed at a lower capital gains rate. Moreover, sellers are generally responsible to a lesser degree for future liabilities, such as product liability claims, contract claims, employee lawsuits, pensions, and benefit plans.

A stock sale is often simpler to implement and more tax efficient for the selling shareholder, particularly if the selling shareholder is a foreign resident where the applicable tax treaty either exempts or reduces the amount of capital gains tax payable. In addition, a stock share will allow any net operating loss of the target company to be preserved and there is no step-up of basis in the assets transferred.

The seller is required to pay a securities transaction tax at the rate of 0.5% of the sales price.[29] In addition, a purchaser acquiring more than 50% of the target company is subject to a deemed acquisition tax in respect of land and certain other assets owned by the target company, since the purchaser is deemed to acquire the underlying assets of the target company.[30]

(b) Financing

A flotation process involves submission of a securities registration statement to the

26 See pages 108–152 of "Korean Taxation 2012", available online at www.nts.go.kr/eng/korean/korean_01.asp?top_code=K001&sub_code=KS01&ssub_code=KSA1.
27 *Id.*
28 *Id.*
29 *Id.*
30 *Id.*

Financial Services Commission (FSC). In addition to filing this statement, an applicant should file a prospectus with the FSC when the securities registration statement becomes effective. Matters to be stated in the securities registration statement include information about the applicant, such as a brief statement about the company and matters necessary for investor protection. Also, there are continuing obligations applicable to listed companies: listed companies are required to make timely and accurate disclosure of material corporate information that may have an effect on the stock price or trading volume, so as to protect the investors.

In South Korea, the patent disclosures were changed from mandatory to voluntary after South Korea's Capital Market and Financial Investment Business Act was amended on February 3 2009. Further, two major stock exchanges, KOSDAQ and Korea Exchange, do not require a listed company to report its procurement of an interest in a patent.

Companies listed on KOSDAQ or Korea Exchange are required to file annual reports which include the company's contingent liabilities. 'Contingent liabilities' include important lawsuits (potentially, patent infringement lawsuits) where the company or its affiliates are parties to the lawsuit or where the lawsuit is brought against assets of the company or its affiliates and such lawsuit has a potential to substantially influence the company's business. Violators of this disclosure requirement may be subject to imprisonment of up to five years or an administrative fine of up to KRW200 million.

(c) *Joint ventures*

Companies engaging in a joint venture to develop technologies in South Korea often contribute their background IP, technology or human resources to enable the joint venture to conduct development work. Thus, the technical contribution of each party should be precisely agreed upon. Furthermore, ownership of the patents and/or know-how developed by the joint venture, as well as the possible licensing thereof, should be considered. The scope of non-competition with shareholders/affiliates and the consequences of joint venture termination including ownership of IP rights should also be contemplated.

As indicated in sections 1.2 and 1.3 above, the ownership of an invention or know-how developed by an employee belongs to the employee/inventor in South Korea unless a prior invention assignment agreement exists and certain statutory notification steps are taken by the employer after receipt of an invention disclosure from the inventor employee. Ownership transfer of the invention or know-how triggers a liability on an employer in a joint venture to pay an inventor remuneration in a reasonable amount determined in view of the profits earned or to be earned by the joint venture (eg, royalties from licensing of the patents/know-how developed by the joint venture to the shareholders/affiliates).

(d) *Divestment*

Another key strategy is to identify IP assets which are no longer required, or which do not provide the required degree of protection to the overall IP strategy of the company, and to find opportunity for the sale or license of these assets. IP rights that are no longer of significant benefit can still hold a significant residual value, because others in the industry may be interested in acquiring those rights, obtaining a straight licence or granting a cross-licence in an exchange of similar classes of rights. Any liabilities and potential defects to IP rights, such as inventorship, ownership (eg, third-party claims) and formalities, must be resolved prior to a decision to divest the asset. Moreover, tax implications and the regulatory requirements of a public company (eg, reporting to the MFDS or the Korea Fair Trade Commission) should be considered.

(e) Consortium agreements

In setting up a consortium or joint (non-corporate) enterprise for a research project, the technical contribution of each party should be precisely set forth. The contribution will include the tasks that each party intends to carry out and the technical resources (such as human resources, equipment, facilities and information) that each party intends to make available. Many of the issues discussed in section 2.2 above, including the issues related to an in-service invention, would also apply in a consortium endeavour.

In addition to a basic confidentiality clause, the commercial provisions in a consortium agreement mainly specify how to divide ownership and rights to use the results of the cooperation (ie, so-called 'foreground' IP rights). In general, the common principle is to ensure that jointly made inventions are jointly owned. In the event of an invention being the work of a single party to the cooperation contract and solely the result of its intrinsic skills rather than shared knowledge, the inventor may be the exclusive owner of the results. In most cases, however, ownership of a joint invention in a consortium context is in effect equivalent to an irrevocable non-exclusive licence.

2.3 Particular business transactions

(a) Material transfer agreements

A material transfer agreement (MTA) is a contract that governs the transfer of tangible research materials between two organisations, when the recipient intends to use the materials for research purposes. Biological materials, such as reagents, cell lines, plasmids and vectors, are the most frequently transferred materials, but MTAs may also be used for other types of materials, such as chemical compounds. As well as the usual contractual terms (eg, confidentiality requirements), an MTA should contemplate terms relating to the following:

- *Definition of materials:* The materials provider should propose a clear definition of 'materials', where the definition refers not only to the original materials but also to modifications or derivatives made from the materials.
- *Use of materials in sponsored research projects:* A typical MTA should contain precise language that outlines the specific scope of sponsored research.
- *Delay in publication:* To protect patentable inventions from public disclosure prior to patent filing, a provider company should demand that the investigator seek its permission and review before publication or public presentation.

- *Control of IP rights:* The importance of the ownership rights over IP generated by the sponsored research project cannot be exaggerated. If an MTA pre-empts ownership rights, investigators may be restricted in their ability to interact with a future sponsor or may have conflicts with obligations to the current sponsors. Conversely, a badly worded licence by the provider to the investigator may result in third-party liability that must be dealt with in case of a downstream transaction.

(b) Commercial research and development agreements

There are two approaches to commercial R&D agreements: non-collaborative arrangements and collaborative agreements.

A party may consider entering into a non-collaborative agreement to access the opposing party's proprietary equipment, facilities or capabilities. In such an arrangement, among other concerns, the using party must ensure that it retains the entire right, title and interest to any invention resulting from the use of the providing party's technology. Conversely, the providing party must impose adequate confidentiality and indemnity terms to perpetuate the value of its proprietary assets.

A collaborative agreement works to protect the interests of each member of the collaboration, including interests in intellectual property, distribution of monies and information sharing. Research, development and collaboration agreements take a wide variety of forms, including: collaborations between commercial and academic organisations; collaborations between commercial organisations; and consortium agreements (see section 2.2 above).

The most contentious aspects of research, development and commercialisation agreements often relate to intellectual property ownership and exploitation. There is no one-size-fits-all approach to intellectual property ownership. It is often necessary for one party to grant to the other a licence to use its so-called 'background' IP to enable the other party to conduct R&D work. If both parties are conducting research and development, cross licences may be necessary. Further, the agreement should unambiguously define both the 'background' intellectual property brought into the project and the ownership of any new intellectual property resulting from the collaboration.

(c) Contract manufacturing agreements

Global companies often execute contract manufacturing agreements with contract manufacturing organisations (CMOs) to manufacture (eg, to prepare and formulate the active ingredient in) a drug. Some key contract provisions that the parties to such an agreement should consider are as follows:

- *Restriction on the use of the intellectual property:* Often, manufacturing a drug requires the company to share its know-how with the CMO. Today, it is not uncommon for global companies to hire the same set of CMOs for manufacturing their products, often in the same field and requiring similar technology. Therefore, the agreement should include specific terms prohibiting the CMO's use of the company's know-how for a competitor's benefit.

- *Compliance:* To comply with the applicable current good manufacturing practice (cGMP) regulations, the agreement should bind the CMO to properly maintain its documents and laboratory controls, to set compliant specifications for raw materials, and to conduct required sampling and testing.
- *Product-specific terms:* The agreement should provide specific terms related to the particular product, including product/component specification, and defined manufacturing operations (eg, batch numbering processes, storage and shipment, and lot disposition).

(d) Product distribution agreements

A distribution agreement is a contract between channel partners that outlines terms on distribution, ownership and title to products, intellectual property rights, minimum performance standards, product liability and recall, and restrictive covenants such as non-disclosure and non-competition agreements. For example, foreign companies that do not have a distribution network in South Korea would execute distribution agreements with local distributors.

The Korea Fair Trade Commission (KFTC) recently issued the Guidelines for Procuring Fairness in Transactions in Pharmaceutical Industry (hereinafter the Guidelines) to address unfair trade practices within the pharmaceutical industry. The Guidelines are not legally binding but in practice they serve as the KFTC yardstick when assessing the fairness of a distribution agreement. The Guidelines are primarily intended to police terms that are unduly favourable to the supplier (and thus less favourable to the distributor) and that are thereby considered an abuse of market dominance and/or unfair business practice. Under the Guidelines, risky terms include:

- restricting the distributor's activities related to competing products during the term of the distribution agreement (eg, prohibiting from distributing other drugs with the same indication(s));
- limiting R&D activities during the term of the contract;
- triggering immediate termination as a result of a failure to meet the minimum purchase amount or minimum sales target; and
- requiring the unconditional and royalty-free transfer of improvements to the supplier that have been developed by the distributor.

(e) Co-promotion agreements

A co-promotion agreement is a joint marketing agreement for a company typically to 'hire' another company's salesforce (sometimes, in addition to its own) to promote the same brand or range of brands. Although there is some distinction between co-promotion agreements (two companies promoting the same product side-by-side, which is typically the method used in the United States), and co-marketing arrangements (two companies promoting the same product under two different names), the industry often uses these terms interchangeably to describe a relationship where one party engages the services of another to help sell a product that it has already developed, where profits on sales of the product are shared as opposed to a commission on sales being earned (as in a sales representation agreement) or purchase and resale of the product occurring (as in a distribution agreement).

A well-written co-promotion agreement should include provisions on the following:

- *Joint marketing committee:* This committee should be set up and composed of individuals from both companies, with the purpose of determining the desirable level of marketing spend and the form the promotion should take. The committee usually includes an even number of people from each company, with a dispute resolution provision that provides a timely and inexpensive method for breaking deadlocks.
- *Regulation and governance:* Governing regulatory responsibilities need to be defined, including obligations for obtaining approvals, procedures for dealing with product complaints and recalls, and mechanisms for ensuring that all promotional materials are reviewed and approved by both parties prior to use.
- *Exclusivity:* To avoid one co-promoting company from offering its own or another's competing product, the agreement should contain exclusivity provisions.

Co-promotion agreements involve many of the same Fair Trade Law issues as implicated in distribution agreements in South Korea. For example, 'reverse payment' provisions – requiring the originator company to compensate the generic company for delaying launch of its generic version – may violate the fair trade laws as they do in some other jurisdictions (see also section 3.8 below). Furthermore, provisions related to resale price maintenance or the imposition of unfair restrictions and aggressive promotions (eg, improper kickbacks to medical facilities through the local co-promotion company) based on superior bargaining position would be viewed as improper trade practices.

(f) ***Clinical trial agreements***

Clinical trial agreements are executed between a sponsor and an institutor or investigator, and should consider:

- compliance with good clinical procedure;
- respective indemnification responsibilities in case patient compensation is needed;
- the provision of materials relating to investigational new drug (IND) approval, as only local entities may be IND applicants;
- the allocation of rights as to intellectual property resulting from the clinical trial;
- the allocation of the costs of the clinical trial; and
- the allocation of indemnification responsibilities with respect to victim compensation.

3. Issues associated with patent transactions

3.1 Regulatory law and compliance

Pursuant to the free trade agreement executed in 2007 between South Korea and the United States, South Korea is implementing a patent-approval linkage system similar to that contained within the US Hatch-Waxman Act, and doing so in several phases up to March 2015.[31]

The patent-approval linkage system is intended to streamline the introduction of generic drugs upon expiration of a patent. At the same time, the linkage system permits a generic manufacturer which believes a patent is invalid or is valid but about to go off patent protection to file a generic application seeking approval of its generic alternative. If the patentee believes otherwise, it may file a patent infringement lawsuit against the generic company. Once the generic company wins the right to produce the generic form of a drug and begins marketing, it will have one year of exclusivity during which no other generic company may enter the market.

Additional regulations related to the full implementation of the patent-approval linkage system are still to come into force. The Ministry of Food and Drug Safety will enact, among other things, the rules for determining generic exclusivity by March 2015. Exclusivity will be awarded to the first generic drug company which successfully maintains Paragraph IV-like certification regardless of whether the patent is invalid or will not be infringed by the generic product. (In US law, 'Paragraph IV certification' asserts that a new product will not infringe an already approved product's listed patents or that such patents are invalid or unenforceable.) The current regulations, however, do not include specific rules on exclusivity – for instance, the mechanisms for triggering the exclusivity clock, 'patent-based' or 'product-based' exclusivity, and rules regarding the waiver or relinquishment of exclusivity have yet to be determined. According to the proposed regulations, generic exclusivity is awarded to one who has filed the first generic approval application, has filed a trial or litigation before filing its generic approval application, and has received a favourable decision in the trial or litigation as a direct party.

It should be noted that South Korea is becoming a hotbed of biosimilar innovation. Its legal framework and regulatory pathway for biosimilars, primarily adopted from the European regulations, were enacted in 2009. Since implementation of the biosimilar pathway, two products – Remsima™ (infliximab) in 2012 and Herzuma™ (trastuzumab) in 2014, both developed by Celltrion – have been approved for use in South Korea.[32]

3.2 Competition rules

South Korea's Fair Trade Law prohibits an act of unfairly coercing or inducing

---

31 See Article 18.9.5(b) of the free trade agreement (available at www.ustr.gov/trade-agreements/free-trade-agreements/korus-fta/final-text).

32 See Celltrion, *What's New*, available online at www.celltrion.com/en/company/notice_view.asp?idx=425&code=ennews&intNowPage=1&menu_num=&align_year=all (accessed May 1 2014). Herzuma™ was approved on January 15 2014 in South Korea. Genentech's Herceptin®, the original product of Herzuma™, generated sales of more than USD6.3 billion globally in 2012.

customers of competitors to do business, or engaging in conduct that unfairly restricts or disrupts business activities. In particular, where coercing or disrupting acts are accompanied by dissemination of facts, statements or assertions that are not objectively proved or that are exaggerated, the Fair Trade Commission or the court is more likely to find that such an act is an unfair trade practice.

In entering into a patent licence arrangement, the Fair Trade Law protects a licensee from potential misuse of a patent right by the patent holder or licensor, when it is considered an unfair trade practice. Examples of such unfair trade practices include: imposing royalty payments after the patent expires; prohibiting a licensee from making improvements to the licensed patent; and imposing obligations to transfer improvements to the licensor for free. In addition, a patent holder's unfair refusal to grant a licence to block market entry may be an unfair trade practice.

3.3 Securitisation of patent rights

South Korea permits patents to be used as collateral for loans but security registration cannot be made based on a patent application.

The main form of security that can be taken over patent rights is a pledge. The Patent Act requires registration of a pledge over patent rights, among other security interests, for claiming against a third party. A pledge over a patent must be recorded in KIPO's patent register to be valid and enforceable.

3.4 The effect of insolvency

Registration does not ensure that a licensor's bankruptcy administrator will not decline to perform obligations under a licence agreement or revoke a licence agreement under the applicable laws of bankruptcy. However, assuming that the administrator does not so decline or revoke, patent licence registration will be beneficial if the licensed patents are transferred to a third-party purchaser in connection with the licensor's bankruptcy. In other words, if a licence is registered, a purchaser cannot acquire the licensed patents free of the licence.

3.5 Taxation[33]

(a) *Main taxes payable by a licensor on the licensing of a patent*

When a licensee in South Korea pays royalties to a licensor outside South Korea, taxes must be withheld. Under South Korea's Corporation Tax Law the licensee must generally withhold tax at a rate of 22% (including local income tax, levied at 10% of income tax) on the gross amount of royalty payments. However, if there is a tax treaty between South Korea and the country where the licensor resides, the withholding tax rate may be reduced – for example, to 16.5% for patent licensors in the United States.

33 See "Korean Taxation 2012", available online at www.nts.go.kr/eng/korean/korean_01.asp?top_code=K001&sub_code=KS01&ssub_code=KSA1.

(b) *Main taxes payable by a seller on the disposal of a patent*

If a seller of a patent is a South Korean tax resident, any gain from the sale is taxable income and is subject to income tax, inclusive of local income tax.

For a non-tax-resident corporate seller, income from the disposal of a patent is classified as royalty income under the Corporation Tax Law. When a buyer pays consideration to the seller, the buyer must withhold tax on the amount. Nonetheless, depending on any tax treaty, income from the disposal of a patent may be classified as capital gains and be exempt from taxation in South Korea.

3.6 Hot topics

The South Korean Supreme Court recently ruled that a reverse payment settlement may violate the fair trade laws.[34] While this decision – the first 'reverse payment' case in South Korea – is couched in generalities, it provides a glimpse into how reverse payments would be judged in the future. First, patent rights do not shield these settlements from the scrutiny of the fair trade laws; instead, the settlements should be analysed under the 'rule of reasonableness', whereby a court must weigh the settlement's possible pro-competitive benefits against its potential anticompetitive effects. Secondly, reverse payments are not unlawful per se – that is, the KFTC needs more evidence than merely the existence of such an agreement; for example, collusion and anti-competitive effects must be proven. The court explained that each settlement will be judged case by case, by considering the totality of the circumstances – for example, the settlement duration, the value of the economic benefits provided under the agreement, litigation costs, and any justifiable reason for providing the economic benefits.

Including the Supreme Court ruling and the implementation of the patent linkage system, the rules of engagement between originators and generics over the South Korean pharmaceutical market are being drawn up. It is predicted that patent disputes – and, potentially, settlements – between originators and generics may become more prevalent.

---

34 See Supreme Court Decisions 2012 Du 24498 and 2012 Du 27794, both rendered on February 27 2014.

Sweden

Jennie Klingberg
Odd Swarting
Setterwalls

1. Patents, SPCs and know-how: foundations for transactions

Swedish intellectual property law is territorially limited but several international conventions are applicable, such as the Paris Convention (1883) in the field of industrial property rights and the 1994 Agreement on Trade-Related Aspects of Intellectual Property Rights (TRIPS). Furthermore, most of the Swedish intellectual property legislation has been developed as a result of close Nordic or European cooperation.

1.1 Patents and supplementary protection certificates

The main regulation regarding patents in Sweden is the Patent Act. A patent is only granted after application. Instead of filing an application for a patent with the Swedish Patent and Registrations Office (PRO), the applicant may choose to make an international patent application according to the Patent Cooperation Treaty. Another alternative is to apply for a European patent under the rules of the European Patent Convention (EPC); such a patent will be valid in the countries designated by the applicant.

An application through the PRO must meet both formal and material requirements and must include:

- a description of the invention, including necessary drawings;
- a distinct specification of what protection is sought by means of the patent claims;
- a summary of the description and claims, merely as technical information; and
- information about the inventor.

The description must be clear enough to enable a professional in the field to implement the invention. Furthermore, the invention has to be new in relation to what was known before the filing date. 'What was known' in this context means everything made available to the public, regardless of whether this was done in writing, in lectures, by public use or otherwise.

The application procedure leads to disclosure of the invention and the patent holder receives the exclusive right to professionally use the innovation. This means that no one else may use, make, offer, import or put on the market products based on the patent. As a general rule, the exclusive right applies only to actions taken in Sweden. Non-commercial use of the invention, such as experiments, is excluded from the exclusive right.

The scope of the patent protection is determined by the patent claims, according to Swedish law. When construing the patent claims, the description may serve as a guide. The decision as to whether an infringement has occurred is mainly based on a literal interpretation of the patent claims. Nevertheless, modifications, including improvements, to the invention that are easily available among professionals in the field are also considered to be within the scope of the patent holder's protection (the doctrine of equivalence). If someone infringes the exclusive right of the patent holder, and has done so intentionally or by gross negligence, he may be fined or sentenced to imprisonment for a maximum of two years. Such a person should also pay reasonable compensation for use of the invention, as well as compensation for additional loss resulting from the infringement.

The maximum period of patent protection in Sweden is 20 years from the date the patent application was filed. The patent protection may cease as a result of any one of the following circumstances:

- the term of the patent has expired;
- the annual fee has not been paid to the PRO;
- the holder of the patent desires that the patent ceases;
- a formal objection to the patent has been upheld; or
- the patent has been outlawed by a court.

It is possible to obtain a supplementary protection certificate (SPC) for pharmaceutical innovations for an additional maximum period of five years. SPCs are applicable to pharmaceutical innovations that require marketing authorisation in accordance with Sweden's Medicinal Products Act. The exclusive right obtained through the SPC is limited and only includes the right to offer the innovation for sale.

1.2 Who owns the patent?

Ownership of a patent is officially recorded in the patent authority's Patent Register. Registration is not mandatory and has no legal effect on ownership. It does, however, affect the possibility of pledging the patent. Ownership is transferred under agreement and is not subject to any formal requirements.

The inventor or his successor in title may acquire the exclusive right to exploit the patent commercially pursuant to the Patent Act. The inventor may be one or more humans, and thus the inventor's right may only belong to a human. But even though a juridical person (such as a company) cannot be the inventor, it may act as an applicant for a patent and thereby become the patent holder.

The requirement to prove the applicant's right to the invention does not exist any longer in the Patent Act; instead, an applicant that is not the inventor must state the grounds for its right to the invention.

Under certain general and special conditions, a compulsory licence can be granted by Stockholm District Court. The holder of a compulsory licence may only exploit the innovation under the terms set out by the court, and the licence does not prevent the patent holder from using the invention or granting another party the right to use the invention commercially. One example of a reason to grant a compulsory license is where an innovation is not used to a reasonable extent in

Sweden and there is a risk of patent blocking (when, for instance, a foreign company applies for and is granted a patent in Sweden without exploiting it) or it causes a risk of suppression of the patent (when, for instance, a financially strong company buys patents only for the purpose of protecting its own production facilities and market position).

1.3 Inventor compensation

As a starting point, an inventor who is an employee has the same right to his invention as other inventors; however, depending on the connection between the invention, the employer's sphere of activity and the employee's tasks within his employment, the employer may have a right to take over the invention or acquire a right to implement it in his business, pursuant to Sweden's Right to the Inventions of Employees Act.

An employer has the right to utilise the invention of one of its employees if the invention:

- arises as a result of the employment and falls within the scope of the employer's business;
- falls within the scope of the employer's business but has not arisen as a result of a work assignment; or
- falls within the scope of the employer's business but has arisen outside the work environment.

The employee has the right to be compensated for both costs in connection with the invention and for the right he is forced to assign. Compensation should be reasonable and based on the value of the invention, the extent to which the employer has acquired its right to the invention, and the significance the employment might have had in the development of the invention. Except for the compensation regulations, most rules in the Right to the Inventions of Employees Act are non-mandatory, and the mandatory regulations are only valid in relation to an employer. This means that if someone else acquires the employer's right to the invention, the successor has no economic obligations towards the inventor.

If a third party benefits from a patent and commits a patent infringement, such person should pay reasonable compensation for use of the invention as well as compensation for the additional loss resulting from the infringement. When determining the amount of compensation, particular account is taken of:

- lost profits;
- profits realised by the party committing the infringement;
- damage to the reputation of the invention;
- non-pecuniary loss; and
- the interest of the patent holder in preventing infringement from taking place.

Damages are intended to compensate for actual loss, but tend to be set by the courts at the lower end of what is reasonable. Punitive damages are not allowed under Swedish patent law.

1.4 Know-how and trade secrets

Know-how is not formally recognised as an intellectual property right in Sweden but is protected under Sweden's Trade Secrets Act if it can be defined as a trade secret. It is therefore customary to use non-disclosure agreements as supplements to the legislation. There is no statutory law regulating non-disclosure agreements but case law confirms the need to make clear in the agreement that the information will be treated in a confidential way, unlike the Trade Secrets Act which does not define any requirements as to what has to be done in order to keep information in the form of a trade secret.

A 'trade secret', according to the Trade Secrets Act, means information concerning the business or industrial relations of a person conducting business or industrial activities which that person wants to keep secret, and the divulging of which would be likely to cause damage to him from the point of view of competition. This definition is only valid for the Trade Secrets Act and in other legislation the term is defined differently. The information may comprise information documented in some form (eg, drawings, models and technical prototypes) or knowledge held by individual persons about specific circumstances. Personal knowledge and skills of employees is, however, excluded from the act.

1.5 Employees, consultants and protecting know-how

The freedom to communicate information, as well as freedom of speech in general, has a strong protection under the Swedish Constitution. The Trade Secrets Act only distinguishes between acts committed wilfully or through negligence; the rules are not applicable if someone exploits or divulges a trade secret that he or someone before him received knowledge of in good faith.

If anyone is found guilty of trade espionage and unauthorised tampering, criminal prosecution and compensation liability for the damage caused are applicable sanctions. The general starting point when it comes to damages in Sweden is that the person who suffered is compensated for his loss, but not more so. Punitive damages are not part of Swedish law.

It can be difficult to decide what belongs to an employer and what to an employee when that employee has left his position in the business. Nonetheless, according to the Trade Secrets Act employee liability is limited to extraordinary circumstances, such as the systematic planning and collecting of information.

In addition to the obligation to pay damages according to the Trade Secrets Act, normal sanctions are applicable arising from a breach of an employment contract. Such sanctions are normally termination of the employment contract or dismissal. Co-ownership is not expressly mentioned in the Trade Secrets Act, nor has it been addressed by case law. Nothing prevents the co-ownership of trade secrets, but this would fall under general civil law principles.

2. Context and features of patent transactions

2.1 The principal dealings in a patent

(a) *Patent assignments*

Patent sales can be driven by having patents where:

- there is no longer any interest in that area of the business;
- there is the potential to have better proceeds through the sale of the patents;
- there is a shift in the business or technology area where the patent is being used;
- there is a need to reduce costs and expenses;
- there are strong liquidity requirements;
- the business is likely to be closed because of bankruptcy; or
- a strategic decision has been taken to sell the patents rather than enter a licensing programme because of the time and resources that the latter requires.

Under Swedish law, both an application and a granted patent can be assigned by contract or inheritance. By selling a patent, the owner surrenders all rights to the invention. Despite the importance of patent transactions, there is only a limited statutory basis in Sweden. The legislature's assumption is that private contracts alone should guide technology transfers. If the parties want, the transfer can be registered with the PRO.

The lack of statutory law means that a transfer agreement must be carefully written to express fully the agreement of the parties and to leave as little as possible to future interpretation by courts or arbitration panels. In the event of a dispute, the court will interpret the wording of the contract but also all factors in the negotiation will be taken into account and balanced against each other. This includes subjective factors relating to the parties' technical knowledge and the awareness of the other party's technical purpose.

Even if a sale has been made on condition that the goods are sold 'in their present condition', there may still be a defect in the purchase object under Sweden's Sale of Goods Act. There is an ongoing debate about whether the parties can rely on analogies from the Sale of Goods Act in a patent transaction. The dominant view is that such analogies cannot be made since the differences between a transaction of patents and a transaction of goods are too great. With this in mind, to limit implicit warranties it is normal that the parties state that the Sale of Goods Act shall not apply.

Transfer agreements are rarely entered into without a prior investigation of the company or the patent and without guarantees from the seller(s). Technical statements, test results or similar documents are often attached and warranted. Some legal experts argue that the patent specification in itself may create implicit warranties, since it creates an expectation that the results mentioned in the description are attainable. It should, however, be remembered that the grant of a patent does not grant any certain qualities of the invention: a general praise of the invention during negotiations is unlikely to impose any liability. Normal guarantees include that the seller represents and warrants that:

- the patent has been duly maintained and monitored;
- the patent does not infringe any rights belonging to a third party;
- any infringement of these rights by a third party has not occurred; and
- trade secrets related to the operation of the patent have been properly protected against access and use by an unauthorised third party.

(b) *Patent licences*

Granting a licence may allow the patent holder to increase profitability compared with the situation where the patent holder exploits the patent on his own. By granting a licence, the licensor refrains from (parts of) the exclusive right of the patent and the licensees receive authorisation to use that patent according to the licensing agreement. There is almost no statutory law regulating patent licence agreements in Sweden. A licence agreement usually contains provisions on the scope and field of use, royalties, territory and term. If the parties want, the licence can be registered with the PRO.

A licence can be a simple, exclusive or sole licence. A simple licence does not give the licensee an exclusive right to the patent – ie, it does not prevent the patent holder from entering a licence agreement with someone else. Under an exclusive licence, the patent holder cannot grant a licence to another party. A sole licence means that the licensor has the right to exploit the innovation together with the licensee, such that the licensor is allowed to compete with the licensee but not to enter into additional licence agreements.

If nothing else is stated in the agreement, it is presumed under Swedish law that the licence is simple and that the licensee is not entitled, without the owner's permission, to transfer or sub-license his right to others. The licensor is not obligated to provide any know-how, unless otherwise is agreed.

The licensee is required to pay reasonable compensation for his use of the invention if royalties are not regulated. Otherwise, royalties often consist of a fixed fee and/or a variable fee that is calculated on the licensee's use, production or sales. Sometimes certain events create milestone or stage payments.

2.2 Patent rights in corporate transactions

(a) *Mergers and acquisitions*

Acquisitions of non-listed companies are not subject to any specific rules, with the exception of the general rules in the Swedish Companies Act and any specific provisions in the articles of association of the acquiring company or in shareholder agreements. There are no requirements as to the form of an agreement to acquire shares or a company's assets. However, it is customary to set out the terms and conditions of the transaction in a written agreement. A share transfer must be registered in the share register of the target company. In order for the purchaser to be protected against the seller's creditors, the share certificates must be duly endorsed and transferred to the purchaser. It is common to draft pre-contractual documents, such as a 'letter of intent', which will address issues relating to the negotiations, exclusivity and confidentiality.

In a share sale, it is important to examine the target company's accounts and obligations and/or receive guarantees from the seller(s), since tax liabilities, outstanding guarantees or pension obligations, for example, may follow. Such obligations will not be taken over by the purchaser of an asset; nevertheless an asset sale is often more complicated since the status of all assets needs to be controlled (eg, are they leased or pledged?). In an asset sale, agreements must also be reviewed since they may contain undertakings not to assign or subcontract. In principle, any intellectual property right which the seller has in its possession may be assigned, unlike assignments of licences which often require the licensor's permission.

(b) Financing

Rules and regulations serve an important purpose in sustaining confidence in the financial market, to enable a common framework for listed companies and to protect minority stakeholders. On the Swedish Stock Exchange (the Exchange), which is the NASDAQ, there are requirements in order for a company to become listed, as well as ongoing requirements that need to be fulfilled while the company is listed.

If the company and the Exchange agree to initiate a listing process, the Exchange appoints an Exchange auditor who makes an assessment as to whether it would be appropriate to list the company. Within this process the company must prepare and publish a prospectus. An attorney must also undertake a legal examination of the company, which should at least cover legal and tax risks and material arrangements. For example, important assets such as intellectual property rights, and risks such as patent expirations or disputes, must be described.

Listed companies have far-reaching obligations to disclose key information and to make public announcements. All information that is likely to materially influence the valuation of a listed company must be published in a way that ensures that it reaches all market participants simultaneously. The requirements might in some cases affect the wishes of the companies involved to keep transactions, researches or similar confidential.

(c) Joint ventures

A joint venture can generally be described as a form of business association between two or more independent organisations to undertake a common project or to achieve a certain goal. More specifically, the parties to the joint venture share risks and each contribute with their intellectual capital towards technology, research and development (R&D), production, marketing and commercialisation. The regulations can be set out either by a specific contractual arrangement or in a joint venture agreement setting up an independent legal entity – a joint venture company. The contractual model is normally used for collaborations that are narrow in scope and of short duration, while the entity model is usually chosen for longer-term partnerships with a broader scope.

As there is no statutory law regulating joint ventures, it is fundamental that the agreement between the parties clearly define the ownership of the intellectual property to be created by the joint venture. Pursuant to Swedish law, the rightful owner of a patent is the registered owner with the PRO; but if the invention is a result of several parties' efforts, all inventors have to be stated in the patent. There is no specific rule in the Patent Act, nor any case law, regulating the situation with several parties having moral interests in the patent; however, it has been assumed in legal doctrine that the general rules on joint ownership are applicable to the situation. The Joint Ownership Right Act is not applicable to the intellectual property rights, but may be applied by analogy.

It is also necessary to agree on:

- the initial contribution of each party to the joint venture;
- each party's respective responsibilities and obligations;
- what happens when new partners enter the joint venture or when partners exit or become insolvent;
- the parties' financial contribution in terms of intangible assets;
- how so-called 'background' intellectual property should be treated; and
- an allocation of costs.

(d) Divestment

Companies considering commercial transactions need to understand the value and risks associated with the transfer. When assessing whether a patent should be transferred or licensed, several questions should be considered and the process often starts with a due diligence exercise where risks are analysed. The risk analysis often includes the company's business plan, the value-creating potential of each patent, competitors' patent portfolios, and potentially infringed or underutilised assets.

A patent can be valued by a number of different methods, involving as their basis a calculation of:

- prospective revenues generated over the patent's lifetime;
- revenue generated by similar technologies; or
- the cost of developing or acquiring reasonable alternatives to the patent/portfolio.

It may also be important to assess elements such as the pool of skilled employees, whether the market can be commercially exploited, know-how, research facilities and other commitments.

(e) Consortium agreements

An effectively drafted consortium agreement should fulfil multiple purposes within a project that will help to ensure the process and the achievement of the project aims as efficiently as possible.

The precise content of a consortium agreement will vary according to the nature and scope of the project. There is no statutory law applicable but it is important to include some means of regulating assignments and warranties, background rights, and foreground rights (including joint ownership, the granting of additional access rights to third parties, and the payment of royalties). In broad terms, the types of issues that it would be normal to expect a consortium agreement to address would include provisions to ensure the technical implementation of the project, the management structure, the allocation of funding, and an allowance for changes in project partner membership.

2.3 Particular business transactions

(a) Material transfer agreements

There are no mandatory rules as to the form of a transfer agreement for research materials, substances or samples for the purposes of research. It is customary to set out the terms and conditions of the transaction in a written agreement, and it is recommended that the agreement defines the rights of the provider and the recipient with respect to the materials and any derivatives.

Where a dispute arises, the general so-called 'specification principle' (which has emerged from case law) will most likely apply. This principle means that the contract will be given a narrow interpretation and only the research materials, substances or samples that are actually specified will be considered as transferred.

(b) Commercial research and development agreements

Research and development projects are often associated with heavy investments, both in time and resources, and are normally carried out over a long period of time. Therefore it may be important that external relationships, such as financial investments, involved in the project are well documented in contracts and that the legal structure is correct from the beginning. No matter whether it is a collaborative or a non-collaborative arrangement, it should be ensured that intellectual property rights are protected and allocated to the right party.

(c) Contract manufacturing agreements

There is no specific legislation governing contract manufacturing agreements, and parties are free to negotiate their own terms. However, Sweden's Product Safety Act sets out provisions concerning liability for defective products. The act imposes strict liability on sellers, importers and manufacturers for personal injury and damage to property used by individuals when it has been caused by an unsafe product.

(d) Product distribution agreements

There is no specific legislation governing distributors, and parties are free to negotiate their own terms for product distribution agreements. However, in some cases the Swedish courts have applied the provisions of the Commercial Agents Act by analogy where the relationship between the principal and the distributor has closely resembled an agency relationship. The act contains mandatory regulations regarding compensation on termination.

Furthermore, the Product Safety Act provides that the marketing and sale of products or services may be restricted or prohibited by a public authority for safety reasons. According to that act, the producer is obligated to recall dangerous products without delay from the distributor's possession, but the distributor also has independent contractual obligations and liabilities to its customers.

Key clauses in a product distribution agreement are thus warranties and liabilities, but also the regulation of resale prices, exclusivity, support, advertising and promotion, and delivery terms.

(e) *Co-promotion agreements*

The marking of patented products is not required. False patent marking is not prohibited under patent law, but it can be expected to be so under marketing law. Sweden's Marketing Act prohibits unfair marketing in general, as well as specific marketing practices such as various forms of misleading marketing and unfair or misleading comparisons between products or businesses. The Marketing Act also includes specific regulations governing the marketing of certain products, such as tobacco, alcohol and pharmaceutical products. These restrictions may make the sales and marketing process of certain products time consuming and more costly compared with that in other industries. Therefore – and, for example, in the life sciences industry – it is common to sub-license the rights to market and sell the products to a partner that has the necessary resources. A co-promotion agreement might therefore be involved. Another reason to enter into a co-promotion agreement is in order to maximise a product's presence in the marketplace.

Key clauses in co-promotion agreements are usually marketing rights, payments clauses (eg, how payments are triggered, rebate pricing, additional territory, priority delivery), non-compete clauses, performance levels, antitrust considerations and liability issues. In addition, inventory monitoring, advertising activity or approval, and quantity guarantees may be vital.

(f) *Clinical trial agreements*

In Sweden, all medicine and medical devices must be approved for use by the Medical Products Agency or the European Medicines Agency. With the exception of naturopathic and homeopathic preparations, the safety and effect of a medicine is always to be based on clinical trials performed by the responsible pharmaceuticals company.

Clinical trials must be performed in accordance with the latest versions of the ICH's Good Clinical Practice and a series of other rules such as OECD's Principles for Good Laboratory Practice, the WHO's Good Clinical Laboratory Practice and EU Directives 2004/9/EC and 2004/10/EC. Furthermore, the Swedish Work Environment Authority has issued regulations on laboratory work with chemicals and therefore indemnification clauses in a clinical trial agreement may be important. It is common that the parties agree to maintain a policy or programme regarding insurance sufficient to support this obligation.

Confidentiality clauses would be expected to contain regulations regarding the extent of confidential information shared (publication of research/study results, pre-publication review and comment period) and the financial liability for disclosure, as well as other conditions regarding intellectual property rights.

3. Issues associated with patent transactions

3.1 Regulatory law and compliance

Intellectual property law in Sweden should not be seen as isolated from its legislative surroundings. There are considerable connections to competition law, marketing law and other intellectual property laws.

The Medical Products Agency is the Swedish national authority responsible for regulation and surveillance of the development, manufacturing and marketing of drugs and other medicinal products. The regulations in Sweden are harmonised with EU directives and published in the Code of Statutes. Regulations are primarily laid down in the Medicinal Products Act and the Medicinal Products Ordinance, which states among other things that medicinal products must receive a marketing authorisation before those products are put onto the market for sale.

3.2 Commonly disputed matters

All patent cases (both infringement and invalidity) are tried by the same department of the Stockholm District Court, and so this effectively makes the court a specialist patent court. On appeal, patent cases are tried by the Svea Court of Appeal in Stockholm. These two courts are composed of both legally trained judges and experts in the specific technical field, who are well versed in patent law. To date, few cases have been tried by the Supreme Court.

The most commonly disputed patents in Sweden are pharmaceutical and chemical patents, and this is a trend that has emerged during the last few years; patent licences are also disputed to some extent. Most cases end with a judgment of the court and settlements are exceptionally rare. There are approximately as many judgments for infringement as there are judgments for annulment, since action for nullity becomes the first defence when an action for infringement is brought. It is not unusual in a case of infringement that the parties settle the dispute with a licensing deal.

3.3 Competition rules

The Competition Act in Sweden is based on prohibitions and sanctions which follow the same applicable principles within the European Union. The Competition Act contains two main prohibitions: that of anti-competitive cooperation and that of companies with a dominant position abusing their market power.

Cooperation between companies must affect competition to an "appreciable" extent in order to be prohibited, where the assessment is based upon market share and the size of the cooperating companies. Cooperation between small or medium-sized companies, where the products affected by the agreement comprise a small proportion of the relevant market, normally falls outside the prohibition. Specialisation agreements, joint production agreements, exclusive agreements or patent licensing agreements may limit the output onto the market and may be prohibited if they are deemed to be particularly anti-competitive.

The classic form of abuse consists of a patent holder trying to exploit its strong position to impose on the contracting party a number of conditions that go far beyond what is reasonable – for example, purchase restrictions or lifelong contracts without termination clauses.

3.4 Securitisation of patent rights

Securitisation normally refers to the pooling of different financial assets and the issuance of new securities backed by those assets. In principle, these assets can be any claims that have reasonably predictable cash flows, or even future receivables that are exclusive. Thus securitisation is, for example, possible for future royalty payments from licensing a patent.

Collateralising commercial loans and bank financing by granting a security interest in patents is possible but not very common in Sweden.

3.5 The effect of insolvency

Swedish legislation and case law are limited and inadequate to some extent when it comes to intellectual property rights and insolvency, and they only contain some isolated regulations. There is no case law to date, and the answer to whether a transfer or licence agreement is protected from third-party claims must be sought on general principles and legal doctrine.

According to the legal doctrine, agreements on intellectual property rights are protected from claims made by third parties or creditors upon the signing of the agreement. Under Swedish law, it is prohibited for an official receiver to terminate a licence agreement on bankruptcy grounds, unless it is stated in the agreement or unless a material breach has been committed by the other party. If such is the case, the other party may claim damages, which will have the character of a non-prioritised claim in the bankruptcy.

Patents which have been granted or which have effect in Sweden may also be pledged. A pledge of a patent or patent application arises by registration of a written contract pledging the property. The application for registration is made with the PRO. Where a registered pledge has been transferred to another party, this transfer must be recorded upon request in the Patent Register or the Register of Patent Applications.

3.6 Taxation

From a Swedish tax perspective, patent transactions are treated as sales of fixtures. The consideration received upon a sale represents a part of the company's overall result, which is currently (2014) taxed at a flat rate of 22%. Provided that the patent has been reported as an asset in the balance sheet, the patent's tax-residual value is deducted as a cost. There are no additional stamp taxes.

3.7 Hot topics

The hot topic affecting Sweden in 2014 is the streamlining of EPC procedures, and other work to unify the approach on patents across Europe.

As mentioned earlier in the chapter, it is possible to apply for a European patent under the rules of the European Patent Convention. The patent applicant must, however, pay a patent fee and provide a translation (in most cases) in each country in which the innovation is intended to be protected. Moreover, disputes regarding European patents have to be tried in each country where the patent is valid. This is a rather cumbersome set of arrangements, and in order to mitigate the situation the European Parliament decided in December 2012 to adopt new regulations on a unitary patent system. The new rules (likely to be implemented in 2015 or 2016) will make it possible to validly file for a patent in almost the entirety of Europe with just one application.

The European Council has decided on the establishment of a specialised patent court, namely the Unified Patent Court, and in due course this will create a more uniform set of case law on patents within Europe. The court will have exclusive jurisdiction for litigation relating to European patents under the European Patent Convention, as well as unitary patents – see the preceding paragraph and for further details see the earlier chapter, "Looking ahead to the unitary patent", in this publication. The court will include a Court of First Instance and a Court of Appeal. The Court of First Instance will act through a central division seated in Paris, with two sections in London and Munich. Moreover, several local divisions for single countries or regional divisions will be established. Together with Denmark, Estonia, Finland, Latvia and Lithuania, Sweden has negotiated for a Nordic-Baltic regional division, and in March 2014 Estonia, Latvia, Lithuania and Sweden signed an agreement for that division to be set up. According to the Swedish minister of justice, this regional division guarantees important accessibility for patent holders in the region, as well as facilitating judges and counsel to retain and develop their knowledge concerning patent law. The Nordic-Baltic regional division will be seated in Stockholm with English as its official language.

Switzerland

David Broenimann
Christoph Heiz
Simon Holzer
Kilian Schärli
Andrea Sieber
Meyerlustenberger Lachenal

1. Patents, SPCs and know-how: foundations for transactions

1.1 Patents and supplementary protection certificates

(a) *Patenting in Switzerland*

A patent is a protective right for a technical invention. 'Invention', in this legal sense, means a solution to a technical problem. First, to be patentable, an invention must be novel. An invention is considered novel when it is not part of the prior art. Knowledge which belongs to the prior art is any information which has been made accessible to the public in any form anywhere in the world before the date of the patent filing. Secondly, the invention must not be derived from the prior art in an obvious way by a person skilled in the art.[1] Unexpected characteristics of products or surprising effects of processes are indications that this criterion has been satisfied. An inventive step may also exist if the solution to the technical problem overcomes a prejudice – "things just don't work that way" – or satisfies a pre-existing need. Thirdly, the invention must be commercially applicable, implementable and reproducible.

Switzerland and Liechtenstein form a unified territory for patent purposes and have the same patent regulations based on a bilateral patent treaty. This bilateral treaty provides common legislation in the field of registration and validity of patents for Switzerland and Liechtenstein. A Swiss patent is valid in Liechtenstein in the same way that a patent in Liechtenstein is valid in Switzerland. Designation of one country under the European Patent Convention implies the other; they cannot be designated separately.[2]

One possible way of filing for a patent that is valid in Switzerland is by applying to the Swiss Institute of Intellectual Property (the Institute). National patent applications are subject to limited examination only. The Institute does not examine novelty and inventive step as part of the national examination process.[3] However, since these two criteria are essential for patentability in Switzerland, an applicant should research the novelty and inventive step himself. As of January 2014 the fees

1 Articles 1 and 7 of Switzerland's Patent Act.
2 Supplementary Agreement between the Swiss Confederation and the Principality of Liechtenstein to the Agreement dated December 22 1978 concerning the protection of patents (hereinafter the Patent Protection Treaty; SR 0.232.149.514.0).
3 Article 59(4) of the Patent Act.

for a national Swiss patent are CHF200 (about USD225) for filing and CHF500 for the examination. Annual patent fees will be due from the fourth year after filing, starting with CHF100, and will then increase annually by CHF50.[4]

In addition, Switzerland is a member of the European Patent Convention. So, secondly, patent protection can be obtained through an application to the European Patent Office (EPO). Since European patent applications are examined for novelty and inventive step, this way leads to a fully examined patent in Switzerland.[5]

Thirdly, the Patent Cooperation Treaty provides the possibility of filing a single international application for a patent in any of the treaty member states. This international route consists of a centralised filing and search procedure. The international patent application is the subject of an international search by an authority specialising in this area, the results of which are made available to the applicant in an international search report. The applicant may opt to request an international preliminary examination, which may be considered to be an expert opinion on the patentability of the patent application. The national or regional offices designated in the application are then responsible for granting the patent within their area of jurisdiction. Switzerland may be designated either via the national route or via the regional route (ie, through a European patent application).[6]

(b) (Unitary) European patents

The Unified Patent Court (UPC) will be a court common to all the contracting member states of the Agreement on a Unified Patent Court and Statute, as signed on February 19 2013. The UPC will become operational after ratification of the agreement by 13 countries, including the three countries where most European patents are registered: France, Germany and the United Kingdom. After a transitional period, it will have exclusive competence in respect of European patents and European patents with unitary effect. The UPC's rulings will have effect in the territory of those contracting member states that will have ratified the above-mentioned UPC agreement at the given time.

As a result of an opinion by the European Court of Justice dating from 2011,[7] the UPC agreement is not open to states outside the European Union. Thus Switzerland and other non-EU states belonging to the European Patent Convention may not join the new European patent jurisdiction. European patents with validity in Switzerland and Liechtenstein will continue to fall under the jurisdiction of the newly established Swiss Federal Patent Court that became operational in January 2012. Because of the head start the Swiss Federal Patent Court has on the European patent court and the attractive procedural rules it offers (eg, cases may be presented in English), patent owners of Swiss patents or so-called 'classic' European patents already have the option today to go before a competent court in Switzerland composed of technically as well as legally trained specialist judges.[8]

---

4 "New Fees – Regulation of the Swiss Institute of Intellectual Property", effective from January 1 2014.
5 European Patent Convention, revised in Munich on November 29 2000 (SR 0.232.142.2).
6 Contract on International Cooperation on the Patent Cooperation Treaty (SR 0.232.141.1).
7 Opinion 1/09 of March 8 2011.

(c) *Supplementary protection certificates*

In order to at least partially compensate for the loss in the effective period of patent protection caused by the delay in obtaining an official marketing authorisation for pharmaceutical drugs and plant protection products, the Institute can grant supplementary protection certificates (SPCs) which extend the patent protection period for active ingredients or combinations of active ingredients by up to five years. An SPC is granted for a product if, at the time of the application, the product itself, a process for its preparation or a use thereof is protected by a patent (either a Swiss national patent or the Swiss part of a European patent) and an official marketing authorisation has been granted for the product to be marketed in Switzerland.[9]

The certificate takes effect on expiry of the maximum term of the patent for a period equal to the period which elapses between the date of filing and the date of the first authorisation to place the product on the market as a medicinal product in Switzerland, minus five years. It is valid for no more than five years.[10]

The current law does not provide for a paediatric extension, but this will most likely change in the course of the current revision of the Swiss Therapeutic Products Act. In Switzerland as in other jurisdictions, efforts are being made so that more medicinal products appropriate for children will be authorised and brought onto the market. To offset the additional work involved in the development of medicinal products for children, the new Swiss provisions particularly provide for a six-month extension to the duration of the SPC, which will be made dependent on the submission of studies that have been carried out according to an agreed paediatric investigation plan. In order to align the Swiss level of protection to the one in the European Union with regard to medicinal products for children, the incentives for the development of such products are planned to be added to the Patent Act in the course of the upcoming revision of the Therapeutic Products Act. In particular, it is proposed and planned for 2016 to introduce a new Article 140n to the Patent Act, which grants a six-month extension to the duration of a granted SPC, provided that the authorisation to place the product on the market as a medicinal product in Switzerland contains a statement that the information concerning the medical product reflects the results of all studies that have been carried out in compliance with the paediatric investigation plan considered during the marketing authorisation.[11]

Since Switzerland is a member of neither the European Economic Community (EEC) nor the European Economic Area (EEA), an approval for marketing of the product in the EEC or EEA has no effect on the granting or duration of an SPC in Switzerland. Hence, the copy of the first official marketing authorisation, which must be enclosed with the application for the SPC, must be the first marketing authorisation of the product in Switzerland.[12]

---

8 Federal Act on the Federal Patent Court (Patent Court Act; SR 173.41).

9 Article 140a and following of the Patent Act.

10 Article 140e of the Patent Act.

11 K Schärli, "The supplementary protection certificate for medicinal products, with special consideration of the requirements for protection and the scope of protection in Switzerland and in the EU", Schulthess Juristische Medien AG, Zurich *et al.*, 2013, N 58 and following.

In contrast, Liechtenstein is a member of the EEA and since Liechtenstein (as a basic principle) automatically accepts Swiss marketing authorisations for pharmaceutical products, the Court of Justice of the European Union (CJEU) held in several decisions that a Swiss marketing authorisation triggers the deadline for the application for an SPC in the EEA.[13] However, as a reaction to these decisions of the CJEU, Switzerland and Liechtenstein entered into an agreement on July 1 2005 that delays the automatic recognition of Swiss marketing authorisations in Liechtenstein for new chemical entities. This agreement has been extended several times and remains in force. The automatic recognition of Swiss marketing authorisations in Liechtenstein is now delayed by a time period of 12 months and at least until the grant of a marketing authorisation for the same new chemical entity in the EEA.[14]

Despite the fact that the Swiss legislature decided to adopt the same substantive SPC solution as was chosen in the European Union,[15] there are important differences. First, the key term 'products' is interpreted slightly differently. In Switzerland, the mere existence of identical active ingredients does not necessarily mean that they are the same product. If, based on an already known product, a new invention is made (such as a more effective dosage regime or the treatment of a new indication with a known substance) and if there is a product claim for this new invention in the basic patent, the new invention may constitute the basis for a new, independent product for which an additional SPC can be granted.[16] Secondly, the Swiss and the European granting practices for combination products diverge on the question of whether a product for which an SPC has been applied is protected by a basic patent. For testing whether the product is protected by a basic patent, the testing in Switzerland embraces the scope of protection of the basic patent. Thus, the product has to be examined to determine whether it would fall within the scope of protection of the basic patent (the so-called 'infringement test'). Pursuant to the CJEU, however, an SPC can only be granted for active substances and combinations of active substances which are specified or identified in the wording of the claims of the basic patent. As a consequence, in certain circumstances (eg, combination products) different SPCs may be granted based on the same basic patent in Switzerland and in the European Union.[17]

---

12 "Guidelines for the examination of national patent applications from Swiss Federal Institute of Intellectual Property", dated July 1 2011, N 13.2.2, p108 (available in German and French).

13 Judgment of the CJEU of April 21 2005, *Novartis AG v Comptroller-General of Patents, Designs and Trade Marks for the United Kingdom* (C-207/03) and *Ministre de l'Économie v Millennium Pharmaceuticals Inc* (C-252/03); order of the CJEU of November 14 2013, *AstraZeneca AB v Comptroller General of Patents, Designs and Trade Marks* (C-617/12).

14 For further details see K Schärli, "The supplementary protection certificate for medicinal products, with special consideration of the requirements for protection and the scope of protection in Switzerland and in the EU", Schulthess Juristische Medien AG, Zurich *et al.*, 2013, N 287 and following.

15 "Official Notice on the Patent Act 1993", p711 and following.

16 *Arzneimittel* decision of the Swiss Federal Court dated November 17 1998, 2/1999, 153 and following; *Cyclosporine* decision of the RKGE dated April 30 1999, 4/1999, p449 and following; "Guidelines for the examination of national patent applications from Swiss Federal Institute of Intellectual Property", dated July 1 2011, N 13.1, p106 and following; K Schärli, "The supplementary protection certificate for medicinal products with special consideration of the requirements for protection and the scope of protection in Switzerland and in the EU", Schulthess Juristische Medien AG, Zurich *et al.*, 2013, N 73 and following.

Swiss courts are not bound by the decisions of the CJEU. Nevertheless, with regard to SPCs the Swiss Federal Administrative Court held that a Swiss legal provision that was copied from European law on purpose must be interpreted in conformity with European law, subject to the Swiss methodology of legal analysis.[18]

As from January 2014 the fees for an SPC are CHF2,500 for registration. Annual renewal fees will be due from the first year after filing, starting at CHF950 and increasing annually by CHF50.[19]

1.2 Who owns the patent?

For national Swiss patent applications and patents, the rules regarding ownership are set forth in the Federal Act of June 25 1954 on Patents for Inventions.[20] Pursuant to Article 3(1) of the Patent Act, the inventor, his successor in title, or a third party owning the invention under any other title has the right to the grant of the patent.

The inventor is the person who has actually created the invention or, expressed otherwise, discovered the patent's teaching. Given that the actual creation of an invention is decisive, legal entities such as companies cannot be inventors in terms of the Patent Act; indeed, inventors can be natural persons only. When several persons are involved in the creation of an invention, these persons may well be regarded as co-inventors. However, co-inventors have to have contributed brain work to the solution of a technical problem; it is not sufficient that they have simply carried out the instructions of other persons (eg, the inventors) or supported an invention (eg, by funding).

(a) Employees' inventions

Regarding third parties who may own an invention and therefore have the right to the grant of a patent,[21] the most important cases are those where an employee creates an invention in the course of his work for his employer (see Article 332 of the Swiss Code of Obligations).[22] In respect of such inventions (whether patentable or not), it is necessary to distinguish between inventions created in performance of contractual obligations and inventions made outside the scope of those obligations. Whereas the former belong to the employer, the latter remain with the employee unless the employer has reserved the right to acquire such inventions in the employment contract. In this last case, the employee who created an invention is obliged to notify

---

17 *Fosinopril* decision of the Swiss Federal Court dated July 10 1998, BGE 124 III 375 and following; *Panitumumab* decision of the Federal Administrative Court dated October 18 2011, B3245/2010; "Guidelines for the examination of national patent applications from Swiss Federal Institute of Intellectual Property", dated July 1 2011, N 13.2.1, p107 and following; K Schärli, "The supplementary protection certificate for medicinal products, with special consideration of the requirements for protection and the scope of protection in Switzerland and in the EU", Schulthess Juristische Medien AG, Zurich *et al.*, 2013, N 218 and following; Schärli/Holzer, "ECJ: supplementary protection certificates for combinations products? Judgments of the ECJ of 24 November 2011 'Medeva' (C-322/10) and 'Georgetown' (C-422/10)", 2012, p284 and following.

18 *Etanercept* decision of the Federal Administrative Court, September 13 2010, BVGE 2010/46, consideration 3.

19 "New Fees – Regulation of the Swiss Institute of Intellectual Property", dated January 1 2014.

20 Known in brief as the Patent Act, SR 232.14.

21 See Article 3(1) of the Patent Act.

22 Federal Act of March 30 1911 on the Amendment of the Swiss Civil Code, Part Five: The Code of Obligations (SR 220).

his employer, which, in return, has to inform the employee within six months if it wishes to acquire the invention (for a reasonable remuneration) or to release it to the employee.[23]

(b) *Other service contracts*

If an inventor makes his invention under a service contract other than an employment agreement (eg, under a contract for work and labour, or under an agency contract), Swiss law lacks specific rules concerning the transfer of the invention from the inventor to the customer. Without a clear rule in the contract it depends on the circumstances of the individual case whether the invention remains with the inventor or whether it is automatically transferred to the customer.[24] As a rule of thumb, under an agency agreement the invention is often seen to belong to the principal, whereas under a contract for work and labour it often remains with the inventor. Because of the huge uncertainties regarding the attribution of an invention in general, it is crucial to provide clear rules in the respective contract, whatever its legal nature might be.

(c) *Joint inventorship*

Where several inventors have made an invention jointly, they own the right to the patent jointly.[25] Swiss doctrine disagrees on the legal treatment of joint inventions. While most authors have previously argued in favour of either a co-ownership or a joint ownership similar to the general provisions on property set forth in the Swiss Civil Code of December 10 1907 (the Civil Code),[26] the prevailing doctrine nowadays argues in favour of a patent-specific co-ownership *sui generis*: each co-owner may independently dispose of his share in the right to the grant of a patent and his right to the patent.[27] Since it might often not be practicable to assess the contributions of each co-inventor, the share of each co-inventor in a patent will mostly be an equal share. Furthermore, every co-owner retains the individual right to bring an action for infringement of the patent or to pay the renewal fee (even if not supported by the others). However, other rights may only be exercised with the consent of all other co-inventors, including (but not limited to) the right to apply for the grant of a patent, to grant licences to third parties or even to use the invention on his own.[28] In any event, it is of the utmost importance to agree on rules regarding the rights of co-inventors in a (potential) future invention in the respective contracts, as otherwise considerable legal uncertainty or even stand-offs might arise.

The situation is different where two or more inventors have made the same invention independently of each other. For such cases, the Patent Act sets forth that the person who makes the earlier application or whose application has the earliest

---

23 It is noteworthy that similar rules usually also apply to employment relationships which are governed by public law – eg, those of employees of the Swiss Federal Institutes of Technology in Zurich and Lausanne (see Article 36 of the Federal Act of October 4 1991 on the Federal Institutes of Technology) or of universities which are governed by cantonal law.

24 Article 393 and following and Article 363 and following, respectively, of the Code of Obligations.

25 Article 3(2) of the Patent Act.

26 Article 646 and following and Article 652 and following, respectively, of the Civil Code (SR 210).

27 Article 33(2) of the Patent Act.

28 Articles 33 (2) and 34(2) of the Patent Act.

priority date has the right to the grant of a patent (the so-called 'first to file' system as opposed to any 'first to invent' system).[29] The slower applicant will have no right in the patent at all and might even be prevented by the patent owner from using the invention, unless the invention had been commercially used in Switzerland or special preparations for that purpose had been made before the filing or priority date of the patent application.[30]

1.3 Inventor compensation

In cases where the inventor and the patent owner are different persons, the question arises whether or not an inventor shall be entitled to special remuneration for the invention. In this context, the most important case is the one where employees are creating inventions in the course of their work for their employer. If inventions are created in performance of contractual obligations (eg, often in cases where inventions are created by employees who are working in the research and development department of a company), no remuneration from the employer will be due, unless agreed on in the employment contract, because the inventive activities are considered to be the performance due under the employment contract and therefore covered by ordinary salary payments. In contrast, if the invention was created outside the performance of contractual obligations and the employer decides to acquire the invention based on a contractual right to do so, the employee is entitled to an adequate additional remuneration for the invention.[31] This right cannot be waived by the employee in advance.[32] The amount of the remuneration depends, among other things, on the economic value of the invention, the contribution by the employer (including the involvement of the employer's staff or the use of its facilities) and the efforts made by the inventor.[33]

The issue of inventor compensation might also arise in cases of inventions that have been made under other service contracts – for instance, under agency contracts or contracts for work and labour.[34] In this context, the wording of the respective agreement is crucial. There is no statutory provision that provides for automatic remuneration for such inventions. Therefore, it is of the utmost importance not only to attribute ownership in the invention but also to determine whether remuneration is due and, if so, in what amount.

1.4 Know-how and trade secrets

In Switzerland there is a long tradition in regard to protection of confidential

---

29 Article 3(3) of the Patent Act.

30 Article 35(1) of the Patent Act.

31 Article 332(2) of the Code of Obligations.

32 Article 362(1) of the Code of Obligations.

33 "Guidelines of the German Minister of Labour for the Remuneration of Employees' Inventions in the Private Sector", dating from July 20 1959 and available online at www.verwaltungsvorschriften-im-internet.de/bsvwvbund_20071959_IIa4.htm (last accessed December 23 2013) are sometimes also referred to in Swiss legal doctrine. However, it should be kept in mind that, pursuant to the German Act on Employees' Inventions, dating from July 25 1957 and available online at www.gesetze-im-internet.de/arbnerfg, (last accessed December 23 2013), employees are entitled to remuneration even in respect of inventions created in performance of their contractual obligations.

34 Article 393 and following and Article 363 and following, respectively, of the Code of Obligations.

information. Therefore, different provisions governing confidential information can be found in contract law, in unfair competition law and in criminal law. At the same time, this means that there is no general definition of confidential information that applies in all legal fields in Switzerland.

Article 321a(4) of the Code of Obligations prohibits employees from disclosing or making use of confidential information or trade secrets. The confidentiality of production and business secrets is protected by Article 162 of the Criminal Code of December 21 1937.[35] Espionage relating to such secrets for foreign governments or organisations or for private companies is also subject to criminal sanctions under Article 273 of the Criminal Code. Professional secrets (eg, those of lawyers or doctors) are protected by Article 321 of the same code. The use of confidential information and trade secrets can also qualify as unfair competition under the Federal Law of December 19 1986 on Unfair Competition (hereinafter the Unfair Competition Act),[36] particularly if the information is used by third parties.

Pursuant to the views expressed by courts and legal writers, the definition of 'confidential information' contains three key elements:

- the information must be confidential (ie, it must not be disclosed to the public);
- it must have commercial value because of its confidentiality; and
- the holder of the confidential information should have made reasonable efforts to keep it confidential.

1.5 Employees, consultants and protecting know-how

According to Article 321a(1) of the Code of Obligations, each employee must loyally safeguard his employer's legitimate interests. This includes an obligation not to make use of or inform others of confidential information that comes to his attention while he is in his employer's service. Even after termination of the employment relationship, an employee must continue to be bound to the secrecy obligation to the extent required to safeguard the ex-employer's legitimate interests.[37]

Similarly, an agent's general liability for faithful performance includes the same obligations as those lying upon the employee under an employment contract.[38] Further, in the case of an agency contract, the agent may not exploit, or inform others of, the confidential information with which he has been entrusted, or of which he has obtained knowledge in the course of his agency relationship. These obligations shall apply even after termination of the contract.[39]

Swiss civil law contains a number of sanctions where confidential information is published or used in breach of the implied obligations of confidence. There is the right to file for preliminary injunctive relief and permanent injunctions, as well as an entitlement to compensation of damages and/or accounting for and payment of profits.

---

35 SR 311.0.
36 SR 241.
37 Article 321a(4) of the Code of Obligations.
38 Article 398(1) and (2) of the Code of Obligations.
39 Article 418d of the Code of Obligations.

2. Context and features of patent transactions

2.1 The principal dealings in a patent

According to Swiss law, the assignment of a patent application and of a patent is only valid if evidenced in writing and signed by both parties. Patents may be transferred without the transfer being recorded in the Swiss patent register; however, until an entry is made, any action provided for in Switzerland's Patent Act may be taken against the former proprietor of the patent. In addition, licensees acting in good faith can still render their contractual performance to the former patent owner.

The patent applicant or the patent owner may grant third parties the right to use the invention (the grant of licences). Where a patent application or a patent is owned by two or more persons, a licence may not be granted without the consent of all entitled persons. Licences of patents are not subject to particular formalities; however, it is advisable to have a written licence agreement. Licences of third parties not recorded in the Swiss patent register are invalid against persons who have acquired in good faith the rights to the patent. Any person who holds an exclusive licence, irrespective of the registration of the licence in the Swiss patent register, is entitled to bring legal action against a potential patent infringer provided this is not expressly excluded by the licence agreement.

An 'exclusive' licence means that no other person or entity other than the licensee is allowed to exploit the relevant patent rights. Importantly, the licensor is also excluded from exploiting these rights. A 'non-exclusive' licence means that the licensor remains free to exploit the same intellectual property and to allow other licensees also to exploit the same patent. A 'sole' licence typically means that the licence is exclusive, except that the licensor also reserves full rights to exploit the intellectual property itself.

Unless stipulated in the licence agreement, or where it results from the nature of the licence that sub-licences are inevitable and have been planned from the very beginning, a licensee may not grant sub-licences.

Any person who holds an exclusive licence is entitled to bring legal action against potential patent infringers independently, provided this is not expressly excluded by the licence agreement.

2.2 Patent rights in corporate transactions

Patents, patent applications, licence rights and other intellectual property rights are often a substantial component of a company's overall value and they therefore play a crucial role in merger and acquisition (M&A) transactions. Such M&A transactions can be implemented by way of a so-called 'asset deal' or a 'share deal'. In the case of an asset deal, the purchaser directly acquires legal title to the patents and licence rights along with the title to the target's other assets. In a share deal, the purchaser acquires the shares of the legal entity that holds the patents and licence agreements. Consequently, in an asset deal, legal title to the patents and licence rights passes from the seller to the purchaser; in a share deal, however, only the beneficial ownership of the patents and licence rights changes from the selling owner to the purchaser.

(a) ***Asset deals***

Traditional asset deals are challenging under Swiss law. They require that the title to each asset be transferred individually pursuant to the specific requirements applicable to the transfer of each type of asset. For the transfer of title to patents, an assignment needs to be signed by the seller in favour of the purchaser, which will subsequently apply to register the change of title in the patent register. Under Swiss law, the written assignment is sufficient to legally effect the change in ownership, and the subsequent amendment of the registration in the patent register is only of a declaratory nature and aims at informing the market that the title to the patent has passed to another. In the case of the sale of a portfolio of patents, a written assignment relating to each patent is required to effect the transfer.

In an asset deal, the first step for the sale of a patent consists in the execution of the (global) asset sale-and-purchase agreement; under the Swiss concept of contract, the execution of such an agreement forms the legal obligation for one party to sell and the other party to purchase the assets. The completion of the sale-and-purchase transaction is effected at the closing where the patents are assigned and transferred.

It is advisable in an international transaction that the assignments and transfer agreements meet the requirements set forth by the jurisdiction under which the patent is registered. This structure may lead to a multi-jurisdictional agreement: whereas the asset sale-and-purchase agreement is governed by Swiss law, the assignment and transfer agreements on the patents are usually governed by foreign laws. The international nature of patent rights generally requires the engagement of foreign counsel in each of the jurisdictions where patent applications are pending or patents have been issued. Furthermore, at the closing, the seller typically hands over to the purchaser a written declaration confirming the transfer of title to the patents. This declaration is useful for the subsequent registration of the new owner in the patent registers.

Licence rights may be a substantial part of the business to be sold. Under Swiss contract law, licence agreements may only be transferred with the prior consent of the contractual counterparty, regardless of whether these agreements contain so called change-of-control clauses.

Compared with the traditional asset deal just described, Switzerland's Merger Act[40] provides for a simplified transfer of assets and liabilities.[41] Under this act, the transfer of the assets and liabilities is effected by operation of law and, consequently, without the need for the prior consent of the counterparty.

(b) ***Share deals***

As described in the introductory paragraph to this section 2.2, an acquisition of the shares of a company holding patents and licence rights does not lead to a legal but to an economic change in ownership in the patents (ie, there is a new beneficial owner). The acquisition of patents and licence rights through a share deal is less complex than through an asset deal. However, should any of the licence agreements contain a change-of-control clause, the consent of the counterparty to the licence

40 SR 221.301.
41 Article 69 and following of the Merger Act.

agreement should be obtained prior to the completion of the transaction unless the purchaser is willing to take the risk that such licence agreement will be terminated by the counterparty.

(c) Dealing with the transaction risks

The risks surrounding a transaction involving the transfer of patents, patent applications and licences mean that a due diligence exercise should always be performed in order to safeguard a potential purchaser's position. Particularly when a technology-driven company or business is to be acquired, the potential purchaser should conduct a thorough analysis of the intellectual property rights owned and utilised by the target company or the seller, as part of the purchaser's pre-investment due diligence review. Such due diligence will enable the purchaser to better assess the value of the target company or target business.

In the course of a due diligence review, the focus will typically be on the following main aspects:

- The purchaser has to review whether the seller or the target company to be acquired has legal title to the patents, the patent applications and the priority rights (if divisional applications are planned), as it is well understood that a seller (in an asset deal) or the target company (in a share deal), respectively, has to have title to the patent rights before it can sell them. Further, a patent owner cannot assert its patent rights against a potential patent infringer unless it is the owner of such rights. It is, however, not unusual that in a group company structure the patents relevant for the purchaser are held by several legal entities of the group, which might have their registered offices in different jurisdictions. With such a structure, should the purchaser be interested in acquiring only a selected number of group companies, the purchaser has to make sure under the relevant jurisdiction that such a company holds legal title to the patent. Typically, other complex issues involving chain of title arise due to the fact that a review of documents filed with the relevant patent register may not in fact disclose the actual owner of the patent in question. In practice, many patents are owned by companies that, over time, have gone through several rounds of corporate reorganisation; as a result, patent rights might have been assigned from one corporate entity to another within the same group. The patent assignments related to these reorganisations may well not be filed in the competent patent registers.
- The purchaser has to verify that all maintenance fees of the patent registers have been paid and all formalities for a due and continued registration of the patents are fulfilled. If such fees have not been paid when due, this could render a patent application irrevocably abandoned or even invalidate an issued patent.
- The purchaser should focus on whether there are any indications of patent infringements by the patent holder on the one hand, or whether there are any infringements of the patent of the target company by third parties on the other.
- Where the patent is of fundamental importance for the business and the existence of the business or company to be purchased, the purchaser is well

advised to verify whether the invention which is the basis for the patent fulfils the requirements to benefit from patent protection (see section 1.1 above) in all the jurisdictions that are or may become relevant for the business operations of the purchaser.

In the due diligence, the purchaser also needs to review carefully patent licence agreements to or from the target company. The purchaser should verify whether the target company retains appropriate rights after completion of the M&A transaction (when the target company is the licensor) or receives appropriate rights (when the company is the licensee). In cases where the target company has entered into a joint development agreement or similar agreements with third parties, a careful review of the rights granted and retained under such an agreement is crucial.

If the M&A transaction is carried out as an asset deal, the transferability of licences granted to the company needs to be analysed. As explained above, contracts are not assignable under Swiss law unless the contractual counterparty (ie, the licensor) has given its consent.

A purchaser of patent assets will also necessarily want to make sure that the sale-and-purchase agreement (regardless of whether it is an asset deal or a share deal) contains representations and warranties from the seller that the patents being acquired exist and are valid and are thus worth the purchase price agreed to as part of the M&A transaction. Specifically, they should state that the seller or the target company holding the patent holds unrestricted title and is registered as owner in the competent patent register. Moreover, the seller should represent and warrant there are no infringements of the patents to be sold. Should there be any infringement claims made by a third party against the patent to be sold or should the patent holder infringe any patents of a third party, such infringement claims should be disclosed. In these situations, the purchaser is well advised to insist on specific remedies and indemnities to be set out in detail in the sale and purchase agreement. The purchaser should also seek to obtain representations and warranties of the seller that the patent is valid and that the patented invention is safe from any challenges of its patent protection. In practice, the seller often gives such a representation and warranty only "to the best of its knowledge" and thus limits the scope.

With regard to the material licence agreements entered into by the target company, the sale-and-purchase agreement should, among other things, contain representations and warranties of the seller that these agreements are valid, have not been terminated or threatened to be terminated, and no counterparty has the right to terminate or modify its rights or obligations under a change-of-control or other specific provision in such an agreement as a result of the transactions contemplated by the sale-and-purchase agreement, and that the target company is not in default under or in material breach of any such agreement.

(d) Joint ventures and consortium agreements

From a Swiss law perspective, it is possible to assign patent rights to a joint venture entity or to the members of a consortium. Patents can also be part of a contribution in kind.

If the shareholders of a joint venture or the members of a consortium provide patent rights, often the joint venture or the consortium does not become the new owner, but only receives a licence to use the patented invention. The scope of the licence might be defined in the joint venture or consortium agreement or in a separate licence agreement.

Due to the lack of specific provisions under Swiss law, the parties to a joint venture or consortium agreement must find appropriate solutions at least for the ownership issue and the right to use of inventions and patents that were generated by the parties before entering into the joint venture or consortium agreement (often called 'background IP') and inventions and patents generated by the joint venture entity or members of the consortium during the term of the joint venture or consortium (often called 'foreground IP'). Sometimes parties also provide specific rules on how to handle 'sideground IP', which is generated during the collaboration phase, but not in project related activities.

2.3 Particular business transactions

(a) *Material transfer agreements*

A material transfer agreement (MTA) governs the transfer of tangible research materials (chemical compounds, biological materials such as cell lines, reagents, plasmids and vectors) between a providing and a receiving party, when the recipient intends to use the material for its own research purposes.

MTAs provide specific provisions defining the transferred material, the scope of and limitations to the planned research, confidentiality rules and delay in publication rules to protect potentially patentable inventions. In addition, they often contain language that prohibits the use of the material in research that is subject to licensing or consulting obligations to any third party, including the sponsor of the research project.

No specific Swiss law allocates the rights in and to the invention arising from research based on material provided under an MTA. The parties to an MTA should therefore provide explicit language to allocate such rights.

(b) *Research and development agreements*

Swiss law allows research and development agreements where an order is given to an undertaking to perform research and development for a principal (non-collaborative research and development agreements) and agreements between two or more parties on research and development on an equal level (joint or collaborative research and development agreements). Even in connection with non-collaborative research and development agreements, Swiss statutory law provides no clear answer to the question of whether the principal acquires all rights in and to possible inventions stemming from the research and development. The parties should thus include specific answers to this question in their agreements.

If universities and universities of applied sciences are involved in research and development, public law normally provides guidelines regarding the allocation of IP rights that result from such activities. The same applies if research and development are (partially) financed by the Swiss Federal Commission for Technology and Innovation (CTI). The CTI encourages the transfer of knowledge and technology between higher-education institutions and businesses through applied research projects. Essentially, the CTI finances salaries. The revised Swiss ordinance on the law on funding of research and innovation[42] imposes limits on how the parties can allocate IP rights and requires that the parties agree on the specific allocation of IP rights in a separate agreement.

Although Switzerland is not a member of the European Union, Swiss antitrust authorities consider the principles of the revised European Block Exemption Regulation for Technology Transfer Agreements when examining the lawfulness of provisions in a research and development agreement dealing with the access to the results of collaborative research and development.

3. Issues associated with patent transactions

3.1 Regulatory law and compliance

(a) *Transfers of marketing authorisations*

If a purchaser acquires a set of patents and is interested in distributing the product that is covered by the patents, the authorisations to do so have also to be transferred. This is of particular relevance in the pharmaceutical sector. A marketing authorisation for the distribution of a pharmaceutical product in Switzerland may be assigned to another company upon written request with Swissmedic at least three months prior to the assignment. That request must be supported by various documents, such as the authorisation of the future holder of the market authorisation, a declaration of assignment signed by the previous holder of the market authorisation, and current excerpts from the commercial register of both companies.[43]

A change of the name and/or domicile of a holder of a marketing authorisation (including the change of the legal form of the company) must be requested at least three months prior to the planned implementation, which also requires various documents, such as the authorisation of the company (see below) and a signed communication regarding the planned change.

(b) *Regulatory data protection*

In the regulated pharmaceutical sector, the protection of innovation is not solely left to the patent system. Regulatory data protection provides protection (ie, exclusivity periods) for the technical data generated by innovator companies for the authorisation of their preparations – that is, data generated by drug firms through preclinical and clinical trials to prove the efficacy and safety of the products concerned. By providing exclusive rights to this data, the relevant regulatory

---

42 SR 420.11.

43 See Swissmedic publication, "Manual transfer of a marketing authorisation and change of name or residence of the marketing authorisation holder", dated January 1 2013.

provisions prevent competitors from obtaining marketing authorisations for low-cost versions during the period of exclusivity.

Current Swiss legislation on regulatory data protection was strongly influenced by the legislation of the European Union at the turn of the millennium, when the current Swiss law was enacted. In Switzerland, subsequent applicants who are not in possession of written permission issued by the holder of the marketing authorisation for the original preparation may apply for a marketing authorisation relying on the data submitted by the first applicant, but not earlier than 10 years after the marketing authorisation was first granted. Under Swiss law, data filed in support of authorisation for new indications, new modes of administration, new preparation forms and new dosages of a known substance are protected for three years. If the applicant proves that the new product brings significant clinical benefits compared with existing therapies, Swissmedic might extend the data exclusivity period to up to five years. Although Swiss law does not contain explicit provisions, Swissmedic takes the view that only the first applicant – that is, only the holder of the first marketing authorisation for the relevant chemical entity – is entitled to data exclusivity of three to five years for new indications, new modes of administration, new preparation forms or new dosages. Second applicants, even if they apply for new indications, new modes of administration, new preparation forms or new dosages of a known substance, do not benefit from regulatory data protection.

The Federal Office of Public Health has issued a preliminary draft amendment to the Swiss regulatory data protection system. The plan is to implement the amendment in the course of the current revision of the Law on Therapeutic Products. The new Swiss regulatory data protection system is intended to implement the European Union's '8+2(+1)' system (i.e. eight years of full data protection, two additional years of marketing exclusivity, and one additional year of marketing exclusivity for new indications of significant clinical benefit). Even though the revised Swiss regulatory data protection system is projected to introduce harmonisation with the EU system, the new Swiss regime leaves noteworthy loopholes. For example, compared with the European law, the proposed revision does not consider the required data protection for new drug combinations or cases of a change of classification of a medicinal product. In addition, the definition of the term 'original preparation' by Swissmedic is still much stricter than the reference products that enjoy regulatory data protection under Directive 2001/83/EC.[44]

3.2 Disputed matters

If Swiss courts have jurisdiction over a dispute in connection with patent transactions, this dispute can be brought before the cantonal courts or the new Swiss Federal Patent Court, which commenced operations on January 1 2012. However, questions of validity and infringement of a patent must be decided by the new Federal Patent Court. Where the nullity or infringement of a patent is to be adjudicated before a cantonal court on a preliminary question or defence basis, the

44 See S Holzer, "Regulatory data protection of medicinal products from a Swiss perspective", *Bio-science Law Review*, Lawtext Publishing Ltd., vol 12 issue 5, 2012, pp184–191.

cantonal judge shall grant the parties a reasonable period to file the validity or infringement action before the Federal Patent Court.[45] The cantonal court shall stay the proceedings until a final and binding decision has been issued on the action. If no action is filed before the Federal Patent Court within the specified timeframe, the cantonal court will resume the proceedings and the preliminary question or defence will be disregarded. No recent case law has been published concerning specific questions regarding patent transactions.

3.3 Competition rules

As a general rule, the Federal Act on Cartels and other Restraints of Competition (hereinafter the Cartel Act)[46] does not apply to effects on competition that result exclusively from the legislation governing IP rights. However, the competitive effects of IP rights which do not arise out of mere existence, but (for example) from the abuse or the licensing of IP rights, may be reviewed by the Swiss competition law authorities. There is no specific legislation dealing with limitations imposed by antitrust law on patent licensing practices and only limited case law is available in this respect. The general rules of the Cartel Act, dealing with anticompetitive agreements (Article 5) and the behaviour of dominant undertakings (Article 7) are applicable.

Other jurisdictions face the problem that drug makers have been able to sidestep competition by offering patent settlements that pay generic companies not to bring lower-cost alternatives to market. These 'pay-for-delay' patent settlements can block generic drug competition for a significant number of branded drugs. However, to date there have occurred neither proceedings nor decisions regarding pay-for-delay cases in Switzerland.

3.4 Securitisation of patent rights

Lending partly or wholly against intellectual property assets is a more recent phenomenon, in particular if the concerned asset generates sufficiently sustainable cash flow. Normally, it is not the patent which is securitised, but rather the income it generates through licence agreements. The royalties stemming from the patent can be pooled and transferred to a special purpose vehicle, which then issues securities in the capital market.

3.5 The effect of insolvency

Swiss bankruptcy law does not deal explicitly with IP rights; the general bankruptcy rules apply. For example, in the event of insolvency, the total assets of the debtor including its IP rights form a single mass.[47]

Article 211 of the Swiss Bankruptcy Law can have a considerable impact on licence agreements, as it provides for the conversion of contractual claims into monetary claims of corresponding value. As a general rule, this leads to the

---

45 See article 26 para. 3 Federal Act on the Federal Patent Court, SR 173.41.
46 SR 251.
47 Article 197 of the Swiss Federal Act on Debt Enforcement and Bankruptcy (SR 281.1).

termination of the contract. However, the administrator of a bankrupt estate is given the choice to prevent conversion by fulfilling a synallagmatic contract. So if the owner–licensor of a set of IP rights falls into bankruptcy, typically the licensee would lose its rights under the licence as these are converted into a monetary claim. If however, the administrator of the bankrupt estate chooses to fulfil the licence contract, the contract remains in force under its original terms.

If the administrator of a bankrupt estate decides to fulfil the licence contract and then sells the IP rights to a third party, legal writers have suggested that the third party acquires the IP rights together with all the obligations resulting from the licence(s) granted. Consequently, the new owner would be bound by the licence agreement. However, no case law exists on this question.

Furthermore, Articles 82 and 83 of the Code of Obligations grant the creditor the possibility of holding back its own contribution in the event of the insolvency of the other contractual party. If the licensor becomes bankrupt, the licensee may retain the licence fee payments, except if there is sufficient surety that the licensee can continue to exploit the licensed rights as before the bankruptcy. If, alternatively, the licensee goes bankrupt, then the licensor may enjoin him from using the licensed rights, unless the licensee gives sureties that the licence fees will continue to be paid.

3.6 Taxation

Switzerland has always been a preferred location for IP holding activities. Besides the generally low tax rates, another reason from a tax perspective is the relationship between taxpayers (and their advisors) and the tax authorities, which is based on mutual trust and cooperation. The Swiss tax-ruling practice is unique in the world: it is best practice to discuss beforehand any critical or uncertain tax-related transaction or structure with the competent tax authorities and to agree on the tax consequences in a written and binding tax ruling. As long as the fact pattern remains unchanged, such a ruling provides legal certainty and the tax authorities adhere to the tax consequences outlined in the ruling.

From an international tax perspective, Switzerland is in a very competitive position for IP holding companies as there is no withholding tax on royalty payments based on domestic law. Thus, based on the principle of reciprocity in many of the more than 80 double-taxation treaties that Switzerland has concluded with other jurisdictions, a full refund of withholding tax is granted to a Swiss IP company.

Other IP-specific tax considerations include the full tax deductibility of R&D cost (also against operating income), tax credits for withholding taxes suffered abroad, and the possibility of applying for a tax holiday for up to 10 years, provided that significant investments are made in Switzerland.

3.7 Hot topics

One of the hot topics in Switzerland is currently the taxation mechanisms.

Switzerland offers highly competitive income tax rates starting at around 12% for corporations. One canton (Nidwalden) applies an 'IP box' regime with an overall effective tax rate of less than 9%.

The IP box regime allows all IP and research and development related activities of a legal entity that has its statutory domicile in the Canton of Nidwalden (in particular, licence income) to be attributed to the IP box within the company. Income generated by the IP box is subject to a reduced tax rate. The definition of licence income is based on the Organisation for Economic Cooperation and Development Model Convention and is defined very broadly.

In other cantons, IP companies usually benefit from the 'mixed company' privilege, resulting in an effective income tax rate of approximately 8–12%. As Switzerland has faced international pressure to abolish the mixed-company privilege, the Swiss government is currently evaluating the possibility of introducing an IP box regime throughout Switzerland. It is likely that such a regime will anyway be introduced by many cantons over the next few years, with the conversion of any mixed-company set-ups in those cantons in a tax-neutral manner, and that the applicable income tax rates will be in the range of today's mixed companies.

Depending on the specific situation of an international group and on the size of the IP operations in Switzerland, tailored solutions may be available which further reduce the effective tax rate. Such solutions include a foreign IP branch of a Swiss corporation or the use of hybrid instruments to finance a Swiss IP company. Also, structuring opportunities are available in order to ensure that no stamp duty is levied upon entry into Switzerland.

United Kingdom

Malcolm Bates
Colin McCall
Louisa Penny
James Stewart
Tim Worden
Taylor Wessing LLP

1. Patents, SPCs and know-how: foundations for transactions

1.1 Patents and supplementary protection certificates

(a) *Introduction to patents*

Patents are a form of intellectual property (IP) that can protect product and process inventions. Once granted, both national patents (for which the publication number is prefixed with "GB") and European patents with a UK designation (publication number prefixed with EP) are registered at the United Kingdom Intellectual Property Office (UKIPO).

The registered proprietor of a patent, or their exclusive licensee, is entitled to prevent third parties from exploiting the product or process technology claimed in the patent for a period of 20 years from the filing date. This is a right of exclusivity to the claimed technology, but not a right to exploit it as such: a patent may fall within the scope of an earlier third-party patent, requiring permission from that third party before the later patented technology can be used.

Typically, a patent has four elements:

- a description – this explains the invention in detail, including the background and context to the claimed technology, and an outline of the 'prior art' (see below);
- examples – these provide details of ways in which the invention has been worked, for example the reactants and experimental conditions for making a compound;
- diagrams – for inventions where illustrations help to understand the nature of the invention (eg, medical devices); and
- claims – these are precisely worded paragraphs that come at the end of the patent and are intended to clearly delineate the scope of the subject matter protected by the patent. There may be one or more claims in a patent. In general, each successive claim in a patent narrows the scope of the invention claimed. The claims are the basis for determining whether a third-party technology is infringing the patent.

(b) *Claims*

Although many patent claims will be easily recognisable as claiming a product or a process, there are forms of claim found in life sciences related patents that are less straightforward. These are as follows:

- *'First medical use' claims:* In English law, claims of this form are understood to mean that the claim protects a product X which is suitable for purpose Y. Such claims are typically seen in the form of a product claim – eg, a pharmaceutical compound for use as a medicament or, more narrowly, for use in the treatment of a specific disease.
- *'Second medical use' claims:* Claims to methods of treatment are prohibited, which was problematic when attempting to claim a newly discovered use of a pharmaceutical (already used for treating a particular disease) for the treatment of a different disease. Either the claims are drafted as product claims limited by use, in which case they are not novel, or they are drafted as process claims for treatment, in which case they are subject to the prohibition. Therefore, in order to claim 'second medical use' for the treatment of a specific disease but avoid the appearance of a method of treatment claim, the European Patent Convention 2000 has sanctioned the use of the claim form "X for use in treating Y" for claims dating from December 13 2007.[1] Prior to this the so-called 'Swiss claim' form of "X in the manufacture of a medicament for treating Y" (a process claim) had been used and patents containing such claims predating the European Patent Convention 2000 are still in force.

However, if a patent claims "X for use in treating Y", how is such a claim infringed? Is it an infringement for an unlicensed third party to manufacture, offer or supply to the market a pharmaceutical X which is not indicated for treatment Y on the patent information leaflet? What if it is subsequently prescribed by a doctor or supplied over the counter by a pharmacist for this purpose? The answer may be that there is an onus on the unlicensed supplier to take steps to alert health professionals and pharmacists that their product is not to be used for an unlicensed purpose. However, an answer to this problem has yet to emerge from the English courts.

(c) Acts of infringement

The scope of the patented technology, as provided for in the claim, is important for determining whether the act of a third party is infringing. Section 60 of the Patents Act 1977 provides that certain acts infringe a patent if they are performed by a third party in respect of a product or patent that shares all the features, or integers, of the patent. It is a critical element of infringement that these acts are performed without the permission of the patent owner (otherwise known as the patentee). The acts are:

- in respect of a product patent: making, disposing of, offering to dispose of, using or importing the product claimed or keeping it for disposal or otherwise;

1 Although the European Patent Convention 2000 form of claim was intended to match as closely as possible the scope of protection of a Swiss claim, the European Patent Office (EPO) Technical Board of Appeal (TBA) decided in case T1780/12 that the Convention form of claim and the Swiss-type claim are not the same: a Swiss type of claim is a purpose-limited process claim; and a Convention type of claim is a purpose-limited product claim.

- in respect of a process patent: using the process claimed or offering it for use in the United Kingdom without permission of the patent holder, providing it can be shown that it was obvious to a reasonable person in the circumstances that its use in the UK without the consent of the owner would be an infringement of the patent; and
- also for process patents: the disposal of, offer to dispose of, use or import of any product that is obtained directly by means of the process claimed, or keeping any such product whether for disposal or otherwise.

The above acts are direct infringements. However there is a further category of indirect, or 'contributory', infringement. This is defined as the supply or offer to supply in the United Kingdom to a party who is neither an owner nor licensee (see next paragraph) to use a patent with any of the means, relating to an essential element of the invention, for putting the invention into effect – provided that it is known, or it is obvious to a reasonable person in the circumstances, that those means are suitable for putting, and are intended to put, the patented invention into effect in the United Kingdom.

All the foregoing acts of infringement require the absence of permission from the proprietor of the patent. Therefore, permission to do these acts is a complete defence to infringement. A licence is essentially an agreement between the patent proprietor (the licensor) and a third party (the licensee) to permit some or all of these acts.

(d) ***Enforcing a patent***

The exclusivity provided by a patent is of little benefit unless it can be enforced against infringers. There are no criminal proceedings available for patent infringement in the United Kingdom and so enforcement requires the patentee, or its exclusive licensee,[2] to bring an action for relief in the civil courts. The remedies that may be awarded to a patentee if they are successful in claiming infringement in proceedings include:

- obtaining an injunction to restrain the infringer from further infringing acts (this can be preliminary or final);
- an order for the infringer to deliver up or destroy any patented product that infringes the patent, or any article of which that product is inextricably a part;
- damages in respect of the infringement or an account of profits derived by the infringer from the infringement; and
- a declaration that the patent is valid yet infringed.

(e) ***Legal requirements for a patent***

The Patents Act 1977 implements the European Patent Convention (the Convention) in English law and sets forth a number of requirements that must be satisfied for a claimed product or process technology to be eligible for patent protection. These requirements are examined further below, but in essence are that the claimed technology:

---

2 Section 67(1) of the Patents Act 1977.

- must be new;
- must involve an 'inventive step' (be non-obvious);
- must be capable of 'industrial application';
- must not fall within the list of excluded subject matter; and
- must be specified in the patent claim in such a way as to disclose the invention in a manner that is clear enough and complete enough for the invention to be performed by a person skilled in the art ('sufficiency').

The invention must be new: To satisfy novelty, the features of the claims in the patent must not have previously been disclosed to the public in any way anywhere in the world before the date at which the application for the patent is first filed (the 'priority date'). The priority date is also the reference date by which the 'state of the art' of the field of technology in question is assessed. The state of the art is taken to consist of all knowledge that is available to the public anywhere in the world.

For this purpose, a disclosure of information needs only to be made to a single person who is not under a duty to keep it confidential in order to amount to being available to the public. The disclosure of the claimed technology to a single person without obligations of confidentiality before filing the patent application is likely to destroy novelty and prevent the grant of a patent. Such disclosures must therefore be avoided – careful attention must be given in particular to managing the publication of research by groups, such as those in universities, if it is likely to be subject to a patent application by the group or a commercial partner.

The invention must involve an 'inventive step': The 'inventive step' in the subject matter being disclosed by the patent must not have been obvious to a person skilled in the art at the priority date. The 'person skilled in the art' is a theoretical creation of patent law, by whom objective assessments of obviousness are made by a patent examiner (during prosecution) or a court (in a patent validity challenge). The skilled person is taken to have background knowledge, called 'common general knowledge', of the area of technology to which the patent relates, at the time of the priority date. However, the skilled person is taken to have no inventive skill of his own.

There are certain issues regarding inventive step that have arisen in life sciences cases:

- a product that has long been sought after (called a 'known desideratum') may still be inventive if it was not obvious how to make it. For example, whilst it was highly desirable to resolve the racemate product citalopram to obtain the enantiomer escitalopram, it was not obvious how to do so and a patent on the product was valid as regards inventive step.[3]
- A claim to a single compound or group of compounds will not be rendered obvious by the earlier disclosure of a wider group of compounds that encompasses it, provided the later selection is not arbitrary but discloses a new technical contribution.[4]

---

3 *Generics (UK) Ltd v H Lundbeck A/S* [2009] UKHL 12.
4 *Dr Reddys Laboratories (UK) Ltd v Eli Lilly & Co Ltd* [2009] EWCA Civ 1362.

- The Court of Appeal has recently held that evidence that post-dates the priority date of a patent may be used to show that the claimed invention actually made no plausible technical contribution and must therefore be obvious.[5] An example of how this may occur in the life sciences context is if later clinical trial data shows the technical contribution described for the product or process claimed to be implausible.

The invention must be capable of industrial application: Being 'capable of industrial application' does not mean it must literally be used in industry or that it should be commercially successful. Instead, the requirement is designed to ensure that the patent system is used for subject matter that has a technical application rather than being an abstract idea.

The requirement of industrial applicability has received greater attention recently in the context of life sciences patents. *Human Genome Sciences v Eli Lilly*[6] concerned a patent disclosing the nucleotide sequence coding for a polypeptide called neutrokine-α (a member of the tumour necrosis factor (TNF) superligand family). Human Genome Sciences had determined the sequence using a computational biology technique called bioinformatics; however, it had no experimental evidence of its activity. Hence, the patent specification could only make speculations about neutrokine-α activity based on shared sequence homology with other members of the same superfamily. The Supreme Court set out a number of points of general principle regarding the industrial applicability of such technology in these circumstances. These include the need for the patent to disclose a "practical application" and a "concrete benefit" that is derivable from the description. But, it was held, these must not be merely speculative; instead, the claimed use must be "reasonably credible", an "educated guess" or "plausible" if the technology claimed is to qualify as industrially applicable. This requirement was judged to have been satisfied in this case.

The invention must not be excluded by law: There are a number of areas of subject matter that are excluded from patent protection. Some of these areas are excluded on the grounds of public morality or ethics, whilst others are excluded because they are considered to be more properly the subject of other forms of intellectual property protection.

The ethical questions associated with stem cells continue to make patents in this area a particular focus of the exclusion provisions, both in the English courts and in the Court of Justice of the European Union (CJEU), under Directive 98/44/EC on the legal protection of biotechnological inventions (the Biotechnology Directive), which the Patents Act 1977 implements in the United Kingdom.

In *Oliver Brüstle v Greenpeace*[7] the CJEU ruled that a process which involves removal of a stem cell from a human embryo at the end of the blastocyst stage,

5 *Generics [UK] Limited t/a Mylan v Yeda Research & Development Co Ltd* [2013] EWCA Civ 925.
6 [2011] UKSC 51.
7 Case no C-34/10.

entailing the destruction of that embryo, cannot be patented. This was on the basis of the exclusion from patentability of the human embryo in Article 6(2)(c) of the Biotechnology Directive. The CJEU held that the concept of 'human embryo' must be understood in a wide sense. Hence, any human ovum must be regarded as a human embryo as soon as it is fertilised, if that fertilisation is such as to commence the process of development of a human being. A claimed process is also excluded from patentability if its implementation requires either the prior destruction of human embryos or their prior use as base material, even if the claimed process itself does not refer to the use of human embryos.

A further reference concerning the patentability of stem cells has now been made to the CJEU for a preliminary ruling by the English Patents Court. This is through the *International Stem Cell Corporation v Comptroller General of Patents* case,[8] which seeks clarification on whether stem cells obtained from parthenogenetically stimulated human ova are excluded from patentability as human embryos, given the wide definition in *Brüstle* and the apparent indication by the CJEU in that case that parthenogenic cells are excluded on the (incorrect) assumption that they are capable of developing into a human being. The reference asks whether the technology used by the International Stem Cell Corporation is excluded from patentability, given that the parthenogenetic activation of oocytes does not actually involve the destruction of a fertilised ovum capable of developing into a human being.

The invention must not be insufficient: Section 72(1)(c) of the Patents Act 1977 requires a patent to disclose the invention in a manner sufficiently clear and complete for it to be carried out by a skilled addressee in the art. If a patent claims more than it enables a skilled person in the relevant art to achieve using their common general knowledge, then it is invalid by reason of insufficiency. Put another way, the patent must disclose enough information for the skilled addressee to perform everything that falls within the scope of the patent claims without undue burden. Alternatively, the patent may disclose a principle of general application that can be performed to obtain everything claimed across the scope of the claim.

However, in the case of product claims, the patent does not have to teach more than one way in which to make the product, regardless of whether several methods are possible. Complications can, however, arise in the context of mixed claims in which the product is defined according to a general method by which it was obtained (for example, recombinant technology). In such cases the patent may be insufficient if there are processes for making it that fall within the general process element, but owe nothing to the teaching of the patent.[9]

In 2013, a number of authoritative decisions in the UK courts embraced a significant principle of the European Patent Office (EPO) Technical Board of Appeal case law (from case T939/92 on *Triazole/AgrEvo*), which says that patentability requires the claimed subject matter to make a plausible technical contribution to the art. These decisions have extended the application of this principle beyond its EPO

8 [2013] EWHC 807 (Ch).
9 *Generics (UK) Limited v H Lundbeck A/S* [2009] UKHL 12.

context of obviousness, to include insufficiency cases. As such, a new threshold of patentability has been introduced into English cases, which the court may consider as a first step when addressing inventive step and sufficiency arguments.

(f) Supplementary protection certificates

A system of extending the term of protection for pharmaceutical products after patent expiry, by so-called 'supplementary protection certificates' (SPCs), was introduced in 1993 by European Council Regulation (EEC) 1768/92 (the SPC Regulation). The purpose of SPCs is to remedy the common problem that patent term for many pharmaceutical products can be significantly eroded by the time taken for a marketing authorisation to be granted – limiting exclusivity on the market and therefore commercial return. An SPC can alleviate this problem by providing up to five years of further monopoly protection for specific 'products' (see below) that were covered by the expired patent.

In order to obtain an SPC, there are three key conditions that must first be satisfied:[10]

- The product must be protected by a basic patent in force.[11] The purpose of this provision is to reflect the intention of the SPC legislation to prolong monopoly protection for products that are already patented.
- A valid authorisation to place the product on the market as a medicinal product must have been granted.[12] This must be the first authorisation to place the product on the market as a medicinal product.[13]
- The product must not already have been the subject of an SPC.[14] The purpose of this provision is to prevent further SPCs for modifications of the same product.

It should be emphasised that an SPC does not prolong the term of a patent across its entire claim scope. Hence, many other potential products that fall within the scope of protection of the basic patent will not be covered by the SPC.

Until recently, it was thought that SPCs could only protect products and the products of processes. However, the ruling of the CJEU in *Neurim*[15] has appeared to open up the possibility of obtaining SPC protection for second medical uses of products even when the active ingredient at issue has already been the subject of an earlier marketing authorisation. The reasoning of the decision was based largely on the understanding that the SPC Regulation is intended to protect the investment made in pharmaceutical research to discover new therapeutic uses of products and so should be interpreted to provide such protection. Whether this reasoning is limited only to second medical use remains to be seen.

---

10 Article 3 of the SPC Regulation.
11 Article 3(a) of the SPC Regulation.
12 Article 3(b) of the SPC Regulation.
13 Article 3(d) of the SPC Regulation.
14 Article 3(c) of the SPC Regulation.
15 *Neurim Pharmaceuticals (1991) Limited v Comptroller General of Patents* (Case C130/11).

1.2 Who owns the patent?

(a) *The general rule*

It is the general rule that the deviser (inventor) of the technology claimed (the invention) is the person entitled to the grant of a patent for that invention.[16] However, any one or more persons can file a patent application and, in the absence of a third party establishing the contrary, they will be assumed to be entitled to its grant.[17] However, the entitlement to the invention may be transferred to a third party by agreement made before the making of the invention.[18] The right to a patent may similarly be transferred at any time after the making of the invention – whether before or after the filing of the application or its grant. In any case, the inventor retains the right to be mentioned in the granted patent.[19]

(b) *Employee-inventors*

The above rules can give rise to disputes about proper entitlement to the ownership of a patent. Such disputes can be referred to the Comptroller of the UKIPO for determination, whether before or after the filing of the application or after grant. The Comptroller can make an order to give effect to their determination, such as a transfer of title or the grant of a licence, or refer the matter to the court.[20]

In the employment context, an invention made by an employee-inventor belongs not to the employee but to his employer, provided that:

- it was made in the normal course of the employee-inventor's duties, or otherwise duties specifically assigned to the employee; and the circumstances in both cases are such that an invention could reasonably be expected to result from those duties; or
- the invention was made in the course of the employee-inventor's duties, and that employee had particular responsibilities arising from the nature of his duties to the employer (eg, as a director), giving him a special obligation to further the employer's interests.[21]

Inventions made in circumstances outside those described above will belong to the employee-inventor. However, it is always open to the employer and employee-inventor to enter into an agreement transferring ownership of the patent to the employer.

(c) *Contractor-inventors*

Unlike an employee-inventor, a contractor who makes an invention pursuant to his contract will be entitled to ownership of the invention unless it has an agreement

16 Section 7(2)(a) of the Patents Act 1977.
17 Section 7(4) of the Patents Act 1977.
18 Section 7(2)(b) of the Patents Act 1977. A party other than the inventor may also be entitled by virtue of any enactment or rule of law, or any foreign law or treaty or international convention – see Section 7(2)(c).
19 Section 13(1) of the Patents Act 1977.
20 Sections 8 and 10 of the Patents Act 1977.
21 Section 39(1) of the Patents Act 1977.

with the commissioning party to the contrary. However, even if ownership is not expressly transferred, a licence to use the invention may be implied to the benefit of the commissioning party if they would not otherwise be able to use the commissioned invention for the purpose intended without infringing the patent protecting it. The existence of an implied licence is typically a matter determined by a court and should not be regarded as a substitute for an express licence or assignment.

(d) Co-contributors

Where there is more than one inventor, they are entitled to the grant of the patent as joint inventors, in equal undivided shares.[22] However, questions may arise regarding the extent to which a party has really contributed to an invention. This question was addressed by the House of Lords in *Yeda Research & Development v Rhone-Poulenc Rorer International Holdings*.[23] Referring to the definition of 'inventor' in the Patents Act as the "actual deviser of the invention",[24] the House of Lords held that 'actual' means the person who "came up with the inventive concept" of the claimed technology. A contribution entitling joint ownership must be to the formulation of that inventive concept – not merely an element of a claim, which might be derived from the prior art and non-patentable.

1.3 Inventor compensation

If the invention made by an employee-inventor belongs to the employer, the employee may nonetheless be entitled to compensation. This requires there to be an outstanding benefit to the employer and that compensation is fair and just in the circumstances.[25] Until relatively recently, no employee had successfully claimed the right to compensation for the benefit his work had brought to his employer.[26] The reason for this was the difficulty of satisfying a former requirement of the Patents Act that it is only the patent for the invention that must have the outstanding benefit for the employer, as opposed to the invention. The Patents Act 1977 has since been amended to expressly state that the outstanding benefit can be derived from both. If such outstanding benefit can be demonstrated, the court must secure for the employee "a fair share (having regard to all the circumstances)" of the benefit the employer has derived or may reasonably be expected to derive from the invention or the patent for the invention, including from its assignment or licence.

Difficulty arises in establishing what an 'outstanding benefit' and 'a fair share' is. In *Kelly v GE Healthcare Ltd*,[27] Mr Justice Floyd (as he then was) held 'outstanding' to mean "something special" or "out of the ordinary"; more than merely substantial, significant or good. However, the judge did not think it necessary to demonstrate that the patent's benefit in that case could not be exceeded.[28] The judge further

22 Section 36(1) of the Patents Act 1977.
23 [2007] UKHL 43.
24 Section 7(3) of the Patents Act 1977.
25 Section 40 of the Patents Act 1977.
26 See *James Duncan Kelly and Kwok Wai Chiu v GE Healthcare Limited* [2009] EWHC 181 (Pat).
27 [2009] EWHC 181 (Pat).
28 Compare *Ian Alexander Shanks v Unilever plc* [2014] EWHC 1647 (Pat).

considered that the meaning of 'fair share' depended on the evidence available, but that it should not be thought of in the same way as loss or damage or seek to put the inventor in the same position as an external patentee.

In *Unilever plc v Ian Alexander Shanks*,[29] the Court of Appeal had to decide how an employee-inventor should be compensated if the benefit of his invention has accrued not to his employer but to a third party to whom rights in the invention have been transferred. In *Unilever*, the inventor's employer, Unilever UK Central Resources, assigned the patent on the invention to another group company, Unilever plc. The assignment was made for a nominal sum, but Unilever plc then licensed the patent in exchange for the payment of royalties totalling approximately £23 million. The inventor claimed that had his invention been licensed sooner it would have raised closer to £1 billion in royalties. The question for the court was whether the sum the inventor was entitled to in compensation should be based on the actual amount made by the licence to the patent or the hypothetical amount that the licence could have made. The court held that the inventor should receive a sum based on the amount of £23m that was actually made, not a hypothetical sum.

1.4 Know-how and trade secrets

Patents are registered rights to an invention in product or process technology in which the scope and features of the invention are precisely delineated in the claims. Further detail is provided in the description, experiments and, if appropriate, figures. The proprietor's name is recorded on a register, together with the inventor and any exclusive licensee. By contrast, the terms 'know-how' or 'trade secrets', the uses of which tend to depend on context,[30] refer to information requiring a high degree of confidentiality that is a vitally important part of a life science company's intellectual property assets. However, know-how or trade secrets cannot be registered, or even easily described by their features and scope. The difficulty of determining whether information falls into this category is underlined by *Printers & Finishers Ltd v Holloway*,[31] in which Cross J comments that it is impossible to provide a definitive list of matters that will qualify as trade secrets. The UK courts have nonetheless so far identified secret processes, chemical formulae, designs and special methods of construction as having the necessary qualities of a trade secret – but this is by no means a definitive list.

In the contemporary world, the protection of know-how and trade secrets is a problem as valuable technical data is generated on a large scale, making this information a difficult asset to control. Know-how is protected as long as its confidentiality remains guarded. However, provided they can be contained and protected by suitable confidentiality terms, the rights to know-how can be transferred. And, because it is often of a technical nature, know-how is commonly dealt with in transactions alongside patent rights.

---

29 [2010] EWCA Civ 1283.

30 Commercial transactions such as assignments and licences in which the relevant information is broadly outlined in the agreement usually refer to it as 'know-how'. The expression 'trade secrets' tends to be used in the litigious context when a party is accused of having breached confidentiality in certain information.

31 [1965] 1 WLR 1.

In recognition of the importance of know-how and in an attempt to harmonise the law in this area across all of the EU member states, the EU Commission announced on November 28 2013 a proposal for a Directive on trade secrets and confidential information. The proposed Directive (full title: Proposal for a Directive on the protection of undisclosed know-how and business information (trade secrets) against their unlawful acquisition, use and disclosure) follows surveys and consultations in 2011/12 which concluded, among other things, that trade secrets effectively fill the gap between copyright and patent protection, but that the levels of protection and the remedies afforded are uneven across the European Union and this impacts business decisions. The Directive therefore seeks to establish across the European Union:

- common definitions of trade secrets and confidential business information, and circumstances when the acquisition, use and disclosure of a trade secret is unlawful;
- measures, procedures and remedies that should be made available to the holder of a trade secret in case of unlawful acquisition, use or disclosure of that trade secret; and
- sanctions in the case of non-compliance with the measures provided; and
- provisions on monitoring and reporting.

The Commission's proposal will be transmitted to the Council of Ministers and the European Parliament for adoption under the ordinary legislative procedure. It is unlikely that this legislation will come into force before 2016.

1.5 Employees, consultants and protecting know-how

The value in protecting know-how against unsanctioned use or disclosure is often only evident in the breach. This is most commonly seen when an employee leaves an organisation, taking valuable trade secrets with him. The obligations the law places on employees whose role involves them knowing and using trade secrets, and also their obligations after they have left that employment, are outlined in *Faccenda Chicken Ltd v Fowler*,[32] which has established that employees and ex-employees are to be treated differently:

- *Employees:* First and foremost, the obligations of an employee on the use and protection of trade secrets should be set out in his employment contract. If there are no express terms dealing with trade secrets, the employee is nonetheless deemed to have an implied duty of fidelity (good faith) to his employer. Whilst the extent of that duty may differ according to the nature of the employment and the contract, as a minimum the duty of fidelity will usually include a duty to keep confidential any information that has been obtained in connection with that employment, regardless of the value of that information.
- *Ex-employees:* When an employee leaves an employment and his contract terminates, the scope of his duty is narrowed but not terminated. An ex-

32 [1987] Ch 177.

> employee retains a duty not to disclose the trade secrets of which he has had knowledge during his employment. A very general summary of an ex-employee's responsibility is that the ex-employee is under no obligation to keep confidential the information that he would naturally hold in his mind, but he must not take information from his employer by copying or memorising for the purpose of using it when he has left that employment.

The application of trade secrets rules to consultants has more recently come into the spotlight in the *Vestergaard Frandsen* series of cases. These all concern secret information used in the manufacture of mosquito nets impregnated with insecticide. Vestergaard, a manufacturer of such nets, had employed as a consultant an insect control specialist, Dr Skovmand, to assist in the development of methods to produce the nets without evaporation of the insecticide. For this purpose, Dr Skovmand had access to a 'fence database' containing information on Vestergaard's techniques of manufacture. He later joined a new company set up to compete with Vestergaard in this area of technology, eventually to be called Bestnet, and there he set about developing a new net called Netprotect. Vestergaard brought an action against Bestnet and its two directors, Mr Larsen and Mrs Sig, who were former employees of Vestergaard Frandsen, for misuse of their confidential information. It was held at first instance[33] that the contents of the fence database were trade secrets owned by Vestergaard. Dr Skovmand knew this but had nonetheless used the information in the database as a starting point for Netprotect. As a consequence, Dr Skovmand was liable to Vestergaard for breach of confidence, as was Bestnet. Mr Larsen, who was aware of the fact that Dr Skovmand was using confidential information from the fence database to develop Netprotect, was also liable. However, on appeal to the Supreme Court,[34] Mrs Sig was held not to be liable for misuse of confidential information on the basis that she had not learned of any relevant trade secrets owned by Vestergaard when employed by it and did not know that any such secrets were being used by an employee of Bestnet.

2. Context and features of patent transactions

2.1 The principal dealings in a patent

The right to exclude third parties from the exploitation of the technology protected by a granted patent, such as making and marketing products, may enable premium prices to be charged and prove highly lucrative if the invention is commercially successful. However, not all patents are owned by manufacturers, and not all patents cover marketable products. Many patents protect inventions that are components of products that others make, and others cover process technologies that are steps to making products for which the owner does not have the exclusive rights. In such circumstances, the patent is of little value if its owner cannot leverage its potential application to make money. The two key ways in which this can be achieved are

33 *Vestergaard Frandsen S/A v Bestnet Europe Ltd* [2009] EWHC 657 (Ch).
34 [2013] UKSC 31.

transferring the ownership of the patent entirely to another party (an assignment), or retaining ownership of the patent whilst contracting with one or more third parties to allow them certain rights to use the technology protected (a licence).[35] It is these rights that are discussed in this section.[36]

(a) Patent assignments

A patent assignment is usually a relatively short and simple agreement to assign the ownership of a patent or patents to a third party. In keeping with basic principles of contract law, the transfer must be in exchange for consideration, unless it is executed as a deed. Consideration is normally a lump sum payment, but the payment does not need to reflect the value of the patent(s) being transferred. Alternatively, the consideration can be deferred and paid in a similar way to royalties under a licence (see (b) below).

An assignment will be void unless it is in writing and signed by the assignor or on the assignor's behalf.[37] However, unless the assignment is registered at the UKIPO, the assignee will not be free of earlier claims to rights under the patent of which it did not know and which were not notified to the comptroller of patents.[38] Assignment of ownership provides the assignee with all the rights of the previous owner in the patent, including the right to license, assign or charge the patent, and bring proceedings for infringement.[39]

(b) Patent licences

A licence is also a contractual relationship between a patentee and a third party. However, in a licence the patentee retains ownership and grants the third party certain permissions to exploit the patented product or process, subject to any limitation prescribed in the terms of that contract. These rights will normally be acts that, without the consent granted by the licence, would infringe the patent and therefore be prohibited.

A patent licence is often a long and complex document, highly tailored to its particular context. This is because of the unique commercial circumstances in each case, the technicality of the subject matter covered, and the process of negotiation by which the terms of a licence are agreed, amongst other matters. However, there are a number of general principles that apply to patent licences in the life sciences context, described next.

Why license a patent?: There are a number of reasons why a patent owner may wish to grant a licence. By licensing a patented product, the patent owner can obtain a share in its eventual sales without the need to invest in the manufacture and marketing of the product itself. In particular, the patented technology may be proved

35 See Sections 30(2)–(7) of the Patents Act 1977.
36 A third way of realising the value of a patent, which is based on assignment and licensing, is securitisation. This is discussed in section 3.4 below of this chapter.
37 Section 30(6) of the Patents Act 1977.
38 Section 33 of the Patents Act 1977.
39 Failure to register has cost recovery consequences for the patent owner or an exclusive licensee if the patent is infringed and a legal action ensues (see Section 68 of the Patents Act 1977).

in principle but require further testing, clinical trials or other development to a marketable form. It may also require a marketing authorisation, the application process for which the patentee has not the experience or resources to undertake.

Alternatively, companies with a wealth of research and development capabilities may have patents in their portfolio covering technology that is no longer a commercial priority for them but which is of interest to third parties. In these circumstances, the patent can be monetised with very little ongoing commitment from the owner to the use to which it is put. Such is the case with 'research tool' patents, covering processes or products needed in the research and development of a third-party product. In other circumstances, it may also be desirable to license patents in exchange for the licensing-in of technology that is needed but owned by a third party (known as 'cross-licensing'), or to pool the patented technology of more than one party in a joint venture (see section 2.2(c) below).

Type of patent licence: When negotiating a patent licence, there are a number of general issues that must be considered and agreed, one of which relates to whether the licence is to be 'exclusive', 'sole' or 'non-exclusive'.

An exclusive licence grants the right to exploit the claimed invention to a single party and no other, including the patent owner. The exclusivity granted under the licence may be for a particular territory or use, entitling the owner to grant exclusive licences for other territories and uses to other parties or to reserve them for itself. For most intents and purposes, an exclusive licensee benefits from the right to work the invention as if it were the patentee; this includes the statutory right to bring proceedings in order to enforce the patent against an infringer.[40] Nonetheless, an exclusive licensor will wish to bind the exclusive licensee to a number of obligations, including measures to ensure that the licensee is making the appropriate efforts to commercialise the invention and provide a return through the payment of royalties (see below). Such licences have the advantage to the licensor that they can enable royalties to be agreed at a premium in exchange for the exclusivity being granted.

A sole patent licence grants the right to work the claimed invention to a single licensee but not to the exclusion of the patentee. The licensee may therefore find that it is exploiting the invention in competition with the licensor. Sole licences do not carry a statutory right to bring proceedings for infringement of the licensed patent.

A non-exclusive patent licence may be granted to any number of third parties, as well as the patentee retaining the right to exploit the claimed technology itself. Because there is no exclusivity in these circumstances, the royalty rate that can be negotiated for such a licence will be correspondingly lower than if the licence had been exclusive. Subject to negotiation, a patent owner is under no obligation to provide a licence to each non-exclusive licensee under the same royalty rate or other terms, and the licensee may have little or no control over other parties that are licensed subsequently. Non-exclusive licences, like sole licences, do not carry a statutory right to bring proceedings for infringement of the licensed patent.

---

40 Section 67 of the Patents Act 1977.

Rights to grant sub-licences: If a licence contains the right to sub-license, the terms of any sub-licence must be consistent with the rights in the head licence – a sub-licence must not purport to grant rights to different territories or uses, or grant exclusivity that the head licensee does not itself have.

Territory and scope of licence: To avoid a potential dispute (see section 3.2 below) it is vitally important that all parties to a licence are clear about what their commercial objectives are and that the rights being granted meet with those objectives as well as, to the extent possible, changes in circumstance that may occur in the future. Essential questions to ask include: What patents will be licensed? Are they applications or granted patents? If applications, what effect might an amendment to the claims have? Which patents to be licensed are subject to royalties? What territory does the licence cover? What activities are licensed to take place in a particular territory – and is manufacture permitted, or only sale? Are there limitations on the purposes that the technology can be exploited for?

Licence duration: The term of a patent licence should be calculated to end with the expiry of the patents being licensed, or for as long as the invention is required by the licensee if that is a shorter term. If the licence term ends before the licensed patents expire, the licensee may not be able to continue its activities with the technology covered by those patents without the risk of infringement. In contrast, if the licence term continues after the licensed patents expire, this may result in an obligation to continue royalty payments unnecessarily. However, it is common to require royalty payments for a specified period of time following the first commercial sale of a product as a way of deferring the consideration for the licence. This is especially so in licences to research tool patents (where the sale of the end product may not infringe any of the licensor's patents, or the licensor's patents may have expired by the time any end product is sold). The parties may continue to deal with each other on the same terms as the licence, after the end of the term, but such periodic licences may be terminated on reasonable notice.

Royalties: The manner in which royalties become payable in patent licences varies in structure and complexity with the nature and circumstances of the deal. Typically, though, royalties are paid as a percentage of sales, or as a predetermined lump sum, or a mixture of both. Basing the royalty on sales rather than profits has the advantage that it avoids complex investigation of the licensee's balance sheet. Sales figures, by comparison, are relatively easily understood and calculated. Furthermore, a badly run licensee may make only a small profit, despite large sales, or, through no fault of the licensor, no profit at all. Accounting issues get more complicated if the licensee sells the relevant product either in combination or bundled with other products, where the licensee can apply profits or sales disproportionately in favour of the products that do not bear a royalty obligation.

Additional or complementary royalty arrangements include the following:

- a down payment – a lump sum paid to the licensor at the time of signing the licence;

- milestone payments – milestone payments are triggered at significant stages of success in the project for which the licence is granted;
- staggering – the percentage rate of royalty payments on sales changes according to the level of sales of the product for which the patent licence is needed;
- stacking – the rate of royalty paid under a licence is decreased according to royalties that subsequently must be paid for licences to third-party patents that are needed in order to develop a product. Typically, this arrangement will be seen in development agreements where several research tool patents are needed to perform steps in a process and it is unclear at the time of completing the agreement how many licences will be needed;
- know-how royalties – it is common for patents not to cover every country in the world and therefore, especially in respect of successful pharmaceutical products, it is likely that commercial sales will be made in territories where there is no patent coverage. Typically, royalties will still be payable in these territories, but at a much lower rate (often half) than that paid in territories where sales benefit from patent coverage.

The royalty rates and lump sums agreed will vary from agreement to agreement, depending on factors such as exclusivity, importance of the technology, alternatives available, projected sales, the financial risks involved and the relative negotiating strengths of the parties.

Royalty rates are usually highly sensitive commercial information and are likely to be kept confidential. It is therefore difficult to obtain comparative rates when attempting to decide on a starting point for a royalty negotiation. However, as a general rule, royalty rates in the life sciences sector tend to be higher than those in other fields of technology, such as telecoms and electronics, and it is not unusual to see rates at around the 10% mark for a product licence – although process licences, such as research tools, are likely to be at a considerably lower rate.

Ownership of improvements: During the course of a research project in which a licensed technology is used, it is quite likely that improvements to that technology will be found. The licence governing the use of that technology must carefully prescribe who is to own such improvements. In addition, the licensor should ensure that it retains all rights to so-called 'background' intellectual property, typically composed largely of the know-how that was licensed with the patent to enable the licensee to work the licensed invention.

Warranties and indemnities: Once all the primary licensing terms are agreed, negotiation is likely to focus on warranties and indemnities. These will be drafted in the licence agreement so as to balance a number of risks between the parties and will differ to reflect the exact nature and circumstances of the licence being agreed. At one extreme, the licensor will not want to warrant any assertions about the patents being licensed and, at the other, the licensee will want a full suite of warranties as well as indemnities in the event of certain breaches. Negotiation will typically produce a result that is somewhere in between the two.

Some warranties will be easier for the licensor to give than others. In particular, a licensor should be very reluctant to warrant that a licensed patent is valid. The reason for this is that it is very difficult to predict the outcome if the validity of a patent is challenged in proceedings. Nonetheless, the licensor ought to be able to warrant that it has ownership of the patent, because it is in a position to know and ownership is a registrable right. Similarly, matters over which the licensor has control, such as the payment of renewal fees, knowledge of current or past infringements by third parties, and conflicting prior rights, are matters upon which warranties should be expected. A licensee will also want a warranty that using the patented technology in the manner being licensed will not infringe the rights of a third party.

A patent should also detail what will happen in the event that warranties are breached. For example, a patent may be revoked because it is held invalid by a court or, if it is an application at the time the licence was completed, it may later never be granted. In these circumstances, the licensee may be able to agree an indemnity in the licence for the repayment of some or all of the royalties it has paid, depending on the extent to which it has already benefited from the licence. The licensee will also wish to be indemnified for the costs and damages payable if it has to defend proceedings, if the licensed technology infringes a third-party right. Alternatively, if a third party infringes the patented technology being licensed, the licensee may want to have the contractual right to fight the proceedings and share or shift the cost to the licensor.

Contracts in English law are subject to the principle of *caveat emptor*. This means that the party taking the benefit of an asset under a contract – in this case a licensee taking the benefit of the rights under a patent – has the responsibility to investigate that asset for potential problems or defects before entering into the contract. There is no duty upon the party offering the asset to disclose these defects. For this reason, a prospective licensee must seek disclosure from the prospective licensor on the above issues. Whilst it is in the interests of the licensor to disclose this information as a qualification of its warranties, the licensee must interrogate it as fully as possible in a process of due diligence so that it is aware of any problematic issues.

Termination: A licensor will wish to retain some control over the manner in which the licensed technology is developed in order to guarantee and maximise as much as possible its return in the form of royalties. In particular, the licensor will want to ensure that it has the right to terminate the licence if the licensee fails to perform certain significant obligations under the licence, such as paying a royalty and marketing a product, and may want to impose certain minima that the licensee has to achieve. At the same time, a licensee will want some flexibility against losing the licence for relatively minor defaults, whilst also having a way out of the arrangement if it proves commercially or practically onerous to perform. It is also common to see a right of the licensor to terminate a licence if the validity of the patent is challenged by the licensee. However, such terms are soon to be made unenforceable (see section 3.2).

Some terms in the licence, such as protection of the confidentiality of know-how or (especially where the licensed technology is a research tool) the royalty obligations, will need to be expressed as surviving the termination of the licence.

2.2 Patent rights in corporate transactions

This section addresses the general patent issues arising for four key activities in which a company may be involved during its lifetime: merger and acquisition; flotation; joint venture; and/or divestment. The section is relevant for UK private companies under English law, including the private-company subsidiaries of quoted public companies. It is not concerned with companies after flotation, where quite different considerations apply and which are outside the scope of this chapter.

(a) Mergers and acquisitions

Mergers and acquisitions (M&As) take many forms, and there can be a clear difference between a true merger – two companies coming together to form a single new entity – and an acquisition – one company being the dominant entity taking ownership of the other. However, the two terms are more often used together as a shorthand description for the sale-and-purchase activity between corporate entities of parts or the whole of their businesses. That is the general sense in which that term is used here.

There are two fundamentally different approaches to a sale and purchase: by acquiring or selling company shares; or, by acquiring or selling the assets of a company. In a share purchase, the target company is purchased as a whole, including all its assets, such as patents and other IP owned by it, as well as all the company's liabilities. All licences in and out of the company being purchased will remain with it, although these must be checked during the due diligence process (see below) for provisions barring change of control. The ownership of patents and other IP is carried across automatically with the shares in the target company and therefore separate assignment is not required.

In an asset sale, the purchasing company buys those assets that it would like from the target company, but it does not purchase the company itself. In this way many assets (and liabilities) remain with the target company. The consequence of this is greater complexity, because all the IP assets being transferred must be assigned separately from the other assets of the company. Although all IP assets can be covered in a single assignment, it is normal to list the registered rights and the most important unregistered rights in a schedule. In a life sciences company, the identification of the patents being transferred is of obvious importance. For registered rights such as patents, the new ownership by the acquiring company must be recorded in the relevant registers.

In an asset sale and purchase, the transfer of licences can be problematic. If licensed-in intellectual property is to be acquired, the consent of the licensor is likely to be needed, creating difficult commercial questions about how and when to approach the licensor to seek permission. In life sciences companies, such licences may be critical to the operation of the business, but there is a risk that the licensor may not wish to license the purchasing business and would rather terminate the licence. Such matters must be checked as early as possible. Exclusive patent licences must also be registered to the purchasing company.

(b) ***Flotation/IPO***

In relation to financing a company's activities, an initial public offering (IPO) is the listing of a company's shares on a public stock exchange. It typically happens at the stage in the life of a company when those investors who have financed the company want to realise some or all of their investment, and it provides an exit route for venture capitalists. The issue and sale of company shares on the market also allows the company to bring in new capital and pay debts. Once listed, the company has the flexibility to raise further capital by rights issues and offers the chance for employees to own a stake in the business.

The IPO process requires the preparation of a prospectus describing the company for potential shareholders. There are a number of obligations on those preparing the prospectus to ensure the accuracy of the information included. This requires extensive and thorough due diligence to ensure that the details given are correct. The due diligence process must encompass IP and, as regards life sciences companies, the value of drugs or potential drugs is very much dependent on the patent applications and granted patents protecting them.

In a sale-and-purchase transaction involving patents, a certain amount of information on the patents can be gathered by the buyer on its own by investigating the EPO, UKIPO and other national registers. But scope is limited because, in respect of know-how and other unregistered IP, no information is available publicly. It is therefore standard practice to require the selling shareholders, the assigning company and/or a potential licensor to give disclosure of the information it has on the patents and any other IP being transferred. If the seller/assignee/licensor has given warranties about the IP, then it is in its interests to make this disclosure because the warranties will be qualified by the matters disclosed: the stronger the warranties the more comprehensive will be the disclosure. The procedure followed should be thus:

- Look at the relevant documents made available by the target company and from an inspection of public sources.
- Seek responses from the target company to a due diligence questionnaire listing appropriate questions relating to patents and other intellectual property rights that are held by the target or licensed to it.
- Provide a draft set of the IP warranties that the third party is expected to agree to as part of the transaction. Together with the due diligence questionnaire, these should be designed to elicit disclosure by the target company of any issues of which it is aware.

When considering purchasing or taking a licence from a target company, the following matters should be the subject of due diligence:

- that the patents and other IP being acquired will give the business the 'freedom to operate' – the ability to conduct the present and future business activities that it envisages without infringement of third-party IP;
- that the target from which the patents and other IP are to be acquired or licensed does actually own them;
- the strength and scope of the IP to which rights are being acquired, including

the status of any current challenges to the validity of any of the IP; and
- infringements and threatened infringements of the patents and other IP being acquired.

For a prospective purchaser or licensee, failing to take these steps could result in significant wasted transaction costs if it later has to walk away from a transaction due to an adverse issue that could have been identified early on. Furthermore, failing to conduct an adequate review and having sufficient warranties could also mean paying for assets that are worth less than was thought. It could even mean paying for assets that have onerous liabilities to third parties attached to them, forcing the business to stop and/or to pay damages, or which are susceptible to invalidation.

Patents are arguably the most important assets a life sciences business has. Therefore, in addition to the due diligence undertaken as a prospective purchaser, a company should also carry out due diligence on its own patents and other IP in preparation for a transaction. It is vital for the business to know and thoroughly understand the following issues in relation to its patents and other IP:

- the quality and quantity of patents and other IP owned, so that a value can be put on them. It should also be possible to identify the IP that is not being used, and the cost of maintaining unnecessary registrations;
- the processes that are in place to capture potentially patentable inventions and other IP created, so that these can be adequately protected and used; further, there should be a regular and routine check for compliance with these processes;
- the chain of title from which ownership of IP is derived and, where applicable, that all IP is properly registered as required – this is to ensure that all the necessary rights are secured in the event that the business is sold or technology is licensed to a third party; and
- infringements or suspected infringements of patents and other IP; moreover, it is imperative to have a process in place to monitor the market for any infringing activity or potentially infringing activity, as soon as it emerges.

For a prospective M&A target, good IP housekeeping practices and a structured approach to IP due diligence will assist not only in maximising the value of its IP assets, but also in ensuring transactions run as smoothly as possible, since it affords the opportunity to regularise any issues discovered.

(c) *Joint ventures*

Circumstances may arise in which two companies wish to pursue the same project, but need the patented technology and know-how of both in order to make this possible. In projects of this kind in which the companies envisage that they will need to work together over the long term, and there is considerable complexity in the assets that need to be shared, the best option may be to form a joint venture. A joint venture is typically approached by forming a new joint venture company into which the patented technology and other assets that are needed are licensed or transferred. The joint venture company can then use the technology to run the project.

Joint ventures raise a number of questions about how the pooled assets are to be owned – eg, whether they are licensed-in or assigned, how they are to be used, and the continued rights, if any, of the joint venture parties to use those rights. Patents are also a particular issue for life sciences joint ventures, because the entire point of a joint venture project is likely to be the creation and development of a new product, which is likely to be the subject matter of one or more patent applications. Questions such as which party will own what proportion of the rights to new patents arising from the joint venture project, and what will be the terms upon which it can be exploited, need to be resolved. Of equal importance will be establishing what the parties are restricted from doing with the newly created technology.

An issue of particular sensitivity that must be agreed between the joint venture parties is what happens to patented technology and other IP when the joint venture comes to an end. Although this may be straightforward in respect of the assets originally contributed by the parties, the rights to new technology developed may be the subject of difficult negotiations.

All of the above matters must be dealt with in the joint venture agreement.

(d) Divestment

The fast-evolving nature of the life sciences sector means that even relatively small companies may wish to change the direction of their business from one area of technology to another. Alternatively, for large enterprises there may be the need to focus on areas of particular expertise and profitability and move away from others. In either case, a company may wish to divest itself of patents covering certain technology areas so that it can realise value for the technology without having to devote resources to its further exploitation or development.

Typically, this can be done by outright assignment of patents, know-how and other assets, or by licence. If it is by licence, then the divesting company will most likely take a royalty and this requires some continuing involvement in ensuring that the licensed technology is being exploited with the appropriate endeavour. If the area of technology is a particularly complex one, such as an entire business area, requiring a number of rights to be assigned or licensed, together with other assets such as staff, the appropriate way to divest may be the creation of a new corporate entity into which the assets are pooled, followed by a sale of the company to a third party.

2.3 Particular business transactions

(a) Material transfer agreements

The developer of biological and/or chemical materials – for example, proteins, cell lines, viruses or pharmaceutical compounds – may wish to transfer them to a third party with the appropriate research interests and expertise so that they can be tested, developed or otherwise used in research. Agreements regulating such arrangements are called 'material transfer agreements'.

For such an apparently straightforward process, these agreements must nonetheless provide for a number of important matters between the transferor and transferee. These include:

- a clear definition describing what the material consists of;
- the limitations for which the material being transferred is to be used – this may be generally for use in non-commercial purposes, or more specifically for a particular analysis or research project (in which case the agreement must carefully define the application that is envisaged);
- restrictions on the publication of research pending applications for patent rights that may arise as a result of that research, or not to publish at all;
- which party will own the rights to patent any invention resulting from the research and know-how (the transferor will most likely want this), and the rights to the material if it is transformed or, as in the case of cell lines and viruses, if further material is cultured from it;
- compliance by the transferee with regulations governing the use of the material, including any restrictions on the transfer of the material to a third party; and
- confidentiality obligations with regard to know-how in the material being transferred.

Thus, the emphasis in material transfer agreements is on preserving confidentiality in the material being transferred and in establishing the rights to any IP that is later developed through research.

(b) Commercial research, development and collaboration agreements

Research, development and collaboration agreements take a wide variety of forms, including:

- collaborations between commercial and academic organisations;
- collaborations between commercial organisations; and
- European FP7 Framework consortium agreements.

The common theme to these agreements is that they involve two or more organisations working together to advance one or more of their existing technologies, or to combine their technologies to produce something new (see also section 2.2(c) above on joint ventures). There is therefore no one-size-fits-all approach to such agreements and the concerns and objectives of the parties in each will be unique to their circumstances. There are, nevertheless, two key topics that the parties to any collaboration agreement should consider: background and foreground patents; and confidentiality and publication. Each is described further below.

Background and foreground patents: Rights to the ownership and the exploitation of patent rights in research and development (R&D) agreements are likely to be particularly contentious, and it is important that clear terms are drafted to deal with this.

In relation to 'background' (ie, pre-existing) patents, it is often necessary for one party to grant to the other a licence under background patents to enable the other party to conduct the agreed R&D work. If both parties are undertaking such work, cross licences may be necessary. Consequently, the value and commercial sensitivity of the background patents should be considered to establish whether any specific restrictions should be placed on their use or disclosure.

For new ('foreground') patents, if there are no provisions in the agreement dealing with patent ownership the default legal position is that a party will individually own the foreground patent that it solely creates, and the parties will jointly own the foreground patent that is jointly created. However, this position can have a variety of undesirable consequences, for example:

- The party owning the background patents will want to control any foreground patents because those patents relate to valuable developments of its initial technology. However, without arrangements to the contrary, the second party will own the foreground patents, allowing the second party to assign or license them to a competitor of the first party without its knowledge or consent.
- If technology arising from the work is jointly owned, agreement needs to be reached about who will take responsibility for patent applications and maintenance for foreground patents covering the jointly owned invention. This must cover all eventualities, including such things as what will occur when one party does not want to file and pay for a patent but the other party does.
- Under English law, joint owners of a patent are permitted to use the patented invention themselves but – in the absence of any agreement to the contrary – they are not permitted to allow third parties to use it or to assign their share to a third party. However, given that any intellectual property developed is in essence a bundle of national intellectual property rights rather than a single intellectual property right governed by the law of a single jurisdiction, and that what a joint owner can or cannot do with its jointly owned intellectual property rights differs from jurisdiction to jurisdiction, dealing with co-owned intellectual property rights can be very administratively burdensome.

Accordingly, co-ownership of intellectual property should be avoided as a default solution in any collaboration context unless the consequences of that co-ownership have been carefully considered. An agreed position should be documented in relation to the prosecution, maintenance, ownership, exploitation and enforcement of foreground patents.

Confidentiality and publication: It is important to ensure that there are confidentiality provisions corresponding to the intellectual property provisions. A common oversight is for the parties to negotiate carefully the patent and other intellectual property provisions but to assume that confidentiality provisions are 'standard' such that the parties fail to tailor them according to the circumstances of the transaction.

For example, an issue that will almost always arise in the context of collaborating with an academic institution is that the institution will require the right to publish the results of the collaboration. Guidance from the Charity Commission in the United Kingdom, which regulates many academic institutions, provides that such an institution needs the right to publish the results "within the appropriate academic time frame". In practical terms, this means a delay of publication of results for up to six months from the date of completion of the research is usually acceptable; however, a delay of any longer will be resisted.

(c) ***Contract manufacturing agreements***

As the development of a drug moves from preclinical development towards approval, the financial risks arising from problems in respect of manufacturing increase substantially and it is essential to put in place agreements even at the preclinical stage in order to minimise the potential exposure. This subject is dealt with further in section 3.7.

(d) ***Product distribution agreements***

Despite sharing features with agency agreements, the legal structure of a product distribution agreement has several key differences. Firstly, the distributor will buy the patented product directly from the manufacturer, apply a mark-up to the manufacturer's price and then sell it on as an independent principal. Once bought from the manufacturer, legal title and (as a result) commercial risk pass to the distributor, and a contractual relationship is forged between the distributor and the customer. As a practical point, therefore, the distributor may require an indemnity from the manufacturer against patent infringement or validity proceedings.

Product distribution agreements are highly regulated by competition law, and any anticompetitive terms – or even the entire agreement – may be rendered void if competition rules are violated. A balance must be struck between requirements for exclusivity that may be expressed in the contract, and stipulations that will affect competition, such as restricting supply within a territory or creating price discrimination between territories. Consequently, close attention must be paid to Chapter I of the Competition Act 1998 in the case of national competition law, and Article 101 of the Treaty for the Functioning of the European Union in the case of European competition law. (See also section 3.3 below.)

(e) ***Co-promotion agreements***

A co-promotion agreement is based on the premise that a single product, under a single brand, can be promoted by two entities in order to achieve greater market penetration. A rarer breed of agreement, it is predominantly employed in the pharmaceuticals and medical instrument spheres.

The basic structure consists of an IP owner sharing promotional rights with a distributor, pharmaceutical company or agent, although it is not uncommon for the assignment to include a conferring of the right to manufacture the product as well. Two IP owners may also make use of this structure in a situation where one has a patented formula and the other a compatible delivery mechanism, and therefore an agreement that they both promote the combined product might be commercially necessary.

Given that there are varying degrees to which rights over the patent may be licensed, there is no set format to which such agreements must adhere, and decisions over whether it extends to manufacturing are peculiar to each case. In fact, the main issue from an IP perspective might not be to do with the patent at all, but with the trademark issues surrounding the brands involved.

(f) *Clinical trial agreements*

When a party engaged in research and development of pharmaceutical or biotechnology products has developed a study drug that has the potential to be developed into a commercial product, it will need to undergo clinical trials in human subjects if it is to progress. However, because of the resources needed to conduct clinical trials, these are now commonly entrusted to a third party in the full-time business of providing clinical trial services. The provider of clinical trials services will be a health authority, academic institution or commercial organisation. It is to regulate the provision of these services that a clinical trials agreement is necessary.

Unlike the agreements described above, a clinical trial agreement regulates activities that would not normally be expected to result in the creation of new patent rights or know-how. Nonetheless, for good order these agreements should govern this possibility and, in this respect, the considerations will be similar to those for R&D agreements (see section 2.3(b) above). However, there will be an emphasis in a clinical trial agreement on the ownership, confidentiality and protection of the patient data produced.

In addition, given the nature of clinical trials, the clinical trial agreement should provide for the division of responsibility for liabilities arising due to injury caused to the trial subjects. The clinical trials services provider will not wish to accept liability if it transpires that the drug is not fit for purpose or it has damaging side effects – it is usual for the company that has developed the drug to provide an indemnity for any liability that arises due to the effects of the drug. However, the provider of the drug will normally expect the services provider to be responsible for any liabilities arising for damage caused by the conduct of the trial itself. In any case, both drug developer and services provider are required to be insured for liabilities arising from conducting clinical trials.[41]

3. Issues associated with patent transactions

3.1 Regulatory law and compliance

Patents provide the exclusivity for product and process exploitation and marketing on which many life sciences business models depend. However, it is in the nature of therapeutic drugs that patent transactions relating to them will also be accompanied by issues of regulatory law. These will be particularly important to consider at the due diligence stage, when the de facto exclusivity protection afforded by some regulations beyond the patent term should be examined; furthermore, compliance with the correct authorisations for placing products on the market must be checked. This section outlines the key regulatory laws applicable.

41 See Directive 2001/20/EC relating to the implementation of good clinical practice in the conduct of clinical trials on medicinal products for human use.

(a) ***Regulatory body in the United Kingdom***

In the United Kingdom, the Medicines and Healthcare products Regulatory Agency (MHRA) is the regulatory body responsible for medicinal products for human and veterinary use. The MHRA is an executive agency of the UK's Department of Health and is responsible for applications made for marketing authorisations through the national procedure. It is also the competent authority when the UK is designated as the reference member state in the mutual recognition procedure or decentralised procedure for marketing authorisation.

(b) ***Clinical trials***

Clinical trials in the United Kingdom are regulated by European Directive 2001/20/EC as implemented into UK law by the Medicines for Human Use (Clinical Trials) Regulations 2004. The MHRA is responsible for enforcing these regulations. Before starting a clinical trial, the 'sponsor' wishing to have the trial conducted must apply for and obtain authorisation from the MHRA. The sponsor, or its legal representative, must be established within the European Union.

In addition to securing authorisation from the MHRA, all medical research involving people in the United Kingdom has to be approved by an independent research ethics committee. The committees are often based at local hospitals and are formed of local people, such as health professionals, patients, lawyers and members of the public. Before the start of a trial, a detailed plan of the proposed research must be submitted to the committee. The committee reviews the proposed research plan to determine whether the anticipated benefits of the clinical trial justify the risks to the participants.

On July 17 2012, the European Commission adopted a proposal for a European Regulation to govern clinical trials in Europe. This would replace the current Directive 2001/20/EC and create uniform rules across Europe relating to clinical trials. It is anticipated that the new Regulation will come into effect in 2016.

(c) ***Marketing authorisations***

Applications for marketing authorisations are first validated for completeness by the MHRA and then the product is assessed for adequate quality, safety and efficacy for the intended use. The application may be referred to an MHRA advisory body, such as the Commission on Human Medicines, which also has a number of subcommittees.

The data requirements for marketing authorisation applications are laid down in the 'Notice to Applicants' published by the European Commission's Committee for Medicinal Products for Human Use. If the application is for a product containing a new active substance, full toxicological, pharmacological and clinical data is required; if it is an active ingredient that has been evaluated before, chemical and pharmacological data, and toxicological and clinical bridging data only are required, linking back the application to the previously authorised active ingredient.

After grant, marketing authorisation holders in the United Kingdom are required to comply with European Directive 2010/84/EU governing pharmacovigilance. 'Pharmacovigilance' is defined as the pharmacological science relating to the detection, assessment, understanding and prevention of adverse effects, particularly long-term and short-term side effects, of medicines. The Directive was adopted in 2010 and came into effect throughout the European Union in 2012. It is implemented in UK law by the Human Medicines Regulation 2012.

The obligations on marketing authorisation holders include:

- an obligation on all applicants for marketing authorisation to submit a risk management plan, and to implement the plan as a condition of any marketing authorisation granted;
- a requirement to establish a Pharmacovigilance System Master File and to audit their pharmacovigilance system at intervals; and
- a requirement to report suspected adverse reactions, not only from the authorised use of a product at normal doses but also from medication errors and misuse of the product.

The pharmacovigilance legislation also allows the MHRA to request that the marketing authorisation holder perform post-authorisation efficacy studies when uncertainty relating to some aspect of the efficacy of a product has been identified.

(d) ***Data exclusivity***

A company that has obtained a marketing authorisation on a product for the first time enjoys a period of data exclusivity, during which their preclinical and clinical trials data may not be referenced in the regulatory filings of another company (typically a generic company) for the same drug substance.

For marketing authorisation applications made before November 2005, the period of data exclusivity varies across the EU member states, and is either 6 or 10 years. However, the rules determining exclusivity changed in November 2005 to follow an '8 + 2 + 1' approach:

- During the first eight years from the grant of the innovator company's marketing authorisation, data exclusivity applies.
- After eight years, a generic company can make use of the preclinical and clinical trial data of the originator in their regulatory applications, but they still cannot market their product for at least another two years.
- After 10 years from the grant of the marketing authorisation, the generic company can market their own product, unless the innovator product qualifies for a further one year of exclusivity.

The additional one year may be obtained, for example, when the innovator company is granted a marketing authorisation for a significant new indication for the relevant medicinal product. In such a situation the generic company can only market its product after 11 years from the grant of the innovator company's marketing authorisation.

(e) ***Orphan drugs***

Drugs treating rare diseases have correspondingly low sales. Therefore, to encourage companies to develop such drugs (so-called 'orphan medicinal products'), European legislation[42] provides certain incentives:

- a 10-year period following the grant of a marketing authorisation, during which no directly competing or similar product can be placed on the market;
- scientific advice and guidance, provided by the European Medicines Agency, on preparing the marketing authorisation dossier;
- possible fee reductions granted by the European Medicines Agency; and
- EU and member state programmes and initiatives for granting funding.

For a product to be classed as an orphan medicinal product, it must be intended for the diagnosis, prevention or treatment of:

- a life-threatening or chronically debilitating condition affecting no more than 5 in 10,000 people in the European Union at the time of submission of the application for orphan designation; or
- a life-threatening, seriously debilitating or serious and chronic condition for which, without incentives, it would be unlikely that the revenue after marketing of the medicinal product would cover the investment in development.

There must also be either no satisfactory method of diagnosis, prevention or treatment of the condition concerned or, if there is such a method, the new medicine must be of significant benefit to those affected by the condition.

An application for orphan medicinal product status is a separate application to an application for a marketing authorisation. Orphan designation may be obtained at any stage before an application for marketing authorisation is made, provided there is proper scientific justification for it. If a drug is successfully designated as an orphan drug, it will be registered on the Community Register of Orphan Medicinal Products.

(f) Medical devices

There are currently three European Directives that govern the requirements that manufacturers must meet before marketing each of the following: active implantable devices; in vitro diagnostic medical devices; and, all other types of medical devices.[43] In the United Kingdom, the provisions of the Directives are implemented by the Medical Devices Regulations 2002. In general, medical devices must meet certain essential requirements before they can be put on the market in an EU member state. If a device is considered to meet these essential requirements, it can be marked with a 'CE' mark.

The European Commission has recently proposed two new draft Regulations, which propose to revise and replace the Directives. This review of the legislation has

---

42 Regulation (EC) 141/2000 of the European Parliament and of the Council of December 16 1999 on orphan medicinal products.

43 Directive 90/385/EEC (the Active Implantable Medical Devices Directive); Directive 98/79/EC (the *In vitro* Diagnostic Medical Devices Directive); and Directive 93/42/EEC (the Medical Devices Directive), respectively.

been proceeding in parallel with both the well-publicised PIP breast implant scandal dating from 2010 and concern over metal-on-metal hip implants. Two of the key aims of the legislation have therefore been to address a perceived need for tougher regulations to protect patients, while still enabling new, life-saving devices to enter the market in a timely fashion. Numerous amendments to the Commission's proposals have been adopted since they were first published, including the radical inclusion of measures for certain high-risk devices that resemble the pre-marketing authorisation requirements for pharmaceuticals. At the time of writing, the proposals are awaiting the results of the European Council's first reading.

(g) ***Advanced therapy medicines***

Advanced therapy medicines are medicines that are based on genes and cells – for example, gene therapy products, somatic cell therapy products and tissue engineered products. Regulation (EC) 726/2004 was adopted to specifically govern the authorisation of advanced therapy medicines, including establishing the Committee for Advanced Therapies (CAT) to assess the quality, safety and efficacy of these medicines.

All advanced therapy products intended for marketing in more than one EU member state are authorised centrally via the European Medicines Agency (EMA). An application for authorised use of such a product is first considered by the CAT, whose opinion will then be considered by the Committee for Medicinal Products for Human Use (CHMP). The CHMP will adopt a recommendation for the European Commission on whether or not the authorisation should be approved. Once authorised, the safety and efficacy of the advanced therapies is monitored by the EMA.

(h) ***Biosimilars***

'Biosimilars' are biological medicinal products which are similar to an authorised biological product and which do not differ significantly from that product in terms of quality, safety and efficacy. The regulation of biosimilars is governed by Article 10(4) of Directive 2001/83 as amended, supplemented by the Annex to the Directive. The Annex in turn refers to a guideline published by the EMA that governs how biosimilars are to be assessed.

There are three EMA guidelines that apply to all biosimilars:

- an overarching guideline that was adopted in September 2005;
- a non-clinical and clinical issues guideline that was adopted in February 2006; and
- a quality issues guideline that was adopted in February 2006.

In 2011 the EMA started a review of the three main guidelines quoted above so that they could be revised to reflect the way that applications for marketing authorisations for biosimilars are currently being considered. At the time of writing, the EMA has adopted draft versions of the revised guidelines. Of these, it is now known that the quality issues guideline will come into effect on December 1 2014.

The three main guidelines are supplemented by a number of product or class-specific guidelines, also issued by the EMA.

3.2 Commonly disputed matters

Section 2.1(b) above described the terms generally needed in a patent licence, or an agreement incorporating a patent licence. One important aspect of patent licence drafting is the need for clarity. This is highlighted by licence disputes that come before the English courts which have arisen from imprecise or ambiguous drafting. This section outlines three key examples of disputes that have been brought before the English courts, concerning: royalty sharing, the scope of the rights licensed, and the timing and period of royalty payments.

(a) *Royalty sharing*

In *Cambridge Antibody Technology v Abbott Biotechnology Ltd*[44] a dispute arose about the royalty-sharing arrangements in a development agreement relating to the anti-TNFα monoclonal antibody drug, HUMIRA. The agreement included a licence for two Cambridge Antibody Technology patents: one for making a 'DNA library' and the other for phage display. Antibody fragments identified as effective against TNFα using the technology in these patents could then be converted by Abbott to fully humanised antibodies and ultimately an anti-TNFα drug. Amongst other risk-sharing clauses, the agreement included a royalty stacking arrangement to offset royalties payable to Cambridge Antibody Technology whenever technology covered by third-party patents was needed as part of the development process. The agreement provided that the offset would be shared for "rights needed to practice or have practiced the technology claimed in the Patents". The dispute was about whether the patents for which third-party royalties would be shared were for technologies similar to phage display and the DNA library or in addition those for the later steps necessary to produce fully human antibodies.

After a lengthy and costly trial considering a large amount of evidence on the meaning of the clause, the judge found in favour of the former interpretation only.

(b) *The scope of the patent being licensed*

Oxonica Energy v Neuftec Limited[45] concerned a dispute over the scope of the patent rights being licensed and therefore the correct amount of royalty payable. Neuftec licensed Oxonica patents and know-how for cerium oxide fuel additives. Oxonica used this technology to develop Envirox, a particulate cerium oxide coated with an alkyl carboxylic acid compound. Royalties were payable on sales of Envirox.

One of the patents was an international application at the time the licence deal was completed. It later matured into the grant of several national patents, but only after the claims in the licensed application had been narrowed in a number of significant countries. Oxonica later produced a product called Envirox 2, which used a different coating compound from Envirox. This new coating compound had been within the claims of the original international application but was not covered by the narrowed granted patents. Hence, Oxonica took the view that it did not owe royalties on Envirox 2 under the narrowed patents. This gave rise to a dispute centred

44 [2005] FSR 27.
45 [2009] EWCA Civ 668.

on the meaning of the provision in the agreement stating that royalties were payable on products.

The Court of Appeal held that the Oxonica construction must be wrong: the licence should be interpreted broadly, to cover the technology being licensed to get Oxonica into the fuel additive business; and it would be unreasonable to have to calculate different royalties for different countries based on the form the claims happened to take when granted there.

(c) ***The timing of royalty periods and payment***

Cambridge Display Technology Ltd v EI DuPont De Nemours[46] concerned a dispute about the meaning of 'minimum royalty', which was specified as "in relation to each Year after 1 January 2003, US 1 million" for the purpose of calculating shortfalls in royalty payable under a clause in the licence.

DuPont argued that the 'minimum royalty' definition meant "each Year commencing after 1 January", and Cambridge Display Technology argued that it should be read as "each Year ending after 1 January 2003" – each party's definition naturally resulting in a more favourable shortfall arrangement for it.

The court held that the natural reading of the shortfall clause taken together with the minimum royalty was that of DuPont.

3.3 National and European competition rules

(a) ***Application of the technology transfer block exemption***

Article 101(1) of the Treaty on the Functioning of the European Union (the Treaty) prohibits agreements which have as their object or effect the restriction or distortion of competition. However, Article 101(3) of the Treaty provides that the Article 101(1) prohibition may be declared inapplicable in certain circumstances.

In the area of technology transfer agreements, the circumstances in which the prohibition will not apply are set out in further detail by the technology transfer block exemption (TTBE) regulation. Any agreement that involves the transfer of rights to a patent is subject to the TTBE. The previous TTBE expired on April 30 2014 and was replaced by a new regulation on March 21 2014, which should be read together with new guidelines on the application of Article 101 of the Treaty to technology transfer agreements. The new TTBE contains a transitional provision, stating that Article 101(1) of the Treaty will not apply before April 30 2015 to agreements that satisfy the existing TTBE but not the new one.

This section focuses on the new TTBE regulation, with reference to the regulation it has replaced (the old TTBE).

General structure: If a technology transfer agreement fulfils the conditions set out in the TTBE, it will be 'block exempted' from the prohibition in Article 101(1) and thus legally valid and enforceable. The old and new TTBEs stipulate that the combined market share of the parties to an agreement must not exceed the threshold

---

46 [2005] EWCA Civ 224.

of 20% for competing and 30% for non-competing undertakings. The new TTBE contains a list of 'hardcore restrictions', which are regarded as almost always anticompetitive. A technology transfer agreement that contains a hardcore restriction will fall outside the scope of the block exemption and will be void and unenforceable.

The new TTBE also contains three 'excluded restrictions'. Obligations in an agreement that are characterised as an excluded restriction will not be protected by the exemption, but the rest of the agreement will remain protected if that obligation is severable from it. The excluded restrictions are not automatically anticompetitive, but it will be necessary to undertake an individual analysis of the effects of the excluded restriction in order to determine whether they will, in fact, be valid.

Challenges to validity: Perhaps the largest change in the new TTBE comprises challenges to a patent's validity. Prior to proposals in the new TTBE, it had been generally accepted practice for a licensor to have a right to terminate the licence in the event of a challenge to validity of the licensed rights by the licensee (although banning such a challenge in an agreement has itself long been considered objectionable under competition law). Many current licences have such termination rights, because a licensor cannot be expected to do business with a licensee that is challenging the very basis of their relationship.

As the Guidelines to the new TTBE explain, the right to terminate on challenge to validity of the licensed rights is now excluded from the safe harbour of the TTBE because

> *such a termination right can have the same effect as a non-challenge clause, in particular where the licensee has already incurred significant sunk costs for the production of the contract products. In these cases the licensee may be deterred from challenging the validity of the … right as it would risk the termination of the licensing arrangement and thus face significant risks which go far beyond its royalty obligations.*

However, non-challenge claims in bona fide settlement agreements are generally considered not to offend competition law, and non-challenge and termination clauses solely concerned with know-how are not excluded from the scope of the new TTBE.

(b) *The legacy of the pharmaceuticals sector inquiry and settlement agreements*

Since its inquiry into the pharmaceuticals sector in 2008, the European Commission has continued its scrutiny of settlement agreements in patent litigation, with a particular focus on practices to delay the market entry of generic medicines in breach of EU competition rules, specifically Article 101 of the Treaty. This has given rise to a number of formal 'statements of objection' against pharmaceutical companies, including Les Laboratoires Servier and Lundbeck in two major cases concerning citalopram, an antidepressant, and perindopril, a cardiovascular medicine. The European Commission has also published four reports on its monitoring of patent settlements in the pharmaceuticals sector, the most recent of which is dated December 2013.

The Commission has recently launched an antitrust investigation against US-based pharmaceuticals company Johnson & Johnson and Swiss-based Novartis to assess whether an agreement between Johnson & Johnson and the Novartis's generic branch, Sandoz, may have had the object or effect of hindering the entry on to the market of generic versions of fentanyl, a strong painkiller for chronic pain, in the Netherlands. Pursuant to this, both companies received a statement of objection on January 31 2013.

3.4 Securitisation of patent rights

In appropriate circumstances a patent that generates a significant and stable income stream for a company may be securitised to realise the lifetime value of the patent with a lump sum in the short term – effectively selling the right to receive the royalty for a one-off payment. A particular method of achieving this aim that has been used by life sciences companies is the assignment of ownership in the patent to a special purpose company, which pays for the patent by issuing bonds or shares to investors. The revenue then generated by the patent royalty is paid back to the investors as interest on their bonds or as dividends on their shares, respectively.

The arrangement is not restricted to patents that are licensed to third parties. Instead, the assignor of the patent may be the only entity exploiting the patented technology and needs to continue to do so after securitisation. In these circumstances, the licensor can take an exclusive licence-back of the right to exploit the patent and in return pay a royalty to the special purpose vehicle to which it was assigned.

3.5 The effect of insolvency

In the United Kingdom, a company can become 'cash flow insolvent' – meaning it is unable to pay its debts as they fall due – or 'balance sheet insolvent', where its liabilities exceed its assets. The statutory insolvency framework in the United Kingdom aims to protect the creditors of the company in difficulty while giving the company a reasonable opportunity to be rescued.

There are two options available to a company in financial difficulty:

- *Administration* – which provides the company with a moratorium from its creditors while it can be turned around or its assets reorganised. An administrator is appointed to manage the attempted rescue.
- *Liquidation (or 'winding up')* – which involves the appointment of a liquidator to oversee the realisation of the company's assets and the distribution of the proceeds to all of the company's creditors. Importantly, as regards patent agreements, a liquidator can disclaim onerous contracts, and set aside agreements entered into in a certain period before liquidation.

Liquidators and administrators are referred to generically as 'insolvency practitioners'.

(a) *An insolvent licensor*

Life sciences companies depending on one or more crucial patent licences should consider the impact the insolvency of the licensor would have on their ability to continue to exploit that patent.

Unless the licence expressly provides for termination upon insolvency, the licence will not fall away merely because the licensor has become insolvent. However, there are immediate risks to the licence that the licensee needs to be aware of. For example, the insolvency practitioner may take the view that the renewal fees for the licensor's patents are not worth maintaining, allowing the patents to lapse. If this happens, then the technology being licensed is open to use by the licensee's competitors and exclusivity is lost. A licensee should therefore ensure that it has the right to pay renewal fees should a licensor fail to do so.

Alternatively, the insolvency practitioner may consider that certain licensor obligations are too onerous and therefore that the licence should be disclaimed – the priority of an insolvency practitioner is to obtain the best realisation of the company's assets for its creditors, and this may conflict with the licensor obligations under the licence. Although the licensee may have a claim for loss caused by breach of contract against a non-compliant insolvency practitioner, this claim will rank alongside the other unsecured creditors and hence a full recovery of that loss is unlikely.

If an insolvency practitioner assigns the licensed patent to a third party, it is likely to do so quickly and without any warranties. In these circumstances a prospective assignee may not make enquiries into the licences granted under the patent. Licensees should therefore protect their position by registering their licence interest against the relevant patents at the UKIPO, thus effecting deemed notice to the assignee of its licence and providing that the patent is assigned expressly subject to it. If a licensee has not registered its interest, and an assignee of the licensed IP is not otherwise made aware of it, the assignee can take the patent free of the licence and will not be required to honour it.

In addition to the steps above, a licensee should protect its position where possible by monitoring the financial position of key licensors and, for critical patents and other intellectual property, consider requiring the licensor to place the patents in an 'insolvency remote vehicle' – a company set up solely to hold the relevant patents and without any other liabilities.

(b) An insolvent licensee

Licensors should also consider the possibility of a licensee going insolvent and how they can protect themselves. A key risk for a licensor is that it will not receive important milestone or royalty payments if its licensee becomes insolvent. Claims against an insolvent licensee will only rank alongside other unsecured creditors, reducing the likelihood of recovery. Licensors should therefore consider royalty structures that front-load payments under the licence as much as possible, including lump sums on signing the licence and royalty payments at regular intervals.

A licensor should be able to terminate the licence upon the licensee's insolvency (or on earlier signs of the licensee's financial difficulty, such as a failure to make certain royalty payments). A licensee is likely to resist such a provision on the basis that it removes further value from the company if it becomes insolvent and reduces the likelihood of a rescue.

Licensors should also consider including a provision that a licence cannot be assigned without consent of the licensor, or even include a change-of-control clause, so that the identity of any assignee of the licence can be controlled by the licensor should it be disposed of by an insolvency practitioner upon insolvency.

3.6 Taxation

(a) General

Income received by UK companies from patent transactions (including the disposal of patent rights and the receipt of patent royalties) is typically subject to UK corporation tax.

The rate of UK corporation tax for the financial year 2014/15 is between 20% and 21% depending on the profitability of the company (although the UK Government is currently proposing a reduction in the top rate of corporation tax to 20% by April 1 2015).

Patents which are acquired or created by UK companies on or after April 1 2002 are subject to a special regime which applies to intangible assets (known as the intangible fixed asset (IFA) regime). Under the IFA regime, the tax treatment of patents is generally aligned with their accounting treatment. Patent rights which were created or acquired before April 1 2002 are subject to separate rules.

(b) Withholding tax

The United Kingdom generally levies withholding tax on patent royalties paid by a UK resident at a current rate of 20%. There are a number of exceptions, including where the recipient of the royalties is a UK company (so UK withholding tax is generally only required on royalty payments to individuals and non-UK licensors). Withholding tax can also be levied on the proceeds of sale of UK patent rights by a non-UK resident.

Withholding tax may be reduced (or eliminated) under the terms of an applicable double-taxation treaty or EU Directive.

(c) Value added tax

The sale or licensing of patents to a UK business is generally subject to UK value added tax (VAT; a sales tax). The current (2014) rate is 20% of the consideration for the sale or licence.

(d) Patent box regime

Since April 1 2013, UK companies which own patents and limited other qualifying IP rights have been able to elect to enter the 'patent box' regime to benefit from up to an effective 10% rate of UK corporation tax on profits relating to those rights.

To be eligible for the regime, a company must hold qualifying IP rights or an exclusive licence in respect of qualifying IP rights (licences must give exclusivity covering at least a whole territory). Qualifying IP rights generally include patents and supplementary protection certificates granted by the UK Intellectual Property Office and the European Patent Office, or by certain other approved EU national patent authorities, and UK and European regulatory data exclusivity rights.

The company must also carry out qualifying development in relation to its IP right, by creating or significantly contributing to the creation of the patented invention or performing a significant amount of activity to either develop the patented invention or develop the way in which the patented invention is used or applied.

Broadly, the patent box regime applies to a proportion of the taxable profits of a company's trade. The income included in the profit calculation includes income from:

- the sale of items in respect of which a qualifying IP right has been granted or products incorporating such an item;
- the licensing of qualifying IP rights;
- the sale of qualifying IP rights;
- receipts of infringement proceedings; and
- damages, insurance or other compensation related to qualifying IP rights.

(e) ***R&D tax reliefs***

UK companies which undertake R&D activities can typically benefit from enhanced corporation tax reliefs. The definition of R&D for these purposes follows its meaning for accounting purposes: R&D reliefs can only be claimed in respect of expenditure on activities within a project which directly contribute to achieving an advance in science or technology through resolution of scientific or technological uncertainty.

There are two main schemes for claiming relief, the availability of which depends on whether the company claiming the relief is a small or medium-sized enterprise (SME) or a large company. The definition of SMEs and large companies follows the principles of EU law.

Both schemes provide enhanced tax relief set off against taxable profits for qualifying R&D expenditure. These enhanced reliefs can reduce a company's taxable profits or increase a company's losses. It is also possible for a company to surrender a loss arising from qualifying R&D expenditure in exchange for a cash payment from the UK tax authority (HMRC).

3.7 Hot topics

There is one hot topic worthy of detailed analysis under UK patent legislation, namely that relating to contract manufacturing organisations (CMOs). As stated in section 2.3(c), the financial risks arising for a company from manufacturing problems increase substantially as a drug development project progresses. It is therefore essential to minimise the potential exposure to these by putting in place appropriate contract manufacturing agreements as early as the preclinical stage if the company is seeking the benefits of efficiency and expertise that can come with outsourcing manufacture to a CMO. However, because many of the leading CMOs are major companies, they can be relatively inflexible in the terms they will agree to in a contract, particularly with small-to-medium-sized pharmaceutical or biotech companies. Set out below are a number of the important issues that need to be considered when seeking to use a CMO.

(a) ***Limiting liability***

CMOs will typically wish to limit their obligations to producing a contract drug to comply with the customer's specification and in accordance with cGMP (current good manufacturing practice). Although CMOs cannot (by law) exclude their liability for death or personal injury as a result of their negligence, they will seek to exclude liability to their customers for indirect, special or consequential losses or for any loss of profits, revenue, business opportunity or goodwill. With such a limitation, customers will be unable to recover damages for any delay to, or increased costs of, clinical trials, caused by a failure of the CMO to provide a contract-compliant drug. Even if the CMO agrees to replace defective products, it may be reluctant to agree to do so within a fixed time frame. For a biotech company, delays of this kind can be extremely serious financially and at least some guarantees on the timing of remanufacture should be sought.

CMOs will also seek to limit their overall liability to the value of the services they provide, because they do not benefit from any share of the potential upsides to successful clinical development and approval of the drug by way of milestone payments or royalties.

Great care should be taken if any raw materials are provided by the CMO's customer and form a large proportion of the cost of manufacturing a batch of the drug. If there is a failure by the CMO to manufacture the drug to specification or in accordance with cGMP, the customer will be left with the expense of replacing the raw materials. It is not unreasonable for the CMO to take on this risk resulting from their own default, particularly if they have insurance which would cover such costs. If not, the CMO may well seek to increase the cost of the services to cover the cost of insurance or at least resist the replacement of raw materials until the manufacturing process has been tried and tested.

If there is any development component to the manufacturing services, such as the early stages of development, the CMO will be very reluctant to agree to fixed delivery dates for a finished drug. This is likely to be in direct conflict with the need of the company to manage preclinical and clinical development timelines and related costs, and there are limits to how this can be addressed. Nevertheless, negotiating some flexibility in relation to timing can help to mitigate this risk.

(b) ***Partnering issues***

Limitations on funding can require biotech companies to partner with pharmaceutical companies earlier and earlier in the development of a drug, potentially accentuating outsourcing risks. An increasingly common occurrence in partnering agreements is for the pharmaceutical company to financially support the preclinical and clinical development of a drug candidate to the stage at which it can opt to commercialise the drug and transfer it in-house. The biotech company must allow sufficient flexibility in the level of financial contribution that it seeks from the pharmaceutical company to deal with additional costs that may arise in the manufacturing process given its inherent risks.

For small-molecule drugs, a pharmaceutical partner taking over the clinical development of the compound may want to manufacture the drug itself or enter into an agreement with a third-party manufacturer. In some circumstances, particularly as regards complex biological molecules where the biotech company has more expertise, it may be desirable for the biotech company to continue its involvement in the supply of the drug to the pharmaceutical partner for the purposes of clinical trials and potentially even commercialisation.

The risk for the biotech company in this arrangement is that it might negotiate a supply agreement with its pharmaceutical partner which stretches it beyond the supply obligations that it can negotiate with its CMO. Furthermore, delivery of non-compliant batches, or delay, by the CMO could leave the biotech company exposed. Furthermore, if the biotech company is an intermediary in the supply chain and adverse events occur in a trial or the trial drug is found to have a latent defect, the company risks exposure not just to the costly process of determining the cause, but also to any liability to trial subjects that results.

It is important that any agreement with a CMO contains strong auditing provisions, such as a right to a minimum number of audits or inspections each year as well as unlimited additional rights in the event of a quality issue. Additional rights should also be sought to attend any audit or inspection by relevant regulatory authorities. A notification of inspection provision will therefore be required, and provision for any licensee of the drug programme to attend is also desirable.

(c) Transfer to a new manufacturer

At some point, as development of a drug progresses from preclinical studies through clinical trials to commercialisation, a new CMO may be needed. This may be for a number of reasons, including a failure of the original CMO to produce the drug product to specification, delays in supply, an inability of the original CMO to supply greater volumes, or CMO insolvency. However, the standard-form agreements of many CMOs often contain provisions limiting the customer's ability to transfer manufacturing, particularly in relation to biologicals.

If a biotech company is providing the process to the CMO, there will be no technology proprietary to the CMO that is essential to the manufacture of the drug. Therefore, it should be possible as a condition of transfer to the CMO that it will facilitate subsequent transfers to another CMO if required. On the other hand, if the new manufacturer intends to contribute significantly to the scale-up of the manufacturing process or has otherwise contributed substantially to the development of the process, it might seek to negotiate financial compensation for the loss of its mandate in circumstances where it is not in default.

(d) What intellectual property is needed by the new manufacturer?

Generally, CMOs will readily assign rights in all foreground intellectual property relating to the product, or exclusively relating to the manufacture of the product, to the new manufacturer. However, they may want to retain rights to the more general foreground intellectual property developed during manufacture to the extent that it is applicable to the manufacture of other products.

Even so, given that it has paid service fees, the biotech company may wish to own all foreground intellectual property. In practice, however, an exclusive or even non-exclusive licence to use such foreground intellectual property in relation to the product may be sufficient for the purposes of transferring manufacture if this is then sub-licensed to the new manufacturer, ideally along with copies of any documentation in which know-how subsists.

Ideally, to avoid the possibility of any later allegation of infringement, a licence under the CMO's background process intellectual property should also be obtained solely to the extent necessary for the exploitation of the foreground intellectual property.

(e) ***How will the transfer practically take place?***

In the event that the relationship breaks down or the manufacturing agreement is terminated as a result of breach by the CMO or the CMO's insolvency, it will be essential for the customer to be able to transfer the process as quickly as possible to another manufacturer. This will be much more easily achieved if any necessary proprietary know-how and documentation is placed in escrow, for automatic release in the event of breach or insolvency.

Obligations to provide assistance and cooperation with a transfer should be included in an agreement with a CMO in respect of any know-how required for subsequent manufacture by a third party, as well as the training of technical personnel involved.

The authors would like to thank the following for their contributions to this chapter: James Marshall, Edward Vickers, Matthew Royle, Talitha Shkopiak, Katherine Hudson and Paul England.

United States

Chuck Larsen
Anita Varma
Ropes & Gray LLP

Intellectual property transactions are at the strategic heart of most life sciences companies. A key goal of such transactions is generally access – one party possesses intellectual property rights and desires to monetise those rights, while the other desires to obtain access to those rights.

Patent transactions, often accompanied by know-how and other intellectual property rights, are the most common transactions in the life sciences industry. A variety of transactional structures are available, and the parties will generally select the structure that best meets their business objectives. Once the appropriate structure is chosen, the parties will memorialise it in a written contract.

1. Patents, patent term and know-how: foundations for transactions

1.1 Patents and patent term

(a) *The fundamentals of US patent rights*

Patent rights in the United States are very much like patent rights granted in most major jurisdictions: the patentee has the right to exclude others from making, using, offering for sale, importing or selling the claimed invention in the United States. If the claimed invention is a process, the patentee has similar rights with respect to products made by that process even if the process is conducted outside the United States.

Although the current statutory term of a US utility patent is 20 years, the term has varied historically from as low as 14 years (during 1790–1835) to as high as 21 years (during 1836–1860). For more than 130 years after that (1861–1994), the term of a US patent was 17 years from issuance of the patent. The current 20-year patent term was adopted in 1995 to conform to the General Agreement on Tariffs and Trade (GATT) and the Trade Related Aspects of Intellectual Property Rights (TRIPs) Agreement. This 20-year term is calculated from the earliest effective filing date of the patent application – but note that the filing date of a foreign priority application or a US provisional application is not considered when calculating the 20-year term.

The patent term can be adjusted by two mechanisms. The first is patent term adjustment (PTA),[1] which accounts for delays in prosecution of the patent application; and the second is patent term extension (PTE), which accounts for delays in the regulatory approval of drugs.

1 35 USC §154.

The PTA is available by means of the American Inventors Protection Act, passed in 1999. This statute provides that the term of the patent shall be extended by one day for each day the US Patent and Trademark Office (USPTO) delays taking action on the application after an allotted time period. In addition, the statute requires that the USPTO issue a patent within three years after the actual filing date of the application. Again, the term shall be extended one day for each day after the end of that three-year period until the patent is issued. Any PTA accrued, however, is reduced by delays caused by the applicant – including, for example, requests for extensions of time to respond to USPTO Office Actions (see section 1.1(b) below).[2]

The PTE process arises from the Drug Price Competition and Patent Restoration Act (commonly referred to as the Hatch–Waxman Act). PTE is designed to compensate a patentee for at least a portion of the time – up to five years – taken to obtain regulatory approval. The statutes embodied in 35 USC §156 and 37 CFR §1.710 provide the framework for calculating the length of the relevant PTE. The scope of patent rights during the PTE term is limited to the approved product, an approved use of the product or a method of manufacturing the approved product, all of which depend on the claims of the extended patent. A patent benefiting from PTE will have its term extended from the original expiration date of the patent, which includes any PTA. Importantly, only that portion of the regulatory review period occurring after the date the patent is issued is considered in the PTE determination. PTE is analogous to the supplementary protection certificate (SPC) granted in Europe. See also section 3.1(f) below.

(b) Brief overview of obtaining a US patent

US patent applications are examined by an examiner at the USPTO, typically in the order in which they are received. Each patent application is classified based on the claimed technology and is assigned to a particular examination unit known as an 'art unit'. Art units are made up of examiners with educational backgrounds suited for examining particular technologies. Patent office examination occurs only between the examiner and the applicant or the applicant's authorised agent who is qualified to practice before the USPTO.

As an initial matter, patent applications are reviewed for formalities, such as the format of the application (eg, the specification, claims, abstract and drawings), payment of the requisite fees, and provision of required documents (eg, inventor oaths and declarations). Once the formalities are completed, substantive examination takes place. Communications regarding examination are in the form of 'Office Actions' drafted by the examiner. As a first step, the examiner will determine whether the claims are drawn to more than one invention. As a general practice, examiners restrict the application to one invention. The applicant then has the opportunity to decide ('elect') which invention to pursue. This initial review is called a 'restriction requirement' and is analogous to the unity rejections made by the European Patent Office. The applicant has the opportunity to pursue the unelected

2 Note however that accrued PTA cannot extend a patent's term beyond the date in a terminal disclaimer, if such a disclaimer is filed.

inventions in 'divisional applications'. In the next step, the examiner will review the claims for compliance with the various patent statutes.[3] Any claims failing to satisfy a statutory requirement are rejected. An Office Action rejecting a claim will contain reasons for the rejection.

An applicant may respond to claim rejections in Office Actions by either arguing that the rejection is improper or amending the claims to address the examiner's concerns. The examiner and applicant may exchange proposals and arguments several times until either reaching agreement on patentability or final rejection. If agreement is not reached, the applicant has the right to appeal.

An initial appeal is made to the Patent Trial and Appeal Board (PTAB), which consists of a panel of three administrative patent judges. If the PTAB agrees with the examiner, the applicant may then appeal to the Court of Appeals for the Federal Circuit (hereinafter the Federal Circuit).

If claims are patentable, the USPTO will issue a Notice of Allowance. After payment of the requisite fees, the allowed claims will issue as a patent. Any PTA accrued during examination will appear on the face of the patent documentation.

1.2 Who owns the patent?

(a) US federal law

Under US patent law, the ownership of a patent initially vests in the named inventors.[4] Inventorship is based on US federal law, and inventorship assessments can be complex – particularly where multiple inventors are potentially involved. Federal law generally provides that parties who jointly contribute to the conception of an invention are considered to be co-inventors.[5] 'Conception' here requires more than a bare idea; rather, the idea must be definite and permanent such that "it involves a specific approach to a particular problem at hand", which must be "sufficiently precise that a skilled artisan could carry out the invention without undue experimentation".[6] The Federal Circuit has observed that "the line between actual contributions to conception and the remaining, more prosaic contributions to the inventive process that do not render the contributor a co-inventor is sometimes a difficult one to draw."[7] As one court put it, inventorship is "one of the muddiest concepts in the muddy metaphysics of the patent law".[8]

Inventorship is fact-specific and is determined on a case-by-case basis. Factors that argue against a finding of joint inventorship may include:

- contributing prior-art elements or general knowledge;
- merely suggesting a desired result without the means to accomplish it;
- following instructions of the conceivers;

3 The principal statutes are: patent-eligible subject matter (35 USC §101), the written description and enablement requirements (35 USC §112), novelty (35 USC §102), and non-obviousness (35 USC §103).
4 *Beech Aircraft Corp v EDO Corp*, 990 F2d 1237, 1248, 26 USPQ2d 1572, 1582 (Fed Cir 1993).
5 See *Fina Oil & Chem Co v Ewen* (Fed Cir 1997).
6 *Burroughs Wellcome v Barr Labs, Inc* (Fed Cir 1994).
7 *Eli Lilly & Co v Aradigm Corp*, 376 F3d 1352, 1359 (Fed Cir 2004).
8 See *Mueller Brass Co v Reading Industries, Inc*, 352 F Supp 1357, 1372 (ED Pa 1972), aff'd, 487 F2d 1395 (3d Cir 1973).

- merely explaining how or why the invention works;
- adopting information derived from another;
- supplying known components or starting material; and
- reducing to practice another's invention.

Statute 35 USC §261 provides that patents are personal property and therefore can be transferred as a property interest:

Subject to the provisions of this title, patents shall have the attributes of personal property. Applications for patent, patents, or any interest therein, shall be assignable in law by an instrument in writing. The applicant, patentee, or his assigns or legal representatives may in like manner grant and convey an exclusive right under his application for patent, or patents, to the whole or any specified part of the United States.

Thus, the inventors may assign their ownership interest in the patent to another party (an assignee) via a written assignment. As with any transfer of property rights, assignment documents must be properly drafted to ensure that the inventor's entire right, title and interest in and to all inventions described in the patent are transferred to the assignee.

In view of this aspect of US law, employers should ensure that the employment, consulting or other agreements provide for adequate transfer of intellectual property rights to the company. These agreements should include prospective assignments, which can be recorded if ex-employees and consultants later refuse to execute specific confirmatory assignments for each patent, or when it is difficult to obtain relevant confirmatory patent assignments in a timely manner.

The exact language used in the patent assignment provision is important because it determines whether or not the assignment would occur automatically or merely creates an obligation to assign in the future.[9] If an automatic assignment is effective, no further act is required on behalf of employer or employee, and ownership of the patent or patent application occurs by operation of law once an invention is made.[10] Contracts that merely obligate an inventor to grant rights in the future may vest an employer with equitable rights in those inventions once made, but do not on their own vest legal title to patents in the employer.[11] Thus, language such as "I hereby agree to grant", "I promise to grant" and "I will assign my right, title, and interest" do not effectuate an automatic transfer and will require additional acts, such as the signing of an assignment, to ensure a transfer of rights to an employer. In contrast, the automatic assignment of an invention as soon as the invention is made can be effectuated by using phrases such as "hereby grants" and "hereby conveys, transfers and assigns".[12]

(b) *US state law*

In addition to the federal patent laws that govern patent assignments, many states have enacted statutes to protect employees from overreaching by their employers.

9 *Abraxis Bioscience, Inc v Navinta LLC*, 625 F3d 1359 (Fed Cir 2010).
10 *DDB Technologies LLC v MLB Advanced Media LP*, 517 F3d 1284 (Fed Cir 2008).
11 *Arachnid, Inc v Merit Industries, Inc*, 939 F2d 1574 (Fed Cir 1991).
12 *Stanford v Roche*, 131 S Ct 2188, 2192 (2011).

For example, state statutes tend to preclude employers from obtaining rights to inventions made outside the scope of an inventor's employment – ie, inventions that were made in the inventor's own time. California Labor Code Section 2870 limits the assignment of inventions to those developed during the course of the inventor's employment using the employer's facilities. Similarly, the Employee Patent Act (765 ILCS 1060) in Illinois states that assignment provisions in employment contracts that assign all rights to the employer do not apply to inventions that were made in the employee's own time and without use of the employer's equipment, supplies, facilities, etc.

1.3 Inventor compensation

Unlike many international jurisdictions, inventors in the United States are generally not entitled to additional compensation for their inventive contributions. An inventor's salary is considered adequate consideration to effectuate the transfer of rights. The adequacy of the compensation has not been the subject of litigation in the United States.

1.4 Know-how and trade secrets

Know-how is an important aspect of most life sciences transactions and in some instances may be the only proprietary right protecting a company's technology. This is particularly true when working with biologics, as patent protection may not be available for the underlying biologic. The only proprietary right in such cases may be the expertise and know-how of the company's scientists. Companies may use trade secret protection and various confidentiality measures to protect these rights.

Confidentiality provisions should define what is meant by 'confidential information' and should cover relevant non-public information that relates to the actual or anticipated business or research and development of the company. As a general principle, it should cover technical data, trade secrets and know-how. These provisions should also preclude employees and consultants from disclosing or using the confidential information for any purpose other than performing services on behalf of the company.

Trade secrets are protected by state law and also by the Uniform Trade Secrets Act (although New York and Massachusetts have not adopted that act). Trade secret protection is not pre-empted by either the US Constitution or the federal patent laws.[13] Under the Uniform Trade Secrets Act, trade secret misappropriation occurs if a trade secret exists and the offending party acquires, discloses, or uses the trade secret improperly.

A trade secret must provide a competitive advantage from not being disclosed. Trade secrets remain protected under the law for as long as they remain secret – a famous example over many decades is the trade secret for the formula used in Coca-Cola. Secrecy is required for these protections to apply – once the subject matter ceases to be secret, it ceases to be protected.

---

13 *Kewanee Oil Co v Bicron Corp*, 416 US 470 (1974).

2. Context and features of patent transactions

The most basic patent transactions involve a direct transfer of rights from one company to another, either by assignment of title or by licence. Other, more complex transactions include mergers and acquisitions, equity and debt financing, joint ventures, divestment, and consortium agreements. In each instance, the patent protection is fundamental to the valuation or structure of the transaction. In some circumstances, life sciences companies enter commercial agreements that are not driven directly by underlying patents but involve patent rights indirectly. Examples of such agreements include employee, consulting, material transfer, manufacturing, distribution, and co-promotion agreements. The following discussion explores these transactional structures in further detail.

2.1 The principal dealings in a patent

Patent assignments and licences are perhaps the most fundamental transactions involving life sciences intellectual property. Assignments and licences share similarities but fundamentally differ according to how the patent's title is handled. In general, patent assignments transfer that title from one party to the other, while in a licence one party grants to the other party only some of the rights in the patent but retains the patent's title.

Transactions with large price tags are becoming commonplace in the life sciences industry. These include mergers and acquisitions, financings, collaboration and joint ventures, and divestments. Often, the principal value driver in these transactions is the intellectual property rights protecting the target company's technology. Comprehensive due diligence is often conducted to ensure that the opposing party possesses the appropriate assets or rights and, where necessary, can transfer these rights to the purchaser/acquirer without impediments, and to ensure that the underlying agreement is drafted to effectuate the terms of the agreement.

Assignments and licences are each explored in further detail below.

(a) Patent assignments

Patent assignments occur in many different contexts in life sciences transactions. They can occur between an inventor-employee and the employer, between consultants and companies, or even between two companies as part of more complex transactions, such as asset purchase agreements. Under an assignment structure, the patent owner (the assignor) transfers an interest (either full or partial) in the patent's title to a third party (the assignee). The transfer occurs by a written contract. Upon execution of the assignment contract, the assignee takes title to the patent and thereafter owns all rights (or a joint right) in the patent, including the right to maintain, enforce and monetise the patent by licensing-out or by transfer.

The form of the written contract can be brief but it must contain the components necessary to create an effective assignment. At a minimum, the assignment must recite the two parties involved, provide a granting clause that transfers the patent's title to the assignee, and refer to the consideration for the transfer. The contract is typically signed by both the assignee and assignor. The effectuation of the assignment is critical to an assignee's ability to later sue a third party for patent infringement under the acquired patents.[14] In addition to the minimum components, if the parties intend the assignee to obtain the right to seek and obtain retroactive damages for past infringement, that right must be set forth explicitly in the contract.[15]

Another important component is to transfer the underlying inventions disclosed in the patent (and continuations and foreign counterparts based on those inventions) in addition to transferring the patent itself. That practice was validated in a recent case by the US Court of Appeals for the Federal Circuit,[16] which held that an assignment of the "inventions and discoveries set forth in a [patent application]" assigns not just the patents that issued from or are related to that application, but also another patent if it claims "inventions and discoveries" set forth in the patent application. In that case, the patent at issue was based on disclosure included in the patent application that was assigned, where "[t]he specification of the [transferred patent application] supports the claims of the patent, meaning it is within the 'inventions and discoveries' in the [patent application] and was therefore assigned."

An assignment contract is typically recorded in the USPTO Assignment Database to notify the public of the ownership change. In some circumstances, transfers not properly recorded can result in a loss of title to the patent or a loss of lien priority to the patent.[17] This principle is commonly known as the 'bona fide purchaser' rule.

(b) Patent licences

The parties to a licence agreement include the patent right holder (referred to as the licensor) and the party receiving the licence (the licensee). The parties have complex but generally aligned interests in the use of the patent(s) being licensed. In a licence, the agreement transfers a right of use, but not full legal title in the patent or a full undivided share of the entire patent. Licences can be structured in multiple tiers – for example, the licensor may permit the licensee to grant sub-licences to third parties.

Licensors commonly include emerging companies and universities, and organisations that have access to patented life-science-based technology but lack the resources to fully develop and commercialise the technology. These organisations seek to monetise the technology by licensing it to other entities that have the resources. Venture capital or private equity financing generally take companies through early product safety and preclinical studies, but at the clinical development stage most companies partner with large pharmaceutical or medical device companies.

Licensing contracts are commercial contracts and, as such, can be considered as extensions of credit. That is, the licensor provides intellectual property rights to a licensee in reliance on the licensee performing certain obligations. The licensee typically bears the bulk of the financial risk in commercialising the technology covered by the licensed intellectual property rights. Accordingly, both parties seek

14 See *Abraxis Bioscience Inc v Navinta LLC*, 625 F3d 1359, 1367 (Fed Cir 2010) cert denied; and *APP Pharmaceuticals LLC v Navinta LLC*, 132 S Ct 115.

15 See *Abraxis Bioscience Inc v Navinta LLC*, 625 F3d at 1367.

16 *MHL Tek LLC v Nissan Motor Co* 655 F3d 1266 (Fed Cir 2011).

17 See 35 USC §261: "An assignment, grant, or conveyance shall be void as against any subsequent purchaser or mortgagee for valuable consideration, without notice, unless it is recorded in the Patent and Trademark Office within three months from its date or prior to the date of such subsequent purchase or mortgage."

contractual protections against potential problems that may be caused by the other party's failure to perform, and those contractual protections form the basis of the written licence agreement. Although the form of the agreement may vary, most licence agreements include at least the topics set out next.

Scope of rights: The granting clause and related provisions define the scope of rights in the licence. As the principal component of the licence agreement, the granting clause specifically identifies the particular patents being licensed and the extent of the licence rights (ie, whether the licence includes the right to make, have made, use, sell, offer for sale and import,[18] or only a subset of those rights). Importantly, the scope of rights also defines exclusivity provisions and other limits imposed on the parties' right to practise the patents.

The exclusivity provisions are structured according to the purpose of the licence, which generally includes at least one of two main goals. The first is to provide the licensee with 'freedom to operate', which is non-exclusive immunity from an infringement suit under the licensed patents. A non-exclusive licence granted to the licensee still allows the licensor to license the patents to other parties without breaching the agreement with the licensee. Non-exclusive licences are often preferred by licensors who own platform or enabling technologies as they can be extremely lucrative, and examples of such platform/enabling technologies include Columbia University's Axel patents, Genentech's Cabilly patents and the Cohen-Boyer PCR patents. A second goal is to provide the licensee with substantive commercial or development rights. Those rights are often exclusive, at least with respect to certain fields and geographies. An exclusive licence typically precludes the licensor from practising the licensed intellectual property within the scope of the licence granted to the licensee.

In addition to exclusivity provisions, the scope of rights may be limited by other terms, including, for example, field-of-use restrictions (eg, therapeutic areas or disease indications), limitations on the products authorised under the licence, and geographic limitations. Geographic limitations allow the patent owner to subdivide the global market for the patented products. One benefit is that the licensor can select local partners who are most familiar with the regulatory and commercial aspects of the foreign markets. Limitations based on indication allow a patent owner with patents directed at pharmaceutical formulations for a new chemical entity to grant separate licences to multiple parties under the patents according to specific indications. Although licensors may have incentives to finely parse their portfolios by granting licences to multiple parties under the same patents in separate fields, geographies or other parameters, licensees need to exercise caution because the licence may limit the licensee's ability to either list the patented products in the US Food and Drug Administration's so-called 'Orange Book' (which contains a list of approved drug products with therapeutic equivalence evaluations) or enforce those patents against generic companies or other competitors.

---

18 The enumerated rights in most licences follow the US law set forth in 35 USC §271(a): "Except as otherwise provided in this title, whoever without authority makes, uses, offers to sell, or sells any patented invention, within the United States, or imports into the United States any patented invention during the term of the patent therefor, infringes the patent."

The scope of rights provisions may also identify reservations of rights by the licensor. For example, university licensors will routinely reserve the right to continue using the licensed technology for research purposes. Other common examples of reserved rights include rights to manufacture and supply the product to be sold, rights to co-market the product with the licensee, or rights to fully exploit the patents outside the licence scope. Federal law allows a patentee the discretion over whether and to whom to issue a licence, and the scope of that licence.[19]

It is necessary to keep in mind that patent licences are licences to intellectual property, and therefore a licensee receives special protection under US bankruptcy law in the case of a licensor becoming bankrupt. Under the US Bankruptcy law the patentee/licensor is deemed a debtor, and the licence is an executory contract under which the debtor is a licensor of a right to intellectual property. If the patentee's bankruptcy trustee or debtor-in-possession rejects the licence as an executory contract in the bankruptcy, the licensee under the agreement may elect to retain its rights to use the intellectual property (including its exclusivity provisions) as such rights existed immediately before the commencement of the bankruptcy case, for the duration of the agreement.[20]

Covenants: Most patent licence agreements include one or more express covenants. An express covenant is a promise by a party to take certain actions or refrain from certain actions for the benefit of the other party. Examples include covenants of confidentiality, non-competition and non-solicitation, use of diligent efforts to conduct product development or commercialisation under the licence, and manufacturing and supply commitments. When conducting due diligence and drafting a sub-licence, particular attention must be paid to the covenants entered into by the primary licensee. Covenants that might be accepted by emerging company licensees might not be acceptable for large pharmaceutical companies. For example, emerging companies pursuing a programme primarily in a specific indication often agree not to compete in other indications. But a large pharmaceutical company wanting to later acquire the emerging company may not be able to abide by such a covenant.

In addition, licence agreements may be subject to implied covenants under state laws. For example, the laws of most states generally require the parties to exercise good faith and fair dealing with each other. Unless explicitly stated otherwise in the agreement, this requirement can impose a duty on the licensee to exercise significant efforts to develop and exploit the licensed technology. For example, an exclusive licensee has an implied duty to use reasonable efforts or even best efforts to exploit the exclusive licence when that licence is granted in exchange for a promise to pay royalties to the licensor.[21]

19 See 35 USC §271(d): "No patent owner ... shall be ... deemed guilty of misuse or illegal extension of the patent right by reason of his having ... (4) refused to licence or use any rights to the patent."

20 See 11 USC §365(n)(1)(B).

21 See *Wood v Lady Duff-Gordon*, 222 NY 88, 118 NE 214 (1917); see also *Perma Research & Development Co v Singer Co*, 402 F Supp 881, 896–97 (SDNY 1975) (Where an exclusive licensee and assignee of the licensor's device could have done "any number of reasonable things ... to perfect the device without unreasonable cost or effort" but it did none of them, the licensee has violated its implied duty to use best efforts).

Implied covenants may also impact the licensor. For example, a licensor that grants exclusive rights to a licensee is subject to an implied covenant not to undermine the scope of those rights. This is applicable where the product sales are intended to generate royalties for the licensor, as is often included in pharmaceutical licensing agreements.[22] The actual language of the exclusivity grant and related limitations is important. As some courts have found, a "covenant [not to exploit the licensed intellectual property] cannot be implied if the parties have either expressly dealt with the matter in the contract or have left the agreement intentionally silent on the point."[23]

The standard for implied covenants may differ among states. The courts will consider various factors, including the express language of the agreement, the payment structure and fairness of the agreement, when assessing the existence and scope of an implied covenant. Indeed, a court will imply a duty on the part of an exclusive licensee to exploit the subject matter of the licence with due diligence where such a covenant is essential as a matter of equity to give meaning and effect to the contract as a whole; but no implied efforts obligation is required where a substantial upfront royalty is provided, the contract is silent on best efforts, and a merger clause is incorporated into the contract.[24]

Royalties and other financial terms: Not surprisingly, the financial terms are often among the most heavily negotiated provisions in a licence. The provisions usually include one or more of upfront payments, royalties and milestone payments, and may include equity. Upfront payments are often made in consideration for technology access or to compensate the licensor for previous investments in intellectual property or product development. The level of payment may vary according to the level of risk in the technology being licensed – for example, licensees of early-stage technology will often seek to defer payments until regulatory and commercial risks are resolved.

Royalties are commonly structured so they are paid according to the sales of the underlying product. Royalties may be reduced according to generic entry and competition. In pharmaceutical collaborations, the royalties are often tiered according to the successful sales of the product. That structure may decrease the royalty rate according to product sales bands, such that a lower royalty rate applies to sales that exceed predetermined levels. Milestone payments in pharmaceutical collaborations are typically one-time payments made when development, regulatory, or sales milestones are reached. Regulatory approval is a common milestone used to trigger payment. Equity is usually reserved for acquisitions and venture financings,

---

22 See *Parev Prods Co v I Rokeach & Sons, Inc*, 124 F2d 147, 150 (2d Cir 1941) ("a person who parts with his formula for exploitation [by a third party also creating other products] should be given protection from a court of equity ... a court would hardly have permitted [a defendant exclusive licensee] from the inception of this contract to lock up the [licensor]'s formula in a vault and freely market [a competing product].") But see *Bloor v Falstaff Brewing Co*, 601 F2d 609, 613 note 7 (2d Cir 1979) (a licensee "having the exclusive right to sell the [licensor's] product may sell a similar product if necessary to meet outside competition, so long as he accounts for any resulting losses the [licensor] can show in the sales of the licensed product." citing *Parev*, 124 F2d at 150–1).

23 *So Good Potato Chip Co v Frito-Lay, Inc*, 462 F2d 239, 241 (8th Cir 1972) (citing *Glass v Mancuso*, 444 SW2d 467, 478 (Mo 1969)); *Cruz v FXDirectDealer LLC*, 12-1252-cv, slip op at 15–16 (2nd Cir June 19 2013).

24 See *Permanence Corp v Kennametal, Inc*, 725 F Supp 907, 909, 11 USPQ2d 1860, 1863 (ED Mich 1989).

although it is occasionally used as consideration in collaborations, for example in university spin-off arrangements.

One particular issue involves the legality of requiring royalty payments after the related patents expire. In general, a patentee's use of a royalty agreement that extends beyond the expiration of the patent(s) is unlawful as misuse of the patent – but there are exceptions.[25]

Patent licences often include trade secrets or know-how in addition to the patents. The payments should be allocated between patent and non-patent rights, for example by a reduction in royalty rate when the patent expires, in order to avoid being construed as an illegal tying arrangement.[26]

Intellectual property: Intellectual property provisions can be complex and subject to extensive negotiation. They typically involve the structure for licensing of core technologies, ownership of improvements that arise from the dealings between the parties, and rights of enforcement. Such issues are discussed below in the context of a simple licence agreement. More complex issues involving the ownership of improvement patents are discussed below in the context of joint venture and collaboration agreements (see sections 2.2 and 2.3 below).

A frequent question that arises in licensing agreements with regard to licensing the core technology is whether there is any meaningful difference between patent rights granted via an affirmative licence as compared with rights given under a 'covenant not to sue'. When considering this question in the context of a patent exhaustion issue, the Federal Circuit held that "[t]he difference [between a covenant not to sue and a licence] is only one [of] form, not of substance – both are properly viewed as 'authorisations.'"[27]

Federal courts also have found that licences and covenants both run with the patent in the event of a sale of the patent: "An assignee [of a patent] acquires title subject to prior licenses of which the assignee must inform himself as best he can, and at his own risk."[28] See also *Medtronic AVE Inc v Advanced Cardiovascular Systems*,[29] which established that a covenant not to sue would pass to a patent assignee when the assignee acquired substantially all the assets of the assignor (including the settlement agreement in which the covenant was contained). Further, it was found in *V-Formation, Inc v Benetton Group SPA and Rollerblade Inc*[30] that a covenant not to sue runs with the patent where the agreement expressly provides that the agreement was binding on the parties' assignments.

25 See *Brulotte v Thys Co*, 379 US 29 (1964); but see also *Bayer v Housey Pharmaceuticals, Inc*, 228 F Supp 2d 467 (D Del 2002), which established that a licence agreement under which a patentee received royalties for 10 years after the expiration of a patent on pharmaceutical products that were developed during the life of the patent, using a method licensed under the patent, was not per se patent misuse.

26 See, eg, *Pitney Bowes, Inc v Mestre*, 701 F2d 1365, 1372 n.12 (11th Cir 1983), cert denied, 464 US 893 (1983); *Meehan v PPG Industries*, 802 F2d 881, 886 (7th Cir 1986), cert denied, 479 US 1091 (1987) ("it is true ... that parties can contract for trade secret payments to extend beyond the life of a patent, [but] there must be some provision that distinguishes between patent royalties and trade secret royalties"); *Boggild v Kenner Prods*, 853 F2d 465, 469–70 (6th Cir 1988) (hybrid agreement unenforceable in the absence of language allocating royalties between patent and non-patent rights).

27 *TransCore v ETC*, 563 F3d 1271, 1275 (Fed Cir 2009).

28 *Keystone Type Foundry v Fastpress Co*, 272 F 242, 245 (2d Cir 1921).

29 247 F3d 44, 58 USPQ2d 1596 (3rd Cir 2001).

30 02 CV 02259 PSF CBS, 2006 WL 650374, at *4 (D Colo, March 10 2006).

Non-exclusive licences and covenants not to sue have also been treated equally in assessing their assignability – see *General Mills v Kraft Foods*[31] (a non-exclusive patent licence is not assignable without the permission of the patent holder) and *Hilgraeve Corp v Symantec Corp*[32] (a covenant was not a transferrable licence where there was no permission to assign). However, in some cases additional implied covenants can apply to covenants not to sue. The Federal Circuit found that a patentee was legally estopped from asserting a patent against a third party, where the patent issued after the patentee granted a covenant not to sue to the third party and the later-issued patent was broader than, and necessary to practise, the patents in the covenant.[33] The court's position was that the covenant would have been meaningless if the patentee were allowed to block its use via a later-issued, broader patent.[34] The parties should consider specifying the patents that are included in the covenant and those that are excluded. It is unclear whether similar implied covenants would apply in the case of a licence.

In relation to grant-back provisions for improvements to the patent product or process, a licence agreement, possibly in its scope of rights section, will often establish a process for allocating rights in improvements and downstream intellectual property that may arise from subsequent activities by the parties. Unlike various international jurisdictions, including Europe, the United States does not preclude or restrict agreements that require the parties to include future improvement patents in the licence or provide a grant-back licence under the improvements.[35] Large pharmaceutical companies that acquire emerging companies should assess the scope of grant-back licenses to determine whether their own technology would also flow back to the initial licensor.

The right of enforcement for a licensed patent may also be negotiated in a licence agreement. As a matter of first principles, the patentee, by statute, has the right to enforce the patent, a right which is called its 'standing'. See 35 USC §281, which provides: "A patentee shall have remedy by civil action for infringement of his patent." That right is also available to successors in title to the patent, because 35 USC §100(d) provides: "The word 'patentee' includes not only the patentee to whom the patent was issued but also the successors in title to the patentee."

Whether or not a licensee can also have the right to enforce is a matter of federal law and depends on the proprietary rights which that party holds in the patent at the time of the infringement. To have standing, a party must have an exclusive proprietary interest in at least a part of the patent.[36] Moreover, an exclusive licensee

---

31 487 F3d 1368, 1373–4 (Fed Cir 2007).

32 265 F3d 1336, 1346 (Fed Cir 2001).

33 *TransCore v ETC*, 563 F3d 1271 (Fed Cir 2009).

34 See *General Protecht Group v Leviton*, 2011-1115 (Fed Cir 2011) (subsequent continuation patents were included in the covenant).

35 See *Transparent-Wrap Mach Corp v Stokes & Smith*, 329 US 637, 72 USPQ 148 (1947), reh'g denied, 330 US 854 (1947) (provisions in a patent licence requiring the licensee to offer to assign any patentable improvements developed by the licensee back to the licensor are not per se illegal or unenforceable).

36 See *Ortho Pharmaceutical Corp v Genetics Institute, Inc*, 52 F3d 1026, 1032–4 USPQ2d 1444, 1450 (Fed Cir 1995), cert denied, 116 SCt 274 (US 1995) (licensee must have beneficial ownership of some of the patentee's proprietary rights to exclude others from making, using, or selling the invention; non-exclusive licensee has no standing to join in a patent infringement suit).

who has the right to exclude others from practising the claimed invention within a licensed field has proprietary rights in the patent and has standing to sue third parties who operate within the licensed field.[37] In some cases, the patent owner must be joined as an indispensable party to the case.[38] A non-exclusive licensee holds no legal right in the patent, and therefore has no standing to bring a claim or join in a claim to enforce the patent irrespective of whether that right is conveyed in the licence.[39]

Similarly, a licensee's grant of an exclusive sub-licence to all of its licence rights can eliminate the licensee's standing to sue for infringement irrespective of the enforcement terms of the agreement. Also note that co-owners of a patent must join in an infringement suit against a third party. In *Ethicon Inc v US Surgical Corp,*[40] Ethicon was unable to prosecute an infringement claim because the other co-owner of the patent had granted a licence to the defendant. This case highlights the importance of ensuring that inventorship of a patent has been appropriately assessed. The co-owner in the *Ethicon* case was an inventor who had not been named as an inventor on the patent in the first instance. The importance of inventorship is also highlighted in the *Yeda Research v Imclone Systems* case,[41] where the district court found that the originally listed six inventors had not been properly listed and found three inventors from the Weizmann Institute to be the proper inventors, resulting in a change in ownership of the patent covering the Erbitux drug. The incorrect listing of inventorship impacted ownership, and thereby all downstream licences granted by the original owner.

Patent marking: Patent licence agreements will often include a provision that requires the licensee to 'mark' the licensed products if those products are covered by a licensed patent. US law does not require a party to mark its products as being patented, but companies are incentivised to do so. 'Marking' notifies the public that the products are patent protected. Proper notice (via properly marking the product) is required to obtain damages for infringement when the patent is enforced. Statute 35 USC §287(a) governs the marking requirements and states:

> *Patentees, and persons making, offering for sale, or selling within the United States any patented article for or under them, or importing any patented article into the United States, may give notice to the public that the same is patented ... In the event of failure so to mark, no damages shall be recovered by the patentee in any action for infringement, except on proof that the infringer was notified of the infringement and continued to infringe thereafter.*

---

37 See *Independent Wireless Telegraph Co v Radio Corporation of America,* 269 US 459, 475 (1926) (standing granted to exclusive licensee to sue third party for using patented invention for commercial radio purposes); *Intellectual Property Development, Inc v TCI Cablevision of California, Inc,* 248 F3d 1333, 58 USPQ2d 1681 (Fed Cir 2001) (standing accorded where licensee received "exclusive licence to make, use, and sell the inventions"); *Hill Phoenix, Inc v Systematic Refrigeration, Inc,* 117 F.Supp.2d 508, 57 USPQ2d 1228 (ED Va 2000) (exclusive sub-licensee had standing).

38 See *Intellectual Property,* above at n37, at 1688.

39 *Western Elec Co v Pacent Reproducer Corp,* 42 F2d 116, 118 (2d Cir), cert denied, 282 US 873 (1930) ("A non-exclusive sub-license grants to the licensee merely a privilege that protects him from a claim of infringement by the owner of the patent ... He has no property interest in the monopoly of the patent.")

40 135 F3d 1456 (Fed Cir 1998).

41 *Yeda Research v Imclone Systems, Inc,* 443 F Supp 2d 570 (SDNY 2006).

The fine for false marking is USD500 per offence.[42] The America Invents Act established that parties can have standing to sue for false marking only if they can prove that they have a competitive injury by the false marking.[43]

IP warranties and indemnities: Warranties and indemnities are important provisions in many licences. Licences typically include warranties about the patentee's ownership of the intellectual property. Examples may include warranties that all intellectual property relevant and/or necessary to the business objective has been included within the scope of the licence and that the patentee has the right to the intellectual property. Other warranties may include the absence of infringement of third-party patents and threats of such infringement, such as litigation, cease-and-desist letters or invitations to license.

Indemnities are structured to address general risks. Parties may also seek special indemnification for specific issues identified during due diligence and negotiation of the agreement. Those indemnity requirements should be carefully drafted as they might be construed narrowly.[44] If post-closing business activities are to be covered – for example, commercialisation of therapeutics that are in development at the time of the acquisition – the parties should consider making the point explicit. In *Owens-Illinois v BTR plc*,[45] the court refused to require the seller (BTR) to indemnify the purchaser (Owens-Illinois) for breach where BTR had warranted that it "owned or had licenses to all the intellectual property necessary to operate the business 'as currently conducted'". At the time of the transaction, a product was being developed but was not being sold. Post-acquisition, the purchaser launched the product and was sued for patent infringement. The court held that in order for the indemnity to apply, it would have to interpret the "as currently conducted" limitation so as to exclude it from the agreement. In most life sciences transactions, at the time of the transaction the products are still in development. This case illustrates why some parties try to obtain more specific IP representations.

Some companies carry insurance to cover liability for adverse legal judgments, such as a patent infringement judgment. It should, however, be kept in mind that many policies exclude wilful misconduct. In some cases, insurance carriers have refused to cover the defendant's liability where wilful infringement was found, on the grounds that the wilfulness finding disqualified the coverage.[46]

Term and termination: Licences are principally commercial contracts. States generally have a presumption against allowing commercial contracts to extend indefinitely, unless explicitly provided in the agreement. If the term of an agreement is not sufficiently limited in the agreement itself, the law may imply that it has no term and permit either party to terminate at will.[47]

---

42 35 USC §292(a).
43 35 USC §292(b).
44 See *Hooper Associates Ltd v AGS Computers, Inc*, 74 NY2d 487, 491 (1989).
45 2010 WL 2607146 (SDNY 2010).
46 See, for instance, *Davila and Rivera v Arlasky and Mulkerin*, 141 FRD 68 (ND Ill 1991) (denied on other grounds).

In many cases, licensors include the so-called 'Medimmune provisions' that attempt to dissuade a licensee from challenging the validity of the licensed patents. The US Supreme Court has held that licensees need not terminate the licence prior to challenging the validity or enforceability of licensed patents.[48] The Second Circuit has held that a broad patent no-challenge clause in an agreement was unenforceable on public policy grounds. This case calls into question some of the techniques that licensors have adopted following the Supreme Court's *MedImmune* decision to try to restrict licensees' ability to challenge licensed patents.[49]

2.2 Patent rights in corporate transactions

(a) General considerations

The basic principles governing patent and know-how rights in complex transactions are similar to the principles discussed above for assignment and licences. As with licences, parties will often conduct due diligence that considers many of the same issues, such as:

- ownership and scope of the patent rights;
- recordation of prior title transfers;
- scope of inbound licences, financial obligations and penalties, potential for infringement liability, and the need for licensor consent in connection with the contemplated transaction;
- outbound licence agreements and restrictions on the target's patents or business; and
- representations, warranties and indemnification obligations previously granted by the target to third parties.

A party's due diligence may be confirmed or supplemented by drafting an appropriate set of representations and warranties. Representations and warranties provide a powerful tool for ensuring complete disclosure by the target company. Generally, target companies are asked to provide representations and warranties addressing one or more of the following:

- the accuracy and completeness of patents listed in the disclosure schedules;
- the validity and enforceability of listed patents;
- exclusive ownership of patents that the target purports to own;
- sufficiency of all owned and licensed patents/know-how necessary to

47 See, eg, *Trient Partners I Ltd v Blockbuster Entertainment, Inc*, 83 F3d 704 (5th Cir 1996) (licensing contract was terminatable at will under Texas law because its only termination grounds were the occurrence of an incurable default, death of a party, change of control, or insolvency, which were found to be events that "are likely never to transpire"); but compare *Zee Med Distrib Ass'n v Zee Med, Inc*, 94 Cal Rptr 2d 829 (Ct App 2000) (implied duration limit found, so contract not terminable at will, where express provisions stated that termination was permitted in the case of bankruptcy by a party, a party's failing to carry on in the ordinary course of business, and material breach not cured within 30 days); *Haines v City of New York*, 364 NE2d 820, 822 (1977) (New York courts will "inquire into the intent of the parties and supply the missing term if a duration may be fairly and reasonably fixed by the surrounding circumstances and the parties' intent").

48 *MedImmune, Inc v Genentech, Inc*, 549 US 118, 135 (2007).

49 *Rates Technology Inc v Speakeasy, Inc*, 685 F3d 163 (2d Cir 2012).

conduct its business, including the right to make, have made, use, sell, offer to sell and import its products;

- accuracy and completeness of disclosure schedule listings of material inbound and outbound licences;
- measures taken to maintain trade secrets;
- any royalty interest or licence right in the target's patents (except as disclosed);
- non-infringement of third-party patents by the target's business and the target's products; and
- special indemnities – and the parties may also seek special indemnification from the target for certain specific issues identified during due diligence and negotiation of the representations and warranties.

In addition to the general considerations discussed above, each type of transaction may have specific considerations that impact its structure and strategy. There follows a review of four common transaction structures and examples of patent issues that arise in those transactions.

(b) Mergers and acquisitions

Merger and acquisition (M&A) transactions involve one entity acquiring a controlling interest in another company or its technology. These transactions are usually performed by equity sale, merger or asset sale, each of which is described further below.

Many of the provisions discussed above in regard to assignments and licences apply to these transactions and will be considered in both the drafting and due diligence phases. For example, M&A transactions are similar to licences when the purchase prices include so-called 'earn-out payments', which are similar to royalties in that part of the purchase price is paid in the form of a percentage of sales of the acquired business.

Equity sales: Equity or share transfers result in the transfer of a controlling equity from the target to the purchaser. With that transfer, the purchaser assumes control over the target company's assets, rights and liabilities, including its patents and contractual rights relating to patents by virtue of owning the stock in the target. But title to the assets remains vested with the target. The transfer of equity ownership of a company in most cases does not constitute an assignment of any intellectual property licences that the company may hold.[50] Patents used by the target company pursuant to contractual rights may be subject to change-of-control provisions requiring third parties to consent to the continued right to use the patents.

Mergers: In a merger, all assets and liabilities of the target are acquired by operation of state merger laws, which take effect when a certificate evidencing the merger is filed with the office of the secretary of the relevant state where the target was incorporated.

---

50 *PPG Industries, Inc v Guardian Industries Corp*, 597 F2d 1090, 1096 (6th Cir 1979).

The ownership of the patents after the transaction depends on the 'direction' of the merger: in a reverse triangular merger, the patents owned by the target company will continue to be owned by the target entity, whereas in a forward merger the title will transfer to the surviving entity as if it were assigned. Patents used by the target pursuant to contractual rights may be subject to assignment or change-of-control provisions requiring third parties to consent to the continued right to use the patents.

A forward merger occurs where a licensee is merged into a parent company or another entity. It constitutes an assignment of the licence rights by operation of law. In the context of a patent licence, a transfer occurs any time an entity other than the one to which the licence was expressly granted gains possession of the licence.[51]

In theory, a reverse merger should not result in an assignment because the original licensee is the surviving entity and retains title to its assets. Recently, the Delaware Chancery Court ruled in *Meso Scale Diagnostics*[52] that "[g]enerally, mergers do not result in an assignment by operation of law of assets that began as property of the surviving entity and continued to be such after the merger". However, this determination may depend on how the applicable state law handles changes to the form of a business. Based on the proposition that an IP licence is personal in nature, at least one court has held that a reverse merger may constitute an assignment of the licence rights if it results in a "fundamental change in ownership", giving licence rights to the competitor of the intellectual property owner, such that a transfer has occurred.[53]

Asset sales and purchases: In an asset sale, the target assigns to the purchaser all patents (and contractual arrangements relating to the patents) included in the acquired assets, subject to transfer restrictions or other consent requirements. A sale of assets including an intellectual property licence is seen as an assignment of that licence. However, if the licence prohibits transfer of intellectual property without prior written consent, a sale without such consent could be held as being null and void.[54]

(c) Financing

In a financing transaction, a party issues a minority portion of its equity to outside investors. The sale can be done privately or to the public.

Special considerations apply to public financing transactions. In particular, Section 10(b) of the Securities and Exchange Act of 1934, along with its regulation in Rule 10b-5 of the Securities Exchange Commission (SEC), prohibit issuers from making false or misleading statements about their business or registration. Specifically, Item 101 of Regulation S-K requires that an issuer describe the

51 See *Cincom Systems, Inc v Novelis Corp*, 581 F3d 431, 437–9, 92 USPQ2d 1085 (6th Cir 2009) (when the licensee merged with the surviving corporation, the licence granted by the licensor solely to the licensee transferred to the surviving corporation).

52 *Meso Scale Diagnostics LLC v Roche Diagnostics GmbH*, 62 A3d 62, 82 (Del. Ch. 2013).

53 *SQL Solutions*, 1991 US Dist Lexis 21097, at 8–10.

54 *McCullough v Dairy Queen of Michigan, Inc*, 121 USPQ 302 (WD Mich 1959) (the purported transfer [of the trademark] without the prior written consent of the trademark owners was null and void and vested no rights to the trademark).

importance, duration and effect of its patents. Item 103 of the same Regulation requires that an issuer describe any material pending legal proceedings. Item 503 of the Regulation requires that an issuer describe the specific risks facing the operation of the business.

Ultimately, in a public offering the issuer must disclose all material risks of its business, such as the lack of a robust patent portfolio, the risk of infringement litigation, and the inability to obtain appropriate licences. This is true even if such disclosure could affect the marketing of the offering. The *Nathenson v Zonagen* case illustrates the importance of accuracy in the totality of disclosures made by a public company.[55] In that case, a class action was filed against Zonagen for alleged violations of Sections 10(b) and 20(a) of the Securities Exchange Act of 1934 and SEC Rule 10b-5. The company had issued a press release saying its patent covered the drug Vasomax™. In subsequent SEC filings, the company listed the patent but did not state that it covered the drug. Although the listing of the patent was not false or misleading on its own, it was considered potentially inaccurate in view of the press release that was never retracted.

(d) Joint ventures and collaborations

Joint ventures and collaborations share many similarities, including that two parties contribute technology and perhaps other resources to develop a product or business. A key difference between them, though, is that a joint venture operates through a separately formed legal entity, where each party to the joint venture owns equity in the joint venture and the parties jointly develop a product through that joint venture entity. Collaborations are similar relationships in practice, in that both parties contribute technology and other resources and work jointly together, but the parties operate through their existing separate entities. The parties will often create a steering committee that is charged with making key decisions for the collaboration.

In such relationships, each party typically contributes its own set of skills and resources to the relationship. For example, one or both parties may contribute existing intellectual property (known as 'background IP') and research and development expertise, while the other party contributes its manufacturing and distribution capabilities. When contributing background IP to the relationship, the contributor should consider:

- the scope of patent rights to be contributed;
- whether the patent rights should be licensed or transferred; and
- the extent to which the party should retain the right to continue to use the patents outside the relationship.

Two key issues that need to be negotiated in collaborations and joint ventures involve the ownership of patents to improvements developed during the relationship, and the rights to use those improvement patents. In some instances, the parties may agree to vest patent ownership (and rights of use) in the party whose employee invented the rights – although as noted in section 1.2 above, determining

55 *Nathenson v Zonagen, Inc*, 267 F3d 400, 423–6 (5th Cir Tex. 2001).

inventorship can be a painstaking and difficult task. In other cases, the ownership and rights of use are determined according to the subject matter of the improvement patents. Other structures may also be used.

In a collaboration or joint venture, the rights of use in the improvement patents may be more heavily negotiated than the actual ownership of those patents. The rights of use in jointly owned patents are governed by federal law. Specifically, the US Patent Act provides that, unlike certain international jurisdictions, "in the absence of an agreement to the contrary, each of the joint owners of a patent may make, use, offer to sell, or sell the patented invention within the United States or import the patented invention into the United States, without the consent of and without accounting to the other owners".[56] A situation where both parties have and exercise this statutory right is usually undesirable to the parties, as it could preclude either party from having any meaningful exclusivity in the patents. In most cases, the parties insist on aligning the ownership and rights of use so they are consistent with the purposes of the venture or collaboration. To do that, most relationships contain provisions in the agreement that allocate the ownership and rights of use of the patents.

Many different structures are used for that purpose. Examples include assignment or cross-licensing to the respective parties or joint venture entity based on inventorship, derivation, and/or subject matter of the patent:

- *Inventorship:* Ownership of the patent is assigned to the party whose employees developed it, and to both parties jointly if jointly invented. Rights of use may be separately allocated or may follow the rights of ownership. These arrangements can be convenient to negotiate, but as noted above can be challenging to implement in practice as they often require the parties to assess inventorship of a technology arising from a group or team of researchers, which can be difficult. To avoid these complex inventorship analyses, some parties choose to provide for joint ownership of all inventions but allocate rights of use to the parties based on their respective fields of interest, as discussed in the third bullet point below.
- *Derivation:* Ownership of the patent is assigned to the party that owned the intellectual property from which the patent was derived, regardless of which party invented it. Technology-driven allocations may be used when expected inventions could be clearly identified as related to one of the party's foundational technology. These approaches have the advantage of avoiding complex inventorship assessments. For example, where one party provides a drug molecule and another provides a delivery device for the drug, one common form of agreement allocates ownership (or rights of use) in any invention that relates to its core contribution (ie, drug improvements are allocated to the party that brings the drug to the project), irrespective of who invented the invention.
- *Subject matter or field:* Ownership of the patent is assigned according to each party's respective field of interest. Field allocations also help avoid the

56 35 USC §262.

complexities of inventorship analyses. There are numerous possibilities for structuring these. For example, rights of use of a new therapeutic drug can be divided according to specific drug formulations, or in the treatment of specific diseases or disease classes, or in combination with specific delivery devices, or in certain medical procedures. Some agreements work best by allocating the fields geographically; others might allocate rights of ownership according to technology (or inventorship) while at the same time providing rights of use according to geography or field. For instance, one party may have the exclusive right to exploit a new drug formulation in Europe, while the other has the right to exploit it in North America.

The agreement among the parties should provide for rights of use post-termination, including licences to background IP and improvements.

(e) Divestment

In a divestment, a party sells its interest in patents (and possibly other technology) to a third party. Transactions of this type are often done by large life sciences companies that elect to discontinue the development of certain products. A key issue in a divestment is determining the IP assets that will be divested.

Initially, the buyer will conduct a due diligence review of the intellectual property identified as relevant to the transaction to assess the strength and transferability of the rights. Such due diligence may include the following:

- searching public records to identify patents registered to the seller and confirm clear title to those patents;
- considering whether the search of the public records should be expanded to include any affiliates of the seller or any other entities related to or partnering with the seller and the relevant business; and
- reviewing inbound and outbound licence agreements.

Once the relevant intellectual property has been identified, the next step is to determine whether the seller is planning to retain any of that intellectual property and what steps may be taken to ensure access to the retained rights.

The purchase agreement will need to correctly identify the patents to be divested. Both the buyer and the seller will need to consider how to identify the divested intellectual property assets in the purchase agreement itself. There are a variety of different approaches that can be taken; each divestment will require the buyer and seller to tailor the approach to both the type and amount of intellectual property involved in the transaction and the relative bargaining power of the parties.

For more general intellectual property, such as know-how and trade secrets, one approach is to identify the divested intellectual property generically as it relates to the divested business. By way of example, the divested intellectual property may be defined as 'all intellectual property that is primarily used in the divested business'. A more seller-friendly approach would be to include 'all intellectual property that is exclusively used in the divested business'.

A combination of these two approaches may also be considered. This combination often involves first defining the divested assets to include all intellectual property primarily or exclusively related to the divested business. The buyer then provides a representation that all such registered and/or unregistered intellectual property is included in a schedule to the purchase agreement. This provides the seller with additional certainty as to the specific intellectual property to be transferred, without restricting the seller to a specific schedule of intellectual property.

In some cases, the seller must decide whether to transfer its patent rights in the divestment, or retain the title to those rights but grant the acquirer an exclusive licence to those patent rights. Some considerations are as follows:

- Transfer:
  - Advantages: the transfer of patent rights to the buyer creates a more straightforward deal structure with lower transaction costs, and both parties will have fully performed at the time of the sale. This structure may be preferable to potential buyers, who would gain an unencumbered right to exploit the relevant patents.
  - Disadvantages: the seller's right in the event of a breach would be limited as the seller would not be able to regain the rights to the transferred patents by terminating the buyer's licence.
- Licence:
  - Advantages: a licence agreement provides the seller with better remedies in the event of a breach, as the seller would be able to terminate the buyer's licence and regain rights to the transferred patents. A licence agreement may also be more attractive to cost-sensitive buyers if it allows sellers to decrease the initial upfront payment.
  - Disadvantages: this approach is potentially less attractive to buyers as their rights would be less certain. Licence agreements are also contractually more complex than a simple transfer of patent rights, increasing transaction costs.
- Collaboration:
  - Advantages: a collaboration arrangement allows the parties to pool their resources and work together to exploit the relevant intellectual property. Depending on the capabilities of the buyer and seller, this approach may be beneficial to both parties.
  - Disadvantages: collaborations are contractually complex and typically associated with the highest transaction costs of the three legal structures outlined here. A collaboration agreement will also require ongoing interaction and cooperation between the buyer and seller, which may make identifying an appropriate buyer more difficult.

Buyers should also consider whether any related tangible personal property should be divested as part of the transaction, including such things as equipment, finished goods, work in progress, raw materials, books and records, branded collateral and any other person property embodying the divested patents.

2.3 Particular business transactions

Other transactions used by life sciences companies tend to have less emphasis on patent rights and more emphasis on product development and sales. These include agreements relating to material transfers, research and development, contract manufacturing, distribution and co-promotion. Each is described further below.

(a) *Material transfer agreements*

Material transfer agreements typically arise in the early stages of product development. They provide the means for working with a potential collaboration partner with access to the licensor's biological or small-molecule materials – for example, viral vectors, host cell lines, or chemical and biological libraries. This access allows the recipient to evaluate the materials to see whether they are useful for its own research or are of interest for a future collaboration. If the recipient requires additional technology or data in order to use the material for the evaluation, that technology and data can also be licensed through the structures discussed in sections 2.1 and 2.2.

Patents are often not included in a material transfer agreement but reserved for a later, pre-commercial or commercial arrangement.

(b) *Commercial research and development agreements*

After a material transfer agreement (and often in lieu of it), parties will enter into a more structured research agreement to develop specific products or classes of products. These agreements are often intended to last for a longer period of time than the material transfer agreement and therefore are structured more like standard licences or commercial contracts. For example, the agreements may include patent and technology licences, detailed provisions regarding the agreed research programme, procedures for altering the research and assessing its success (or failure), allocation of the right to enforce intellectual property against competitor products, and the split of responsibility for defending infringement actions brought against the product being developed. More extensive commercial terms are also typically included, such as representations, warranties and indemnities, dispute resolution and termination procedures, and economic provisions with royalties and milestone payments (see section 2.1(b) above).

(c) *Contract manufacturing agreements*

Contract manufacturing agreements are commercial contracts that focus on responsibilities for the manufacturing – and often the supply, too – of products. The agreements resemble product development agreements and licences, in that they contain many similar terms identified above. As they emphasise manufacturing, they also typically include detailed provisions specific to manufacturing, including quality control, testing and maintenance. Patent licences may be included in the agreement, but the licences are often focused on rights to manufacture under the patents.

A note of caution when entering into such agreements is required, to ensure that a contract manufacturing agreement is not construed to trigger the 'on-sale bar' of

US patent law. Specifically, statute 35 USC §102 creates a bar to patenting if an invention was on sale more than a year prior to the patent application filing date. The Supreme Court laid out a two-pronged test to determine whether the statutory on-sale bar applies: "First, the product must be the subject of a commercial offer for sale ... Second, the invention must be ready for patenting."[57] The statutory bar does not apply if the offer for sale was for experimental rather than commercial purposes.[58] The courts have held that even a secret offer for sale from a contract manufacturer to the patentee can trigger the on-sale bar.[59]

A conservative way to protect against the on-sale bar is to file a patent application directed at the invention before negotiating any such contract manufacturing agreement. A patentee may also be able to protect against a manufacturer or supplier triggering an on-sale bar by structuring the agreement to cover services instead of sales of products. Nonetheless, see *Enzo Biochem, Inc v Gen-Probe Inc*,[60] where the contract language was judged to render a services agreement an offer for sale.

It is not clear whether the on-sale bar would be triggered under the America Invents Act when the sale occurs via manufacturing agreements. The language of statute 35 USC §102(a)(1) reads: "the claimed invention was patented, described in a printed publication, or in public use, on sale, or otherwise available to the public before the effective filing date of the claimed invention." Some commentators, including the USPTO, have taken the view that the statute applies only to public sales and not confidential agreements between parties. This view has been formed because of the phrase "or otherwise available to the public", which was not included in the statute prior to the America Invents Act.

(d) Product distribution agreements

In product distribution agreements, the licensor is usually referred to as the supplier (or 'OEM', other equipment manufacturer), while the licensee is referred to as the distributor. The supplier supplies the products to the distributor, and the distributor distributes the products through established channels of trade. Unlike contract manufacturing agreements, which focus on manufacturing arrangements between licensor and licensee, product distribution agreements focus on late-stage agreements that obligate the distributor to commercially distribute the products of the other party.

The intellectual property rights in these transactions typically focus on trademarks and branding, with limited or no patent licences. One reason for this is that the rights in the patents are exhausted when the product is sold to the distributor and therefore not needed.[61]

---

57 *Pfaff v Wells Electronics, Inc*, 525 US 55, 64, 48 USPQ2d 1641, 1645 (1998).

58 *Pfaff*, n57 above, at 1644.

59 *Hamilton Beach Brands, Inc v Sunbeam Prods, Inc*, 107 USPQ2d 1901 (Fed Cir 2013) (holding that a confirmation from a supplier for an order of 2000 slow cookers triggered the on-sale bar).

60 76 USPQ2d 1616 (Fed Cir 2005).

61 See *Quanta v LG Electronics*, 553 US 617, 625 (2008) (the longstanding doctrine of patent exhaustion provides that the initial authorised sale of a patented item terminates all patent rights to that item).

(e) ***Co-promotion agreements***

Co-promotion agreements are also commercial agreements and contain many of the same core terms found in licences and collaboration agreements. Co-promotion agreements focus on shared responsibilities for selling the product and are often used when both parties have sales forces and customer bases. For example, large pharmaceutical companies may from time to time enter into co-promotion arrangements for the sale of blockbuster drugs.

Under these agreements, each party participates in the promotion and sale of the product, with economic mechanisms by which each party accounts to the other for the sales made. The responsibilities for selling the products may be divided by geography (eg, the United States as one region and Europe as another), by channel (eg, consumer points of sale as one channel and hospitals as another), or as otherwise agreed.

The economics may include upfront payments, royalties and sales milestone payments, as discussed earlier. The co-promotion might alternatively be structured as a profit-sharing arrangement, where each party shares the costs and proceeds of the programme's sales. The sharing arrangement can vary according to the parties' relative contribution to the costs, or can be based on a fixed or variable percentage.

Like distribution agreements, these agreements typically focus on trademark licences and branding, with limited patent and technology licensing involved.

(f) ***Clinical trial agreements***

In a clinical trial agreement, one party provides a small-molecule drug or therapeutic biologic and the second party provides the facilities, resources and expertise to conduct clinical trials using the therapeutic. These agreements are similar to research and development agreements in scope and complexity, although the research itself is typically more specifically defined to include specific clinical trial activities. Most of these agreements do not focus on intellectual property provisions, although they could and in some cases they probably should. Observations made in clinical trials may lead to new inventions. One example is the drug Lumigan that was being studied for glaucoma. During the clinical trial, it was observed that one of the side effects of the drug was to extend the patient's eye lashes.

3. Issues associated with patent transactions

In addition to patent exclusivity, many small-molecule therapeutics (known as 'new chemical entities', NCEs) and biologics may be eligible for regulatory exclusivity protections. Life sciences companies are often trying to optimise their patent and regulatory exclusivity so as to create effective barriers to market entry for generic companies. The following section provides a brief outline of the available regulatory protections.

3.1 Regulatory law and compliance

In 1984, with passage of the Hatch–Waxman amendments to the Food, Drug and Cosmetic Act, Congress authorised the US Food and Drug Administration (FDA) to approve generic drugs that are 'the same' as drugs previously approved by the FDA, under the 'new drug application' (NDA) provisions of the amended statute. To obtain approval of a generic small-molecule drug, an applicant must submit an abbreviated new drug application (ANDA) to the FDA. In support of such an application, a generic manufacturer may rely on the preclinical and clinical testing previously conducted for the drug product earlier approved under an NDA. The previously approved drug is known as the 'reference listed' drug (RLD).

In order for an ANDA to be approved, the FDA must find that the generic version is identical to the RLD with respect to the active ingredients, the route of administration, the dosage form, and the strength of the drug. At the same time, the FDA must also determine that the generic drug is 'bioequivalent' to the RLD. Under the statute, a generic drug is bioequivalent to an RLD if "the rate and extent of absorption of the drug do not show a significant difference from the rate and extent of absorption of the listed drug".[62]

Upon approval of an ANDA, the FDA indicates whether the generic product is 'therapeutically equivalent' to the RLD in its publication "Approved Drug Products with Therapeutic Equivalence Evaluations" (also referred to as the Orange Book). Physicians and pharmacists consider a therapeutic equivalent generic drug to be fully substitutable for the RLD. In addition, by operation of certain state laws and numerous health insurance programmes, the FDA's designation of therapeutic equivalence often results in substitution of the generic drug without the knowledge or consent of either the prescribing physician or patient.

Under the Hatch–Waxman amendments, the FDA may not approve an ANDA until any regulatory exclusivity for the RLD has expired. The Food, Drug and Cosmetic Act provides a period of five years of non-patent data exclusivity for a new drug containing a new chemical entity. In cases where such exclusivity has been granted, an ANDA may not be filed with the FDA until the expiration of five years unless the submission is accompanied by a so-called 'Paragraph IV certification'. A Paragraph IV certification asserts that the new product will not infringe the already approved product's listed patents or that such patents are invalid or unenforceable.

If a Paragraph IV certification is made, the applicant may submit its application four years following the original product approval. The FDCA also provides for a period of three years of exclusivity if the NDA includes reports of one or more new clinical investigations, other than bioavailability or bioequivalence studies, that were conducted by or for the applicant and are essential to the approval of the application. This three-year exclusivity period often protects changes to a previously approved drug product, such as a new dosage form, route of administration, combination or indication.

(a) Hatch–Waxman patent certification and the 30-month stay

Upon approval of an NDA or a supplement thereto, NDA sponsors are required to list with the Food and Drug Administration each patent with claims that cover the applicant's product or an approved method of using the product. Each of the patents listed by the NDA sponsor is published in the Orange Book. When an ANDA

62 21 U.S.C. § 355(j)(2)(A).

applicant files its application with the FDA, the applicant is required to provide a certification concerning any patents listed in the Orange Book.

Specifically, the applicant must certify with respect to each patent that:

- the required patent information has not been filed;
- the listed patent has expired;
- the listed patent has not expired but will expire on a particular date, and approval is sought for after that patent expiration; or
- the listed patent is invalid, unenforceable or will not be infringed by the new product.

A paragraph IV certification is not required for patents directed at methods of use for which the ANDA applicant is not seeking approval. Such a certification is, though, needed for a 'Section 505(b)(2)' filing, which is not an ANDA filing but relies on studies known in the field.[63] To the extent that the Section 505(b)(2) applicant is relying on studies conducted for an already approved product, the applicant is required to certify to the FDA concerning any patents listed for the approved product in the Orange Book to the same extent that an ANDA applicant would.

If the applicant does not challenge the listed patents or indicates that it is not seeking approval of a patented method of use, the ANDA application will not be approved until all the listed patents claiming the referenced product have expired (other than method-of-use patents involving indications for which the ANDA applicant is not seeking approval).

If the ANDA applicant has provided a Paragraph IV certification to the FDA, the applicant must also send notice of the Paragraph IV certification to the NDA and patent holders once the ANDA has been accepted for filing by the FDA. The NDA and patent holders may then initiate a patent infringement lawsuit in response to the notice of the Paragraph IV certification. The filing of a patent infringement lawsuit within 45 days after the receipt of a Paragraph IV certification automatically prevents the FDA from approving the ANDA until the earlier of 30 months after the receipt of the Paragraph IV notice, expiration of the patent, or a decision in the infringement case that is favourable to the ANDA applicant.

(b) Biologics Price Competition and Innovation Act

The Biologics Price Competition and Innovation Act of 2009 (hereinafter the Biosimilars Act) was signed into law by President Obama on March 23 2010. This legislation was inspired by the Hatch–Waxman Act and provides a new biosimilar pathway for the approval of large-molecule (biologic) drugs. Like Hatch–Waxman, the Biosimilars Act provides pioneer companies that invest in and develop biological drug products with an exclusivity period so that they can recoup their investment.

The Biosimilars Act provides 12 years of marketing exclusivity for new 'biological' molecules. Under this scheme, there are two types of generic biologics: those that are 'biosimilar' and those that are 'interchangeable'. A 'biosimilar' product is defined as one that is "highly similar" to the pioneer product notwithstanding

63 See 21 USC 355(b)(2).

minor differences in clinically inactive components and where there are no clinically meaningful differences in terms of safety, purity and potency. An 'interchangeable' biological product is a biosimilar that is expected to produce the same clinical result as the pioneer in any given patient and, if administered more than once to an individual, the risk of safety and diminished efficacy of alternating or switching between use of the product and use of the pioneer is no greater than the risk of using the pioneer without such alteration or switch. Under the Biosimilars Act, a biosimilar applicant can file an Abbreviated Biologics Licence Application (ABLA) after four years of RLD exclusivity.

(c) ***Paediatric studies and exclusivity***

Under the Paediatric Research Equity Act of 2003, an NDA or supplement thereto must contain data that is adequate to assess the safety and effectiveness of the drug product for the claimed indications in all relevant paediatric subpopulations, and to support dosing and administration for each paediatric subpopulation for which the product is safe and effective. With enactment of the Food and Drug Administration Safety and Innovation Act in 2012, sponsors must also submit paediatric study plans prior to the assessment data. Those plans must contain an outline of the paediatric study or studies that the applicant plans to conduct, including study objectives and design, any deferral or waiver requests, and other information required by regulation. The applicant, the FDA and the FDA's internal review committee must then review the information submitted, consult with each other, and agree upon a final plan. The FDA or the applicant may request an amendment to the plan at any time.

The FDA may, on its own initiative or at the request of the applicant, grant deferrals for submission of some or all paediatric data until after approval of the product for use in adults, or full or partial waivers from the paediatric data requirements. Additional requirements and procedures relating to deferral requests and requests for extension of deferrals are contained in the Food and Drug Administration Safety and Innovation Act. Unless otherwise required by regulation, the paediatric data requirements do not apply to products with orphan designation (see section 3.1(d) below).

Paediatric exclusivity is another type of non-patent marketing exclusivity in the United States and, if granted, provides for the attachment of an additional six months of marketing protection to the term of any existing regulatory exclusivity, including non-patent and orphan exclusivity, which is discussed in (d) below. This six-month exclusivity may be granted if an NDA sponsor submits paediatric data that fairly responds to a written request from the FDA for such data. The data need not show the product to be effective in the paediatric population studied; rather, if the clinical trial is deemed to fairly respond to the FDA's request, the additional protection is granted.

If reports of requested paediatric studies are submitted to and accepted by the FDA within the statutory time limits, whatever statutory or regulatory periods of exclusivity or patent protection cover the product are extended by six months. This is not a patent term extension, but it effectively extends the regulatory period during which the FDA cannot approve another application.

(d) *Orphan designation and exclusivity*

Under the Orphan Drug Act, the FDA may designate a drug product as an 'orphan drug' if it is intended to treat a rare disease or condition (generally meaning that it affects fewer than 200,000 individuals in the United States). A company must request orphan product designation before submitting an NDA. If the request is granted, the FDA will disclose the identity of the therapeutic agent and its potential use. Orphan product designation does not convey any advantage in, or shorten the duration of, the regulatory review and approval process.

If a product with orphan status receives its first FDA approval for the disease or condition for which it has such designation, the product generally will receive orphan product exclusivity. Orphan product exclusivity means that the FDA may not approve any other applications for the same product for the same indication for seven years, except in certain limited circumstances. Competitors may receive approval of different products for the indication for which the orphan product has exclusivity, and may obtain approval for the same product but for a different indication. If a drug or drug product designated as an orphan product ultimately receives marketing approval for an indication broader than what was designated in its orphan product application, it may not be entitled to exclusivity.

(e) *Qualified infectious disease product (QIDP) exclusivity*

Antibiotic products are now eligible for additional regulatory exclusivity under 21 USC §355(f), which was enacted in June 2012. Within this framework, antibiotic products may be designated as 'qualified infectious disease products' (QIDPs) and obtain a five-year extension to any existing regulatory protection. In other words, if an antibiotic is entitled to FDA exclusivity such as NCE exclusivity, clinical investigation exclusivity or orphan drug exclusivity, QIDP designation will extend the available exclusivity for the antibiotic by another five years.

The statute lists a few representative 'qualifying pathogens' but otherwise delegates the task of determining what is a qualifying pathogen to the FDA. According to a June 12 2013 Proposed Rule, the FDA has proposed at least 18 different pathogens for inclusion in the qualifying pathogen list.[64] The FDA's final regulations with the finalised list of qualifying pathogens have not yet been published, but the statute has required that the FDA issue its final regulations no later than July 9 2014.[65] Although the FDA has not issued its final regulations, it has still granted QIDP designation for several drugs over the past year.

Additional benefits to QIDP designation may also be available, including priority review of the drug application by the FDA (which would shorten the FDA timeline for reviewing an NDA for a QIDP-designated drug) and eligibility for fast-track designation (which would expedite certain aspects of development and review).

---

64 See www.gpo.gov/fdsys/pkg/FR-2013-06-12/html/2013-13865.htm.

65 The final rule was finalised 5 June 2014. See: https://www.federalregister.gov/articles/2014/06/05/2014-13023/establishing-a-list-of-qualifying-pathogens-under-the-food-and-drug-administration-safety-and.

(f) *Patent term restoration and extension*

A patent claiming a new drug product may be eligible for a limited patent term extension under the Hatch–Waxman Act, which permits a patent restoration of up to five years for patent term lost during product development and the FDA regulatory review. The restoration period granted is typically one-half the time between the effective date of an Investigative New Drug and the submission date of an NDA, plus the time between the submission date of a NDA and the ultimate approval date. Patent term restoration cannot be used to extend the remaining term of a patent past a total of 14 years from the product's approval date. Only one patent applicable to an approved drug product is eligible for the extension, and the application for the extension must be submitted prior to the expiration of the patent in question. A patent that covers multiple drugs for which approval is sought can only be extended in connection with one of the approvals.

The USPTO reviews and approves the application for any patent term extension or restoration in consultation with the FDA.

3.2 Hot topic

One of the major changes in the US patent system in recent years was the enactment of the America Invents Act in 2011. The act was a step towards harmonising the laws of the United States with the rest of the world. Two key changes should be mentioned here:

- US patent law has been changed to a 'first to file' regime for all patent filings made after March 16 2013. Patents filed prior to that date will continue to be subject to the 'first to invent' system. Thus, both conventions will be in play in the United States for many years to come – potentially until 2034. Parties conducting due diligence on patent portfolios will need to take account of these two systems, as appropriate.
- Some of the advantages previously provided by filing US provisional patent applications have been lost. Under the America Invents Act, international applications will also be accorded the benefit of their filing dates and be relevant prior art.

In addition, the America Invents Act provides more options for challenging the validity of competitor patents before the PTAB, including *inter partes* reviews, post-grant reviews and third-party observations. These post-grant proceedings are similar to European opposition and third-party observation proceedings. Although post-grant proceedings in the United States have not been favoured historically by life sciences companies, preliminary data indicates that Hatch–Waxman litigants are now using this forum. Generic filers that use these new provisions may be able to invalidate patents well before the expiration of the five-year regulatory exclusivity period. One attractive feature of the new post-grant proceedings is a statutorily mandated 12–18-month time requirement for completion of the case. In addition, PTAB proceedings require a lower burden of proving invalidity and potentially broader claim construction than does a US district court.

A review of initial post-grant cases shows some interesting strategic use of these proceedings by generic filers. In *Apotex, Inc v Alcon Pharmaceuticals Ltd*,[66] Alcon had two patents listed in the FDA Orange Book for Vigamox® (moxiflocacin hydrochloride). Teva, the first Paragraph IV filer, challenged the patents in district court and lost. Apotex was the second generic filer. Alcon sued Apotex for infringement in the district court, and Apotex filed an *inter partes* review petition before the PTAB. Apotex relied on the same prior art and same arguments that had previously been used by Teva. The PTAB instituted the proceedings on the finding that there was a "reasonable likelihood" that the patents may be found invalid for obviousness. Because of the lower burden of proof and different claim construction standard, the PTAB was not constrained by the earlier district court decision.

66 IPR2013-00012-00015.

About the authors

Disket Angmo
Associate, Khaitan & Co
disket.angmo@khaitanco.com

Disket Angmo is a member of the IP practice of Khaitan & Co. Her practice mainly focuses on trademark prosecutions, litigation and transactional advisory work involving trademarks, designs and patents. Ms Angmo also works with the corporate M&A team of Khaitan & Co. She has represented clients in several core corporate transactions, domestic as well as global, including mergers, acquisitions and divestments. She has also advised clients on strategic corporate assignments and provides advice on matters in relation to several aspects of company law, labour law, securities law and information technology and privacy laws.

Malcolm Bates
Partner, Taylor Wessing
m.bates@taylorwessing.com

Malcolm Bates is a partner in the Taylor Wessing IP department and head of life sciences in the United Kingdom. His legal specialisations are: the commercial and IP aspects of technology transactions, including spin-outs, start-ups, venture capital investments, trade sales and initial public offerings (IPOs); IP and IT agreements, including patent and software licensing, research and development agreements, collaborations, clinical trials agreements, distribution and agency agreements; and IP and IT contract disputes and patent validity and infringement issues. Dr Bates works primarily in the life sciences and technology industry sectors. He was previously vice president and general counsel at Genset in Paris, a NASDAQ-listed biotechnology and bioinformatics company. Dr Bates speaks regularly on commercial and intellectual property issues. Dr Bates has a degree in molecular biology and biochemistry from the University of Durham and a Ph.D. in molecular biology from Imperial College, London. He is the company secretary of One Nucleus, the United Kingdom's largest biotech network. Dr Bates speaks both French and German, and is recommended in *Chambers* 2014 and 2013 and the *Legal 500* 2012.

Nicole Battistini-Kohler
Partner, Taylor Wessing
n.battistini-kohler@taylorwessing.com

Nicole Battistini-Kohler is a partner and member of the corporate transactions practice area in Taylor Wessing's Munich office. She has specialised in corporate law for 11 years, including all types of transaction in the area of life sciences and multidisciplinary internal reorganisation projects. In 2004 Ms Battistini-Kohler earned an LLM Eur in International and European business law at Ludwig Maximilians University of Munich. She passed the two legal state examinations in 1999 and 2002 and trained as a notary public in Basel, Switzerland. In 2006 she passed the qualifying test for European attorneys in Dusseldorf and was admitted to the German bar. Before joining Taylor Wessing in June 2006, she worked for international law firms in Germany and Switzerland. Ms Battistini-Kohler, a native German speaker, is fluent in English and French, and has a basic knowledge of Italian.

Laëtitia Bénard
Partner, Allen & Overy Paris
laetitia.benard@allenovery.com

Laëtitia Bénard is a partner in the IP and litigation department at Allen & Overy Paris. She leads the Paris IP practice and is co-head of the firm's life sciences group. She has a wealth of experience in all IP areas, in particular in patents, both contentious and non-contentious. Ms Bénard has represented pharmaceutical clients on a number of market-shaping patent litigation cases and is widely considered to have broken new ground in this field. She also advises on regulatory issues and contracts. Her practice covers all life sciences product regulations, as well as regulations and ethical rules applicable to healthcare professionals. She regularly publishes articles on developments in the life sciences sector.

She is recognised as "very impressive" and an "outstanding savvy and very creative lawyer who broke new ground in the field of pharmaceuticals patent litigation" (*Legal 500 EMEA 2013*, Intellectual Property, France).

David Broenimann
Partner, head of tax department, Meyerlustenberger Lachenal
david.broenimann@mll-legal.com

David Broenimann is a certified tax expert in Meyerlustenberger Lachenal's Zurich office. He advises domestic and international clients in all areas of corporate tax law. He focuses on mergers and acquisitions, real estate taxation and the structuring of IP and financing transactions. He is a frequent lecturer, publishes articles on different tax topics and serves as an examiner at the certified tax experts examinations.

Rory Cattanach
Partner, Wildeboer Dellelce LLP
rory@wildlaw.ca

Rory Cattanach is a partner at Wildeboer Dellelce LLP practising primarily in the areas of securities, corporate finance, corporate/commercial, mergers and acquisitions and technology law. He acts for a variety of early stage and mature technology, biotechnology, telecommunications and traditional economy companies (both Canadian and international), and has been involved in numerous public and private financings and secured lending transactions. In addition, Mr Cattanach has acted for several underwriters in public and private equity and debt financings, has extensive merger and acquisition experience and has participated in several cross-border transactions. He has served as a director of a number of private, public and not-for-profit companies and has spoken at conferences on a variety of corporate and commercial law topics. Mr Cattanach teaches securities law at Osgoode Hall Law School at York University as an adjunct professor.

Joanna Chan
Paralegal, Dechert LLP
joanna.chan@dechert.com

Joanna Chan is a paralegal in Dechert's Hong Kong office. She focuses her work on drug regulations and assists on technical issues relating to life sciences arising from client matters, such as identifying the nature and scope of technologies which may be material in a business transaction pursued by clients and interpreting the underlying technologies in patent disputes.

Frederick Ch'en
Of counsel, Hogan Lovells
frederick.chen@hoganlovells.com

Frederick Ch'en is a solicitor (England & Wales), US attorney (California), US patent attorney and a registered foreign attorney in Japan. He trained and worked for seven years in the London office of a Magic Circle international law firm before becoming head of intellectual property in the Tokyo office of another top-tier law firm (2010–2013). There, in collaboration with numerous Japanese law firms, he advised various companies on patent-related matters in Japan. Dr Ch'en has wide experience of advising on a range of contentious and non-contentious IP matters, focusing on multi-jurisdictional patent litigation and also IP aspects of transactions and IP licensing. Before becoming a lawyer, Dr Ch'en was a research scientist at Oxford University, having obtained the top first-class honours degree in molecular and cellular biochemistry from Balliol College and a doctorate in medical sciences from Merton College. Dr Ch'en is a member of AIPPI, LES and Pharma Delegates.

Mikyung (MK) Choe
Attorney, Kim & Chang
mkchoe@kimchang.com

MK Choe is a member of the IP transaction practice group. She specialises in IP transactions, including technology licensing, software licensing, patent & trademark licensing, IP-intensive M&A, due diligence and corporate transactions, foreign-invested R&D center related issues and technology export control issues. She has extensive experience in drafting, negotiating and reviewing IP-related contracts such as trademark or technology licence agreements in corporate transactions like joint ventures. She also advises on IP disputes including patent enforcement litigation, disputes with licensees or distributors and measures against counterfeit drugs.

Ms Choe is a US-licensed lawyer, admitted in the state of Pennsylvania and before the USPTO. She received a BS from Seoul National University and her JD from the University of Pittsburgh.

H. Joon Chung
Attorney, Kim & Chang
hjchung@ip.kimchang.com

H. Joon Chung is a U.S.-licensed attorney specialised in IP practices, including pharmaceuticals, chemical, patent prosecution, litigation and transactional areas. Before joining Kim & Chang, Mr Chung practised as in-house counsel for Vertex Pharmaceuticals, Inc, located in Cambridge, Massachusetts, United States. At Vertex he played an integral role in creating and protecting the worldwide IP portfolio for INCIVEK, a drug for treating hepatitis C (2012 annual US sales of $1.30 billion). He regularly counselled the company on in- and out-licensing and acquisition opportunities, including conducting due diligence. In addition, he oversaw patent portfolios related to clinical candidate drugs in cancer indications, developing and implementing worldwide patent lifecycle strategies.

Mr Chung is admitted to practise in Massachusetts, New York and before the US Patent and Trademark Office. He holds a degree from Oberlin College, a master's in chemistry from Boston College and a law degree from Suffolk University Law School.

Noel Courage
Partner, Bereskin & Parr LLP
ncourage@bereskinparr.com

Noel Courage is a partner with Bereskin & Parr in Toronto. He is a Canadian patent attorney and a US patent agent. His practice focuses on the licensing and patenting of biopharma, chemical and medical device inventions. He also provides opinions on patent validity and freedom to operate. He has experience with regulatory issues in drug and medical device approval, drug advertising, determining Patented Medicine Prices Review Board jurisdiction and Patented Medicine (Notice of Compliance) Regulations strategy.

Noel is a member of the American Intellectual Property Law Association (AIPLA) and the

Canadian Healthcare Licensing Association. He is also co-founder of the award-winning AIPLA International Patent Law Handbook. He was ranked in the *IAM Patent 1000 – The World's Leading Patent Practitioners 2013*.

Philippe de Jong
Partner, IP department, ALTIUS
philippe.dejong@altius.com

Philippe de Jong is a partner in the IP department at ALTIUS. He specialises in IP law, with a particular focus on patents, plant variety rights and parallel import issues. He also has extensive experience in the regulatory regimes for medicinal products and plants and is the head of the cross-practice Plant Law Practice within ALTIUS. Mr de Jong regularly lectures at conferences and seminars.

Mr de Jong has published articles in various leading European legal journals on specific topics of IP litigation. Together with two ALTIUS colleagues, including Christophe Ronse, he is the co-author of an authoritative review of patent case law which is published every three years by one of the leading Belgian business law journals. He worked as a consultant for the European Community Plant Variety Office and is a member of EPLAW.

Fiona Desouza
Senior associate, Khaitan & Co
fiona.desouza@khaitanco.com

Fiona Desouza is a patent attorney practising in the IP team of Khaitan & Co. Ms Desouza specialises in patent-related work in the fields of life sciences, bio-chemistry, chemistry and pharmaceuticals. She drafts patent specifications and prosecutes patent applications in these fields, and is experienced in performing prior art searches and freedom-to-operate searches, giving opinions in these matters and representing clients in patent revocations, appeals and petitions before the Intellectual Property Appellate Board and the Bombay High Court. With regard to the commercial and regulatory aspects of intellectual property, Ms Desouza has advised foreign clients in relation to clinical trial regulations in India, requirements under the Indian biological diversity statute and also on agreements involving research collaboration, technology licences and clinical trial outsourcing. Ms Desouza has written on patent-related developments in India for the website of CNBC TV18's *The Firm* and is a regular contributor to the IP newsletter *Ergo*, released periodically by Khaitan & Co.

Paul England
Senior associate and psl, Taylor Wessing LLP
p.england@taylorwessing.com

Paul England is a senior associate and professional support lawyer who specialises in life sciences patent matters, including related issues of supplementary protection certificates and regulatory protection. Before joining Taylor Wessing, Dr England had many years' fee-earning experience at another leading patent practice, where he advised in a number of complex and high-profile disputes for well-known pharmaceutical companies. For the past four years Dr England has developed a mixed practice of client advice, knowledge management and business development in patent law. This includes writing regularly (with more than 30 articles in peer-reviewed publications). He also speaks on patent matters for clients and at conferences. Dr England is currently particularly active in providing thought leadership on the implications of the Unitary Patent Court and the unitary patent.

In addition to *Patent Transactions in the Life Sciences* Dr England is the consulting editor of *Intellectual Property in the Life Sciences* (also Globe Law and Business) and author of *'Expert Privilege' in Civil Evidence* (Hart Publishing).

Dr England is also the editor of Taylor Wessing's Synapse microsite, a unique international online portal providing legal and commercial updates and analysis for the life sciences industry.

Before becoming a lawyer, Dr England was a scientist and he has a doctorate in biochemistry and molecular biology from the University of Oxford.

Manja Epping
Partner, Taylor Wessing
m.epping@taylorwessing.com

Manja Epping is a partner and member of Taylor Wessing's patents and pharma practice area and co-head of the life sciences and healthcare industry group. She has extensive experience in advising national and international companies from the pharmaceutical, biotech, medical devices and food sectors on complex, mostly cross-border contractual projects, in particular on manufacturing and distribution agreements, research and development collaborations and licence agreements. In addition, she advises such companies on regulatory issues as well as advertising and labeling law. Dr Epping passed the two German legal state examinations in 1993 and 1995. Before that, she obtained a Maîtrise en droit in international law from the University of Aix-en-Provence, France. She was admitted to the Munich Bar in 1996. Dr Epping, a native German speaker, is fluent in English and French, and has a basic knowledge of Spanish.

Kevin Fritz
Partner, Wildeboer Dellelce LLP
kfritz@wildlaw.ca

Kevin Fritz is a partner at Wildeboer Dellelce LLP in Toronto, Canada. His practice focuses on tax law, including corporate and personal tax planning, with a particular emphasis on corporate finance, mergers and acquisitions and business restructurings. He has significant experience in the tax structuring of mutual funds and other investment products. He has acted for a wide array of clients, both domestic and international, including individuals, governments, crown corporations, financial institutions and businesses of all sizes, from start-up, owner-managed firms to large multinational corporations. He is an adjunct professor of law at the University of New Brunswick and has been recognised in *The Best Lawyers in Canada* since 2013.

Uri Fruchtman
Senior associate, Dr Shlomo Cohen & Co
uri.fruchtman@shlomocohen.co.il

Uri Fruchtman is a senior member of Dr Shlomo Cohen & Co. He graduated with distinction from the Tel Aviv University and served as editor of the *Tel-Aviv University Law Review*. He represents many innovative pharmaceutical companies in licensing, regulatory, transactional and corporate matters.

Michael Gaul
Associate, Taylor Wessing
m.gaul@taylorwessing.com

Michael Gaul is a member of the corporate transactions practice area. He advises national and international clients on all corporate law matters. The main focus of his activities is advising on national and cross-border M&A and private equity transactions. Mr Gaul studied law at the University of Bayreuth and passed his first legal state exam in 2008. As a trainee lawyer, he worked in the antitrust/competition team of an international law firm and the legal department of a German DAX-listed company. After passing the second legal state exam in 2010, Mr Gaul was admitted as an attorney at law. Before he joined Taylor Wessing in 2013, he was a research assistant at the University of Bayreuth and member of the DFG graduate school.

Christoph Heiz
Partner, head of capital markets practice group, Meyerlustenberger Lachenal
christoph.heiz@mll-legal.com

Christoph Heiz is a partner in the corporate finance department and heads the firm's capital markets practice group. He practises corporate law and capital markets, M&A, leveraged transactions and corporate finance/banking. He has recently been involved in various national and cross-border acquisitions and acquisitions financing structures for Swiss and international clients. He regularly advises companies and financial institutions on capital market transactions, public takeovers, investment funds and corporate finance. He is a member of the board of directors of several Swiss companies.

Lewis Ho
Partner, Dechert LLP
lewis.ho@dechert.com

Lewis Ho leads the life sciences practice in Asia. He helps life sciences and technology companies and their financial sponsors to capture, manage, risk assess, evaluate and monetise their IP assets. His practice focuses on structuring technology transfer transactions and enforcing IP rights.

Mr Ho has extensive experience of advising on both inbound and outbound technology transfer transactions, and has negotiated more than 50 collaboration, outsourcing, joint venture and licensing deals with various life sciences companies, academic institutes, hospitals and contract research organisations in the past few years.

Mr Ho has been consistently recognised by *Chambers Asia* and *Legal 500* as one of the leading life sciences and IP lawyers in China. *Chambers Asia* noted him for "approaching issues in a 'down-to-earth' manner" and clients described him as "professional and prompt in providing feedback" and as producing "a high quality of work". He was also included in the *IAM Patent 1000 – The World's Leading Patent Practitioners 2012*, where he was recommended in both the licensing and litigation sections, and the *IAM Life Sciences 250 – The World's Leading Life Sciences Patent Litigators 2012*. He is also recommended by *International Who's Who Legal 2013* as a life sciences lawyer for IP, regulatory and transactional advice in China. Mr Ho is frequently invited to speak on life sciences and IP issues at conferences and seminars. Before joining Dechert, Mr Ho served as a resident partner in the Shanghai and Beijing offices of a major international law firm.

Christof Höhne
Counsel, Taylor Wessing
c.hoehne@taylorwessing.com

Christof Höhne is a member of the patents practice area in Taylor Wessing's Dusseldorf office. He advises and represents national and international clients in patent infringement matters and parallel (in)validity proceedings. He is a certified IP lawyer with comprehensive practical experience. The main technical focus of his work is on electronics and medicinal products. Dr Höhne studied law at the University of Constance and subsequently completed an LLM course in IP law in the United Kingdom. He completed his legal training in Dusseldorf and was subsequently awarded a PhD in law from the Heinrich Heine University, Dusseldorf. Dr Höhne is a native German speaker and fluent in English.

Simon Holzer
Partner, head of patent and life sciences practice group, Meyerlustenberger Lachenal
simon.holzer@mll-legal.com

Simon Holzer has been the head of Meyerlustenberger Lachenal's patent and life sciences team since 2012. He specialises in patent and other IP litigation, pharmaceutical law, supplementary protection certificates (SPCs), trademarks and geographical indications. He is an adjunct judge with the Swiss Federal Patent Court. Dr Holzer is ranked and recommended for his work in the legal directory *Chambers Europe 2014*, *Who's Who Legal Switzerland* (Patents and Life Sciences), the *IAM Patent 1000 – The World's Leading Patent Practitioners* and the *WTR 1000 2014 – The World's Leading Trademark Professionals*. Dr Holzer's recent publications include various articles in the IP sector.

Yoon Young Huh
Attorney, Kim & Chang
yoonyoung.huh@kimchang.com

As a member of Kim & Chang's health practice, Yoon Young Huh represents multinational companies in the pharmaceutical, medical device, food and cosmetic industries, advising on regulatory, compliance and corporate law issues. Regarding the pharmaceutical industry, Ms Huh specialises in pharmaceutical regulation, pricing and reimbursement, product liability claims stemming from adverse drug effects, compliance, bribery and competition law issues relating to promotional and marketing practices in the industry, and corporate restructuring and M&A. Most recently, Ms Huh has defended pharmaceutical companies in investigations by competition authorities with respect to marketing and promotional practices, advised on implementing global acquisitions in Korea, and advised companies in obtaining desired health insurance pricing for new products.

Ms Huh earned her BA in political science and economics from Smith College in 2003 and her JD from the University of Michigan Law School in 2007. She is admitted to practice in New York.

Matthias Hülsewig
Counsel, Taylor Wessing
m.huelsewig@taylorwessing.com

Matthias Hülsewig is a member of the patents practice area in Taylor Wessing's Dusseldorf office. He advises and represents national and international clients in patent infringement matters and parallel (in)validity proceedings. The main technical focus of his work is on electronics and medicinal products. In addition, he advises on R&D agreements, as well as corporate transactions involving IP rights. Dr Hülsewig studied law at the University of Wurzburg. He was awarded a PhD from the University of Wurzburg in 2006 and the same year completed his legal training in Dusseldorf. In 2007 Dr Hülsewig completed his LLM in commercial law at the University of Glasgow. He is a native German speaker and fluent in English.

Ji Eun Kim
Patent attorney, Kim & Chang
jekim@ip.kimchang.com

Ji Eun Kim's practice includes diverse IP issues, with particular emphasis on the areas of pharmaceuticals, chemistry and biotechnology. She has assisted clients in filing and managing patent portfolios, including for candidate drugs, representing multi-national companies. She has practised before the Korean Intellectual Property Office and the Korean courts at various stages of litigation. In addition to patent prosecution and litigation, her practice involves disputes related to freedom-to-operate opinions, trade secret misappropriation, licensing agreements, unfair competition and other IP disputes. She also has extensive experience in licensing and other IP transfer transactions.

Ji Eun holds a degree from the College of Pharmacy, Seoul National University and an LLM from Boston University Law School. She is also a member of the Korean Patent Attorneys Association and the Korean Pharmaceutical Association.

Jennie Klingberg
Associate, Setterwalls
jennie.klingberg@setterwalls.se

Jennie Klingberg's main areas of practice are life sciences, corporate law and IP rights law. Ms Klingberg has experience in handling complex litigations within these areas and regularly assists both Swedish and international companies in a number of different industries with contract negotiations, acquisitions and legal investigations. Ms Klingberg has vast experience within the areas of healthcare, pharmaceuticals and medical devices, and regularly advises clients in contract negotiations and the procurement process.

Eiichiro Kubota
Partner, Hogan Lovells
eiichiro.kubota@hoganlovells.com

Eiichiro Kubota specialises in counselling and litigation in the areas of patent law and all other areas of intellectual property. Mr Kubota studied at the Faculty of Engineering at Tokyo University. He qualified as a Japanese lawyer in 1991 and as a Japanese patent attorney in 1993, and between 1991 and 1994 he worked at a well-known Japanese IP firm. In 1994 he established Kubota Patent and Law Office, which merged with Hogan Lovells in 2008. Since qualifying, Mr Kubota's practice has centred on IP litigation, particularly patent litigation, and related matters such as advising on infringement and invalidity. Mr Kubota is a member of the International Association for the Protection of Intellectual Property (AIPPI), the Asian Patent Attorneys Association, the Licensing Executives Society (LES) and the Japan Trademark Association. He prepared the Japanese section of *Intellectual Property in the Global Market Place*, and he has published widely in Japanese on matters including patents for business models, copyright in buildings and trademarks in the internet age.

Chuck Larsen
Partner, Ropes & Gray LLP
charles.larsen@ropesgray.com

Chuck Larsen is a partner in Ropes & Gray's Boston office, and is dually qualified as a US lawyer and a UK solicitor. His work has a particular emphasis on IP strategies and transactions in the United States and Europe. Mr Larsen represents clients before the USPTO and the European Patent Office, conducts patent validity, infringement and freedom to operate studies for both US and European patents, and advises clients in preparing and filing re-examinations, patent challenges before the USPTO Patent Trial and Appeals Board, European oppositions and other contested proceedings. Mr Larsen also performs IP portfolio due diligence for equity investments and acquisitions, and works with clients to structure their transactions to address the IP risks involved. A significant component of his practice also includes negotiating and drafting complex licensing and commercial agreements involving intellectual property and advising clients on US export and embargo issues pertaining to their international technology transactions. Mr Larsen represents major medical technology and pharmaceutical companies in connection with a wide range of technology applications, for example medical implants, orthopedics, healthcare informatics systems, medical diagnostics and signal processing, energy-based therapeutics, neuro-monitoring, organ transplants, pharmaceuticals and drug delivery systems.

Monique Lee
Associate, registered foreign lawyer in Hong Kong, Dechert LLP
monique.lee@dechert.com

Monique Lee focuses her practice on IP litigation, licensing and regulatory issues in Greater China. Ms Lee has represented and advised multinational companies and ventures in various sectors, including electronics, wireless technologies, consumer healthcare, pharmaceuticals, medical equipment and health services, in matters such as freedom to operate, licences and compliance, IP acquisitions and infringement analysis, and anti-counterfeiting investigations.

Ms Lee's experience also includes handling design, copyright, computer software, trademark, unfair competition, antitrust and data privacy matters.

Before joining Dechert, Ms Lee was an associate in the IP practice of a major US law firm. She has also worked in boutique IP firms in Washington, DC, Tokyo and Hong Kong.

Anja Lunze
Partner, Taylor Wessing
a.lunze@taylorwessing.com

Anja Lunze is a partner of the patents and pharma group of Taylor Wessing in Munich. She is a German qualified lawyer and a German specialist litigator in industrial property law, with 10 years' experience in patent litigation. In particular, Dr Lunze specialises in patent infringement litigation and parallel opposition and nullity proceedings before the European Patent Office and the German Patent Court. She has particular experience in dealing with cross-border patent litigation, in particular in the pharmaceutical field. Her technical focus lies on pharmaceuticals and biologics, medical devices and chemistry. For her PhD thesis on the legal consequences of patent invalidity she was awarded the Faculty of Law Doctoral Prize at the Technical University of Dresden in 2007. Dr Lunze regularly publishes on current IP law issues, in particular on patent-related questions, and she is speaker at conferences and seminars on patent law. Dr Lunze, a native German speaker, is fluent in English and French and has a good command of Japanese.

Jose Madan
Executive director, Khaitan & Co
jose.madan@khaitanco.com

Jose Madan heads the IP practice of Khaitan & Co. He is an IP professional with more than three decades of varied and extensive experience in intellectual property. He has received practical training in IP law with a reputed German IP law firm. Mr Madan specialises in patents, especially drafting patent specifications and handling contested proceedings (eg, pre and post-grant opposition and revocation proceedings), including attending hearings, briefing counsel in patent-related court matters and giving opinions. He has extensive experience in handling such matters, gives presentations on IP rights and has co-authored several articles on patents and trademarks. Several leading IP publications have recommended Mr Madan as a leading IP expert.

Klaus Ewald Madsen
Partner, Bech-Bruun
kem@bechbruun.com

Klaus Ewald Madsen is head of the IP and technology department of Bech-Bruun. He specialises in patent law, life sciences, marketing, copyright and design law. Mr Madsen represents Danish and foreign businesses and advises clients on issues concerning IP law, marketing law and the regulatory aspects of medicinal products. Mr Madsen has substantial international experience, in particular in the negotiation and conclusion of licence and other IP agreements. Moreover, Mr Madsen is a seasoned litigator and regularly litigates injunction and court cases within the IP law field. In recent years he has represented clients in a significant number of patent cases, particularly within the area of pharmaceutical law.

Colin McCall
Senior associate, Taylor Wessing
c.mccall@taylorwessing.com

Colin McCall is a senior associate in Taylor Wessing's IP department. His practice focuses on the life sciences, healthcare and technology (with a particular focus on med-tech and clean technology) sectors, and he advises on both non-contentious and contentious matters. Key areas of his practice include IP disputes, including patent and copyright litigation; commercial agreements, including patent and know-how licensing, R&D agreements, collaborations, distribution agreements; contract disputes in the technology sectors; and IP aspects of mergers, acquisitions, joint ventures and investments. Chambers and Partners recently singled Mr McCall out as a key individual for having impressed clients with his "very practical" approach, and with his ability "to give you business like solutions" and clients describe him as "very sharp and intelligent" and a "very good IP all-rounder". Mr McCall has a first-class degree in cell biology from Durham University, holds a diploma in IP law and practice from Bristol University and an MSc in energy, environmental technology and economics from the London City University.

Jongguk Pak
Attorney, Kim & Chang
jongguk.pak@kimchang.com

Jongguk Pak is an attorney at Kim & Chang, focusing on healthcare/regulatory practices related to, among others, M&A, drug pricing, fair trade law and privacy issues. Before joining Kim & Chang, Mr Pak was a commissioned Air Force officer of the Korean military, practising as a judge advocate. He received an award from the commissioner of the defense acquisition program administration for his outstanding services.

Mr Pak holds a degree from Yonsei University.

Louisa Penny
Senior counsel, Taylor Wessing
l.penny@taylorwessing.com

Louisa Penny is senior counsel at Taylor Wessing LLP, specialising in all aspects of EU and UK competition law. She advises across a variety of sectors – pharmaceuticals, patent pools, media and entertainment, sport, telecoms, real estate and life sciences – and was recognised as an Associate to Watch in Chambers and Partners in 2013 and 2014.

Louisa Penny has extensive experience in contentious matters, including both European Commission and Office for Fair Trading cartel investigations, and regularly represents clients before UK and European regulators and courts. She advises on all aspects of defending a competition investigation, including attending dawn raids, leniency and responding to statements of objection. She also advises on day-to-day competition matters, including commercial arrangements, block exemptions and compliance.

Ms Penny has gained clearance for mergers from both national and international regulators, and also advises on state aid issues. She regularly contributes to legal journals and provides training on a range of EU and UK competition matters.

Christophe Ronse
Partner, IP and dispute resolution department, ALTIUS
christophe.ronse@altius.com

Christophe Ronse is a partner in the IP and dispute resolution department at ALTIUS. He is a seasoned commercial litigator, and focuses on patent law and life sciences. He is also the head of the cross-practice healthcare and life sciences team within ALTIUS.

Mr Ronse is the author of various publications and has spoken at many seminars, particularly on IP and patent law issues. Together with two ALTIUS colleagues, including Philippe de Jong, he is the co-author of an authoritative review of patent case law which is published every three years by one of the leading Belgian business law journals. He is also a member of the European Patent Lawyers Association (EPLAW).

Christian Bennet Rosenbæk
Lawyer, Bech-Bruun
cro@bechbruun.com

Christian Bennet Rosenbæk is a member of the IP and technology department of Bech-Bruun. He is specialised in IP law, with primary focus on patent law, life sciences and infringement of intellectual property. Mr Rosenbæk advises and represents Danish and foreign businesses on IP matters and has experience from several patent infringement litigations within the pharmaceutical industry. In addition to his Danish law degree, Mr Rosenbæk has an LLM in law, science and technology from Stanford Law School.

Kilian Schärli
Associate, Meyerlustenberger Lachenal
kilian.schaerli@mll-legal.com

Kilian Schärli is a member of the IP team of Meyerlustenberger Lachenal. He specialises in all areas of IP law (patents, SPCs, designs, trademarks, copyright, licensing, unfair competition), pharmaceutical law and regulatory affairs (medicinal products and drugs). Dr Schärli advises clients from the pharmaceutical, engineering, advertising/media and consumer goods industries. He is a frequent speaker and author on IP litigation and other IP topics. Dr Schärli wrote his doctoral thesis on SPCs for medicinal products.

Carreen Shannon
Partner, IP/IT department, Deterink Attorney and Notaries
carreen.shannon@deterink.com

Carreen Shannon's practice – both litigation and non-litigation – covers all aspects of IP law, such as patents, trademarks, designs and copyright. She focuses on patent litigation and operates in several technical areas ranging from mechanical engineering and pharmaceuticals to electronics and biotech. Ms Shannon also assists clients in exploiting their IP portfolio and has extensive experience in drafting a wide range of IP-related contracts (eg, agreements relating to licensing and assignment of intellectual property and joint research, consortium, manufacturing and distribution agreements). Ms Shannon is also experienced in litigation and counseling in the area of regulatory pharmaceutical law. Before working at Deterink, she worked as an attorney at De Brauw Blackstone Westbroek and Hoyng Monegier (Howrey LLP) and as a senior legal counsel at TomTom. Ms Shannon regularly speaks and publishes on IP-related issues and is co-author of a book on the enforcement of IP rights in the Netherlands.

Andrea Sieber
Partner, head of M&A practice group, Meyerlustenberger Lachenal
andrea.sieber@mll-legal.com

Andrea Sieber is partner of the corporate finance department and the head of the firm's M&A practice group. She specialises in national and international mergers and acquisitions, private equity and capital market transactions, as well as national and international company reorganisations. She also advises Swiss and foreign clients on collective investment schemes and issues related to contract, commercial and corporate law. Ms Sieber has extensive experience representing clients in international and national M&A transactions (sell and buy side), and has represented clients in capital market transactions and in public tender offer proceedings (bidder and target side). As a regular contributor to M&A-related articles, she writes for numerous publications, including *Practical Law Company Multi-Jurisdictional Guide, IFLR, Mergers & Acquisitions, Going Public* magazine and the International Law Office.

James Stewart
Senior associate, Taylor Wessing
j.stewart@taylorwessing.com

James Stewart is a senior associate in the Taylor Wessing tax and incentives department. He works on a range of transactional matters, and advises on various tax related issues, including corporation tax, stamp taxes and value added tax. Mr Stewart also has particular experience advising on the Enterprise Investment Scheme and the Seed Enterprise Investment Scheme.

Odd Swarting
Partner, Setterwalls
odd.swarting@setterwalls.se

Odd Swarting works as an adviser to life science, forestry and other industrial companies, entities and institutions. In addition to all sorts of commercial contracts, his practice largely involves research and development projects and various issues. Mr Swarting also has substantial experience counselling both creditors and debtors during financial crises, and administering bankruptcies and restructurings. Mr Swarting has 25 years' experience providing qualified legal advice in these areas. He is also a regular contributor to life science-related articles and writes for numerous publications.

Luca Trevisan
Senior partner, Trevisan & Cuonzo Avvocati
ltrevisan@trevisancuonzo.com

Luca Trevisan is a founding partner of Trevisan & Cuonzo Avvocati and one of the leading names in Italian litigation in the commercial, IP and product liability fields. During a career spanning more than 30 years, Mr Trevisan has handled a number of landmark cases widely reported in the international press. He also has extensive experience in leading and managing multi-jurisdictional trademark and patent disputes. His domestic and international clients come from a broad cross-section of industries, including pharmaceutical, mechanical, oil and gas, electronics and automotive.

Mr Trevisan teaches on the master's in legal studies at the Il Sole 24 Ore Business School. He is a frequent speaker at conferences and regularly contributes to international legal journals. He is a member of the International Association for the Protection of Intellectual Property, the International Bar Association, the German Association for the Protection of Intellectual Property and the International Trademark Association.

Anita Varma
Partner, Ropes & Gray LLP
anita.varma@ropesgray.com

Anita Varma is co-head of Ropes & Gray's IP rights management practice group and has over 20 years' experience in IP law as a lawyer and as a patent examiner at the US Patent and Trademark Office. She is also a UK solicitor and is qualified to practise before the European Patent Office. Ms Varma focuses her practice on developing, analysing and managing patent portfolios in diverse areas of technology for life sciences companies. She leverages this broad patent experience to counsel clients on transactional matters in connection with financial investments, mergers and acquisitions, and collaborations. Ms Varma combines her patent experience with an understanding of regulatory exclusivities to help clients evaluate target portfolios and conduct worldwide due diligence assessments.

Ms Varma is viewed by competitors as "an extremely skilled attorney with both legal and business smarts" (*LMG Life Sciences*) and clients note that she always "finds a strategic path for clients involved in litigation" (Chambers & Partners independent survey). She has been recognised as a leading lawyer in publications including *Managing IP's* Top 250 Women in IP, *Legal 500*, *Best Lawyers in America* and the *IAM Patent 1000*.

Liad Whatstein
Head of patent practice, Dr Shlomo Cohen & Co
liadoffice@shlomocohen.co.il

Liad Whatstein is head of the patent practice at Dr Shlomo Cohen & Co and has vast experience in the pharmaceutical and life science fields, in both contentious and non-contentious work. Among others, Mr Whatstein has successfully represented many of the innovative pharmaceutical companies in some of the most important patent conflicts in Israel. Mr Whatstein has acted for companies including Sanofi-Aventis,

Merck, Novartis, Eli Lilly, Procter & Gamble, Wyeth, Genzyme, Abbott, Amgen, Lundbeck, Schering-Plough and Pfizer. He has acted in cases involving blockbuster drugs such as Prozac, Fosamax, Allegra/Telfast, Losec, Klacid, Vioxx, Cosopt, Lexapro/Cipralex, Propecia, Lovenox/ Clexane, Viagra, Cerezyme, Januvia, Ritalin and Truvada. In addition, Mr Whatstein acts for the state of Israel in complex patent matters relating to the exploitation of pharmaceutical inventions. Mr Whatstein graduated from the Tel Aviv University and the London School of Economics, both with distinction. He is acclaimed by international directories for his achievements and skills, and is a regular contributor to IP publications and a frequent speaker on IP matters. Mr Whatstein served as the chairman of the IP Committee of the Israeli Bar Association and is involved in the legislative revisions of the Israeli IP legislation.

Steffen Wirsing
Associate, Taylor Wessing
s.wirsing@taylorwessing.com

Steffen Wirsing is a member of the patents and pharma practice area in Taylor Wessing's Munich office. He advises national and international companies from the pharmaceutical, biotech, medical devices and food sectors in all stages of their product lifecycle. In addition, he advises clients on cross-border contractual projects, in particular on manufacturing and distribution, research and development and licence agreements. Mr Wirsing studied law at the University of Wurzburg and passed his first legal state exam in 2010. Since 2011 he had been working for Taylor Wessing in Munich in tandem with his legal clerkship. Before passing his second legal state exam in 2013, he worked for an international law firm in New York. Mr Wirsing speaks German and English.

Tim Worden
Partner, Taylor Wessing
t.worden@taylorwessing.com

Tim Worden is a partner in Taylor Wessing's Cambridge office and specialises in transactional, non-contentious and regulatory intellectual property in the life sciences, IT and hi-tech sectors. He advises on: a range of IP, IT and commercial agreements, such as technology and software licences, collaboration agreements, R&D agreements, clinical trials agreements and a range of services agreements; regulatory issues in the pharmaceutical industry – in particular, in relation to the promotion of medicines in the United Kingdom – and in the medical devices sector; IP, IT and commercial issues arising out of mergers and acquisitions, venture capital investments and IPOs; and IP and commercial contract disputes. Mr Worden was previously legal counsel and company secretary at Eli Lilly and Company Limited, the UK subsidiary of the US pharmaceutical company. He has a degree in natural sciences (chemistry and biochemistry) from Cambridge University and a diploma in IP law and practice from Bristol University.